D1382801

Interpretation and Processing
of Vibrational Spectra

Interpretation and Processing of Vibrational Spectra

MILAN HORÁK
and
ANTONÍN VÍTEK

Translation
KAREL ŠTULÍK

A Wiley-Interscience Publication

JOHN WILEY & SONS

Chichester • New York • Brisbane • Toronto

Published in co-edition with SNTL, Publishers of Technical Literature, Prague.

Library of Congress Cataloging in Publication Data

Horák, Milan.

Interpretation and processing of vibrational spectra.

'A Wiley–Interscience publication.'
1. Vibrational spectra. I. Vítek, Antonín, joint author. II. Title.

QD96.V53H6713 543'.085 78-16741

ISBN 0 471 99504 5

Printed in Czechoslovakia.

Contents

PREFACE

At the present, when methods employing infrared and Raman spectroscopy have penetrated into practically all chemical laboratories, many chemists work independently with vibrational spectra. One of the difficulties they face is a lack of suitable literature. Data required for everyday spectroscopic practice are scattered in original papers, monographs, textbooks and manufacturers' literature; this leads chiefly to a loss of time in finding the appropriate material and sometimes also to ineffective utilization of information stored in the spectra.

We feel that this manual, "Interpretation and Processing of Vibrational Spectra", will prove useful to people interested in the routine application of spectroscopy in chemical laboratories. Rules and methodical notes for procedures employed in the handling and interpretation of vibrational spectra are summarized in the book. Commonly useful definitions, constants, conversion factors, relationships and schemes of computing procedures for numerical treatment of the experimental data are given and the book is supplemented with programs in the FORTRAN IV language applicable to the solution of some common tasks in vibrational spectroscopy using a computer.

The book was written for university-educated chemists; however, we have tried to make as large a part of the text as possible intelligible to graduates of specialized secondary schools. The aim of adding well-tested computer programs with a detailed description of their use is to help in promoting computing techniques in those laboratories where they are not yet commonly used.

CHAPTER 1

Introduction

The initial successes of vibrational spectroscopic methods in analytical chemistry and structural diagnosis were experienced during the second World War. An important contribution of spectroscopy to the elucidation of the structure of penicillin and its progressive role in the identification and structural diagnosis of the components of liquid fuels became important stimuli for the development of infrared spectroscopy and related experimental techniques. Although the cost of infrared spectrometers, the manufacture of which was started at the end of the second world war, was many times higher than that of instrumentation commonly used at that time in chemical laboratories, they rapidly spread all over the world. Further encouraging reports of the successful application of infrared spectroscopy appeared. The evolution in the use of infrared spectroscopy was completed by the designing and manufacture of double-beam spectrometers towards the end of the 1940s; the newly introduced compensation for measuring beam attenuation enabled automated operation and removed tedious manual evaluation of the spectra. During the time when infrared spectroscopy was undergoing rapid development, another method for the study of vibrational spectra of molecules, Raman spectroscopy, was somewhat superseded. However, this situation did not last long. After the introduction of laser excitation sources (at the beginning of the 1960s) the technique for obtaining Raman spectra underwent a basic change which helped to restore this method to its rightful prominence. The technical possibilities of the two methods of vibrational spectroscopy, infrared and Raman, are almost equal today and the advantages of the former or the latter can be utilized according to the task to be solved.

At present, after almost thirty years of systematic use of the methods of vibrational spectroscopy on a wide scale, their advantages, drawbacks and limitations have been thoroughly explored and the value of the information obtained properly appreciated. Now the methods of vibrational spectroscopy can be discussed soberly; thus the fact that they have stood up to competition from newer, often very effective methods and have remained (and, we feel, will remain) among the principal methods employed in the chemical laboratory is all the more noteworthy.

The growing general acceptance of the vibrational spectroscopic methods has brought them to the attention of a progressively wider community of specialists who are not and do not intend to become professional spectroscopists. They have, however, become accustomed to systematic work with spectra obtained using their samples, e.g. from central spectroscopic laboratories. This way of dealing with "supplied" spectra has a number of special features and leads us to specific selection of the material for this book. For example, detailed description of spectrometer design and of the function of the various components has been omitted, provided such knowledge is not required for correct evaluation of spectra; questions of experimental techniques in spectral measurements are discussed only in connection with spectra quality, etc.

The book is intended primarily for chemists and therefore does not contain calculations of spectroscopic and molecular constants which would be attractive for physicists. This omission was also caused by purely practical reasons: explanations required for understanding the principles of calculation of force constants, contributions from bond dipole moments, molecular geometry, etc. would occupy far more room than is available in this small publication and, moreover, many excellent books dealing with this topic have been published. In this book, we propose to concentrate on the problems of obtaining so-called "primary data" from the spectra (wavenumbers and band intensities), i.e. information which is usually employed in the solution of chemical problems.

The book is not concerned with non-trivial methods of obtaining information on the vibrations of molecules, such as non-elastic neutron scattering, photoelectron spectroscopy, etc. Another method, Fourier-transformation spectroscopy (the FTS method), which now appears quite frequently in laboratories, is only briefly mentioned. We consider the information obtained by this method to be equivalent to that obtained from infrared spectra.

Similar to many other branches of science, vibrational spectroscopy has been strongly affected by the development of modern computing techniques. For this reason, this book is also supplemented with a collection of programs for the solution of basic problems in spectral analysis.

In order to attain as high an effectiveness as possible, independent subroutines are given in the appropriate section of the book and can be treated as elements of a flexible building box. In addition to the programs described in the book, other programs, suitable for the solution of particular problems, can be assembled from the subroutines.

The program language, FORTRAN IV, which has been selected, can be used without any modifications for all IBM 360 and IBM 370 computers and for other types after minor modifications.

Considerable difficulty was encountered while writing this book because of the discrepancy between standards and rules dealing with nomenclature, symbols, dimensions and units and contemporary laboratory practice. Our opinions are expressed in the list of symbols on p. 379, to which explanations and notes are added. This list also contains references to parts of the text where the problems of symbols are discussed in greater detail.

Regarding the emphasis placed on the practical aspect of the book, little attention has been paid to derivation or proofs of the presented relationships. However, we have verified most of them during our many years of spectroscopic practice. To make the contents of the book as lucid as possible, we have made generous use of examples; they are generally related to the spectra of methanesulphonyl chloride, CH_3SO_2Cl. This compound has been selected as a model after careful consideration, chiefly because of the relative simplicity of the appearance and interpretation of the vibrational spectra of its molecule.

We would feel particularly rewarded if this book were to become an everyday tool in chemical laboratories, used in the handling and interpretation of molecular vibrational spectra. We would therefore be grateful to our readers for comments on the topic and contents of the book and for critical remarks concerning any errors, formal faults or omissions.

CHAPTER 2

Vibrational Spectra

This book will deal with the problems in handling and interpretation of vibrational spectra of polyatomic molecules;*) therefore, it is necessary in the beginning to briefly summarize the principal concepts in this field of physics and the rules for the formation, description and obtaining of vibrational spectra.

2.1 Basic Spectroscopic Quantities

Generally, a spectrum is considered to be any functional dependence of two quantities, one of them having the dimension of energy. According to this definition, it follows that a vibrational spectrum is the functional dependence of the intensity of radiation absorbed (with infrared absorption spectra) or scattered (with Raman spectra**) by a given substance on the energy of the quanta of this radiation; characteristic absorption or scattering occurs due to quantization of the vibrational (or vibrational-rotational) movement of molecules.

The principal requirement in spectroscopy is that the radiation be monochromatic. Instruments for the measurement of spectra are constructed so that at a given moment radiation with defined quantum energy E is incident on the detector; this energy is described by the well-known formula relating the corpuscular and wave properties of radiation:

$$E = h\nu; \tag{2.1}$$

where h is Planck's constant and ν is the radiation frequency.

---

*) Almost everything true for the vibrations of molecules is also valid for the vibrations of polyatomic ions or radicals; this fact will not be further emphasized.

**) These spectra are named after Sir C. V. Raman who first reported their measurement in 1928.

Due to the historical development of spectroscopy, the wave properties of radiation are used for its description rather than the corpuscular properties. Therefore, quantity E is rarely used during the recording of spectra; the use of the following four interrelated quantities is much more frequent: period, T, frequency, ν, wavelength, λ, and wavenumber, $\tilde{\nu}$. The mutual relations among these quantities are expressed by the following four conversion equations:

$$T = \nu^{-1} = \lambda c^{-1} = \tilde{\nu}^{-1} c^{-1} \quad \text{(s)}, \tag{2.2}$$

$$\nu = T^{-1} = c\lambda^{-1} = c\tilde{\nu} \quad (\text{s}^{-1}), \tag{2.3}$$

$$\lambda = cT = c\nu^{-1} = \tilde{\nu}^{-1} \quad \text{(cm)}, \tag{2.4}$$

$$\tilde{\nu} = c^{-1}T^{-1} = \lambda^{-1} = \nu c^{-1} \quad (\text{cm}^{-1}), \tag{2.5}$$

where c is the velocity of light (in cm per second in a vacuum). It follows from the conversion relationships, (2.2)–(2.5) and equation (2.1) that quantum energy E is directly proportional to frequency ν and wavenumber*) $\tilde{\nu}$, while inverse proportionality holds between energy E and wavelength λ or period T.

Wavenumbers $\tilde{\nu}$ and wavelengths λ are most frequently employed for the description of infrared and Raman spectra. Although the use of either of the two quantities is equally justified from a purely theoretical point of view, wavenumbers are preferred for a number of practical reasons which will be discussed elsewhere (see Section 2.4); the same holds for the present book. Conversion tables between wavenumbers and wavelengths are given on p. 383.

The energy of a quantum of radiation, E, must be distinguished from the radiant energy, Q (also with the dimension of energy), which represents the sum of the energies of all the quanta of polychromatic radiation, passing through the instrument during a certain time. If energy Q is related to a time unit, quantity Φ (with the dimension of power) is obtained, termed the

---

*) It should be pointed out, for the sake of completeness, that the wavenumber depends on the medium in which the spectra are measured. This fact follows from the relationship, $\tilde{\nu} = \nu c^{-1}$; frequency ν is a quantity which is independent of the conditions of the spectral measurement, while the velocity of the radiation depends on these conditions. In a medium with refractive index n, the velocity of radiation, c_{n}, equals $c_{\text{n}} = c_0 n^{-1}$ and the wavenumber is $\tilde{\nu}_{\text{n}} = \tilde{\nu}_0 n^{-1}$. In precise measurements it is thus necessary to recalculate the experimental wavenumber to the wavenumber in vacuo, unless the measurement is performed in vacuo. However, the error thus created is relatively small and hence the correction for the refractive index of air is negligible during work with common spectrometers.

radiant power. The amount of energy passed in a unit time and corresponding to a narrow wavenumber interval, $\Delta\tilde{\nu}$, can be expressed as $(\mathrm{d}\Phi/\mathrm{d}\tilde{\nu})\,\Delta\tilde{\nu}$, where the differential

$$\Phi_{\tilde{\nu}} = \mathrm{d}\Phi/\mathrm{d}\tilde{\nu} \tag{2.6}$$

is called the spectral radiant power. The spectral radiant power can be analogously defined in other scales than the wavenumber scale; the relationships between the thus-defined quantities are

$$\Phi_{\lambda} = \mathrm{d}\Phi/\mathrm{d}\lambda = -\tilde{\nu}^2\Phi_{\tilde{\nu}} \tag{2.7}$$

or

$$\Phi_{\nu} = \mathrm{d}\Phi/\mathrm{d}\nu = \Phi_{\tilde{\nu}}/c. \tag{2.8}$$

In single-beam infrared spectrometers the dependence of the magnitude of the radiant power in a narrow wavenumber interval (which is directly proportional to the spectral radiant power) on the wavenumber, wavelength or frequency is measured; a similar situation occurs during the measurement of Raman spectra.

On the passage of monochromatic radiation through a sample, part of the radiant power is absorbed by the medium, part is scattered and reflected and the remainder passes through. The ratio of the spectral radiant power after passage through the sample, $\Phi_{\tilde{\nu}}$, to that before passage through the sample, $\Phi^0_{\tilde{\nu}}$, is denoted τ,

$$\tau = \Phi_{\tilde{\nu}}/\Phi^0_{\tilde{\nu}} = \Phi_{\lambda}/\Phi^0_{\lambda} = \Phi_{\nu}/\Phi^0_{\nu} \tag{2.9}$$

and called the transmittance. Analogously, the absorptance, α, is defined as the ratio of the absorbed spectral radiant power to the incident power and reflectance ϱ as the ratio of the reflected spectral radiant power to the incident power. It follows from the law of conservation of energy that the three quantities must be interrelated according to the simple equation

$$\tau + \alpha + \varrho = 1. \tag{2.10}$$

With double-beam instruments, the ratio of the spectral radiant power in the measuring beam with the sample inserted (denoted as Φ) and in the reference beam without the sample or with a blank sample (denoted as Φ^0) is measured. This thus basically corresponds to measurement of dimensionless quantity τ, transmittance, defined by relationship (2.9).

If the mechanism of the process of absorption of radiant energy in the medium along the optical path of the beam is independent of the magnitude of the radiant flux density the spectral radiant power after passage through an absorbing layer with thickness d is given, according to the Lambert law, by

$$\Phi_{\tilde{\nu}} = \Phi_{\tilde{\nu}}^{0} \,.\, \exp(-kd). \tag{2.11}$$

Dimensionless quantity τ (transmittance) can then be expressed as

$$\tau = \Phi_{\tilde{\nu}}/\Phi_{\tilde{\nu}}^{0} = \exp(-kd). \tag{2.12}$$

Its decadic logarithm is proportional to d with proportionality constant k

$$\log \tau = -0.4343 \,.\, kd. \tag{2.13}$$

The absolute value of this dimensionless quantity is called the absorbance or internal transmission density D of the measured sample:

$$D = -\log \tau. \tag{2.14}$$

Its magnitude depends on the probability of photon absorption and thus is directly proportional to the length of the optical path in the absorbing medium.

2.2 Origin of Vibrational Spectra

It follows from the laws of quantum mechanics that the vibrational energy of molecules (energy related to the oscillations of the atom nuclei) can attain only certain discrete values, levels. The system of these levels can in the simplest case, i.e. for a diatomic molecule, be represented by a relatively simple term diagram (Fig. 1); when developed according to a single vibrational quantum number, v, it unambiguously describes the vibrational states of a molecule with a single real vibrational degree of freedom. In the term diagram, the vibrational state energy is plotted on the vertical axis and the horizontal segments symbolize the individual energy levels.

The molecule can exchange energy during interaction with a photon; the energy of this quantum of radiation can then be utilized for transition from a lower energy state to a higher energy state or the released energy (corresponding to the opposite transition) can be converted into radiation. The molecule therefore accepts and releases energy non-continuously, in certain quanta; if the energy of the quantum of radiation which was exchanged by the molecule is determined, then the separation of the two energy levels between which the transition occurred can also be found. This is the key to the understanding of the majority of applications of vibrational spectroscopy.

The arrows in the term diagram (Fig. 1) denote transitions between individual energy levels. Transitions from the ground state level (vibrational quantum number $v = 0$) to higher levels ($v = 1, 2, \ldots$) are collectively termed harmonic frequencies.*) Transition $1 \leftarrow 0$ corresponds to the first harmonic frequency, the so-called fundamental frequency, transition $2 \leftarrow 0$ corresponds to the second harmonic (also called the first overtone), transi-

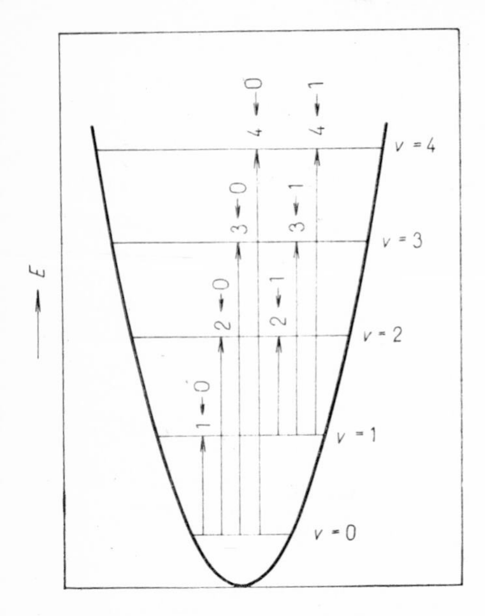

Fig. 1. The term diagram for a vibrating diatomic molecule. The levels denoted by vibrational quantum numbers v are drawn on the curve representing the course of the potential energy in the quadratic approximation. Transition $1 \leftarrow 0$ is fundamental, transitions $n \leftarrow 0$ ($n > 1$) are called overtones and transitions $n \leftarrow 1$ ($n < 1$) are called hot.

tion $3 \leftarrow 0$ corresponds to the third harmonic (the second overtone), etc. There are, of course, also transitions starting from higher energy levels. In the term diagram they are represented by the group of transitions starting from the first excited state level ($v = 1$). These transitions are less pronounced in the spectra, as the number of molecules thermally excited to higher vibrational states is always lower than the number of molecules in the ground state and the level population sharply decreases with increasing quantum number. Since the molecule population in higher vibrational states increases

*) The concept of frequency has two meanings in vibrational spectroscopy; on the one hand it is a measure of the quantum energy ($E = h\nu$) and on the other it is a synonym of the concept of "oscillations". In this book, where wavenumber is used as a measure of the quantum energy, $\tilde{\nu}$ ($E = h\tilde{\nu}c$), the concept of frequency will express the basis of the oscillating movement.

with increasing temperature, the intensity of the bands corresponding to these transitions also increases with increasing temperature. For this reason they are usually termed "hot transitions", corresponding to "hot bands" in the spectra.

We can now explain the relationship between the term diagram given in Fig. 1 and the vibrational spectrum of the corresponding molecule. To emphasize the mutual relations, the spectrum will be replaced by its line scheme (Fig. 2). On the horizontal scale of the spectrum scheme (Fig. 2)

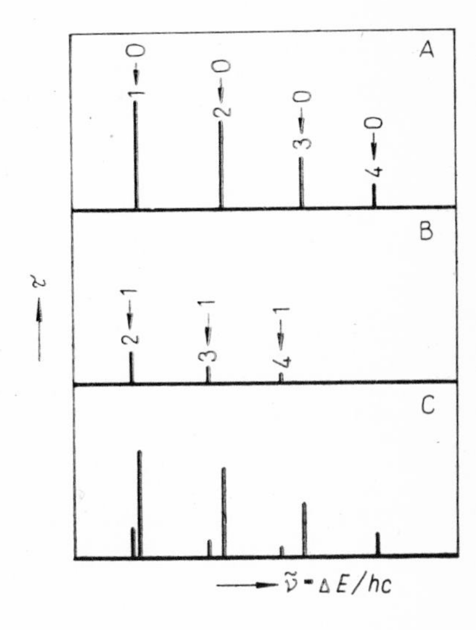

Fig. 2. Line scheme of the vibrational spectrum of a diatomic molecule with marked transitions taken from Fig. 1. A — harmonic frequencies; B — hot transitions; C — the scheme itself with both transition types.

are plotted the wavenumbers $\tilde{\nu}$ of the energy differences of vibrational levels between which transitions take place, with vibrational quantum numbers v'' and v' (for the lower and higher states, respectively),

$$\tilde{\nu} = [E(v') - E(v'')]/hc. \tag{2.15}$$

The position of the lines in the spectra, expressed in wavenumbers, thus corresponds to the separation of the vibrational levels.

Harmonic frequencies are depicted in the first line of the scheme, hot bands in the second, and the third line – representing the vibrational spectrum of a diatomic molecule in this approximation – involves both these transition types simultaneously. The uneven height of the lines should express the differences in the intensities of the vibrational transitions; however, this question is rather complicated and will be dealt with in detail later.

The term diagram and the spectrum scheme discussed above describe a situation in which the nuclei of a diatomic molecule only vibrate (e.g. in the liquid state). If the molecule also rotates – e.g. in the gaseous state – the movement must be simultaneously described by vibrational and rotational coordinates. Since the energy separation of the rotational levels is one to three orders of magnitude smaller than that of the vibrational levels, molecular rotation is manifested in the fine structure of the vibrational

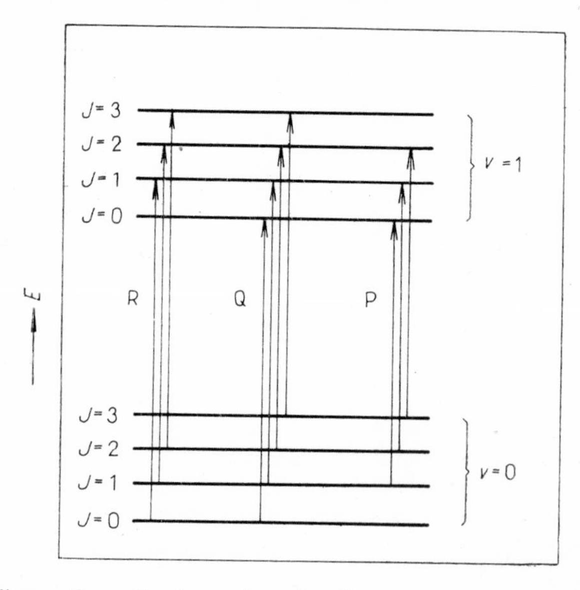

Fig.3. Term diagram for a vibrating and rotating diatomic molecule. Only two vibrational levels (for $v = 0$ and $v = 1$) with fine rotational structure are depicted, the ground ($v = 0$) and the first excited ($v = 1$) levels; J is the rotational quantum number; transitions for selection rules $\Delta J = +1$ are denoted by the letter R, $\Delta J = 0$ by letter Q and $\Delta J = -1$ by letter P.

bands. An energy scheme of a vibrating and simultaneously rotating diatomic molecule is given in the term diagram in Fig. 3. The fact that the rotational levels of a diatomic molecule can be described by a single rotational quantum number, J, is unique for this group of molecules.*)

*) A single rotational quantum number describes the levels in molecules of a so-called spherical rotator (such as the methane molecule), with identical moments of inertia along the three axes. However, diatomic molecules have only two rotational degrees of freedom (as have all linear molecules) and the moments of inertia along the two axes are identical.

Rotational-vibrational transitions are denoted by the arrows in the term diagram in Fig. 3. All transitions take place between identical vibrational levels (excitation from the ground vibrational state to the first excited state) and among various rotational levels. Part of the infrared spectrum of carbon monoxide is given in Fig. 4 and corresponds to the term diagram in Fig. 3. This is a transition between the ground and first excited vibrational states of the CO molecule, which is extensively split due to rotational

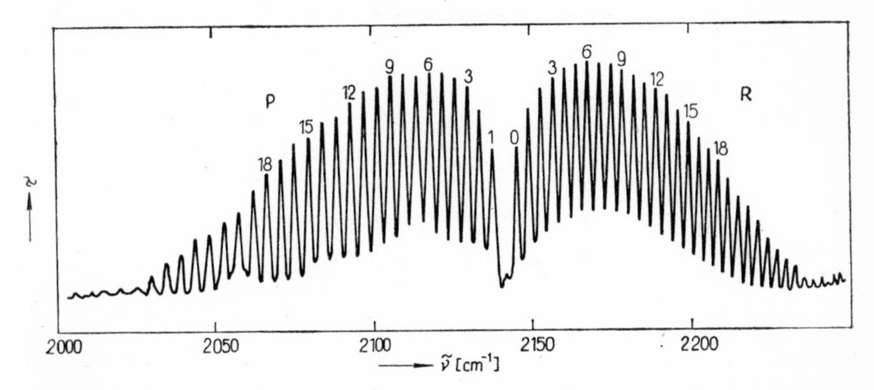

Fig. 4. Rotational-vibrational $v_{1\leftarrow 0}$ band for carbon monoxide CO from the experimental infrared spectrum. The bands in the R- and P-branches are numbered by the rotational quantum numbers of the initial rotational states. The Q-branch does not appear in this band in consequence of the selection rules.

transitions. The great number of components of the rotational-vibrational band is related to the large population of collision-excited molecules in higher rotational states; owing to the smaller separations of the rotational levels, the molecules are excited during collisions to higher rotational states more readily than to higher vibrational states.

Two band groups can be seen in the spectrum in Fig. 4; the R-branch towards higher wavenumbers and the P-branch on the opposite side. In the R-branch are concentrated transitions with selection rules $\Delta J = +1$ and in the P-branch transitions with $\Delta J = -1$. The Q-branch also appears in the infrared spectra of many molecules (transitions $\Delta J = 0$) between the former two branches; however, it is forbidden for the described $1 \leftarrow 0$ transitions in diatomic molecules. Bands in the P-, Q- and R-branches are usually denoted by the value of the rotational quantum number of the level from which the transition started. It is evident from the spectrum in Fig. 4 that the carbon monoxide sample contains a sufficient number of excited molecules with rotational quantum numbers up to $J = 30-35$.

With polyatomic molecules the whole vibrational and rotational-vibrational situation becomes considerably more complicated. While diatomic molecules have only a single proper vibrational degree of freedom, polyatomic molecules have $3N-6$ corresponding degrees of freedom (N is the number of atoms in the molecule). This, of course, is true only for non-linear molecules; linear molecules have one more proper vibrational degree of freedom, i.e. $3N-5$. Therefore, the term diagram for the simplest

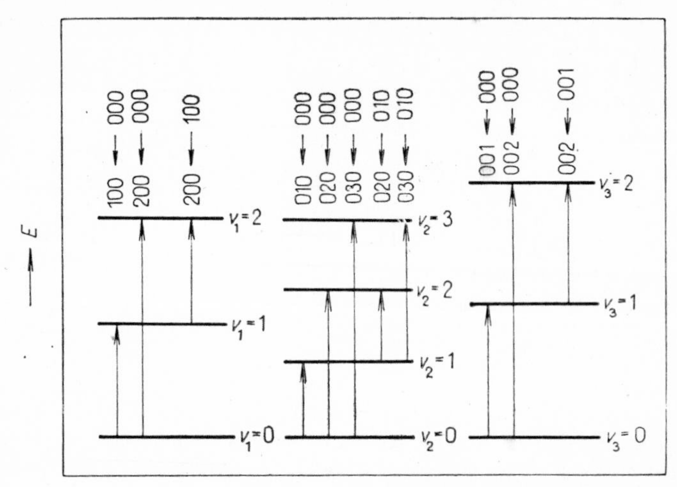

Fig. 5. The term diagram for a vibrating non-linear triatomic molecule, expanded according to three vibrational quantum numbers, v_1, v_2 and v_3. Transitions 100 ← 000, 010 ← 000 and 001 ← 000 are fundamental, transitions n00 ← 000, 0n0 ← 000 and 00n ← 000 are overtones and the rest are hot transitions.

polyatomic—triatomic—non-linear molecule must be developed using three quantum numbers, v_1, v_2 and v_3 (Fig. 5; the same transition types are included as in the diagram in Fig. 2).

A scheme of the spectrum of a triatomic molecule, corresponding to the term diagram, is given in Fig. 6. The first three lines of the scheme contain the harmonic frequencies developed according to quantum numbers v_1, v_2 and v_3 and in an additional line are given the hot band frequencies. The fifth line is related to the so-called combination frequencies, which were not encountered with diatomic molecules. These are transitions in which the photon energy is utilized for simultaneous excitation of two or more higher vibrational levels. For the sake of simplicity,*) only binary

*) Higher harmonic frequencies, as well as the combination frequencies, are not simple multiplets, sums or differences of the fundamental frequencies, due to their anharmonic

sum combinations, $(\nu_i + \nu_j)$, are considered in the scheme, although experimental spectra exhibit difference combinations $(\nu_i - \nu_j$, etc.) and even higher-order combinations (third-order, $\nu_i + \nu_j + \nu_k$, etc.). In the last and sixth line of the scheme are collected all the transitions represented

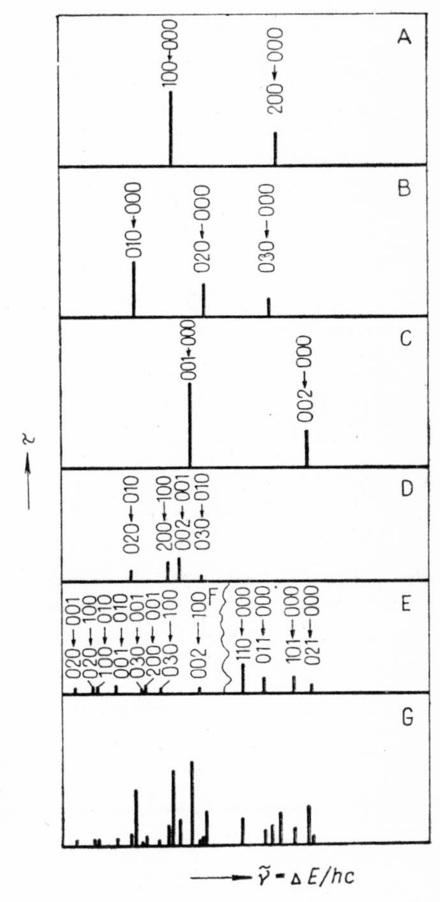

Fig. 6. A line scheme of the vibrational spectrum of a non-linear triatomic molecule with marked transitions from Fig. 5. A, B and C — harmonic transitions according to quantum numbers v_1, v_2 and v_3, respectively, D — hot transitions, E — combination sum frequencies (the corresponding transitions are not given in Fig. 5), F — combination difference frequencies (the same holds as for the sum frequencies), G — scheme of the resultant spectrum.

in the term diagram (Fig. 5) and shown separately in the first to the fifth line of scheme 6. It can be seen that even systems as small as triatomic molecules, moreover in such a simplified approximation, yield a relatively complex spectrum. This complexity inherently contains the high potentiali-

---

character. This fact will be discussed in chapters 4 and 5 and is not considered here in order to simplify the discussion of the origin of vibrational spectra.

ties of using vibrational spectra for the identification of molecules of chemical compounds, because it is improbable that the vibrational spectra of different compounds will be identical.

Similar to diatomic molecules, the simultaneous occurrence of vibration and rotation and the consequent appearance of rotational-vibrational bands in the spectra must be expected with polyatomic molecules in the gaseous state. Provided that the mass of the molecules does not exceed a certain value (given by the spectrometer resolution), the rotational-vibrational bands exhibit a structure with resolved rotational transitions. If this value is exceeded, the value of the moments of inertia increases so that the individual lines in the rotational-vibrational band merge and form an envelope for the rotational-vibrational band. The laws governing the appearance of such unresolved bands are treated in detail in Section 4.3.3.

So far, the energy aspect of molecular vibrations, expressed by term diagrams, has been predominantly discussed. However, in an effort to explain the relationships between these diagrams and spectra lucidly, considerations on the intensities of infrared bands and Raman lines were neglected. Although this topic will be treated in greater detail in Chapter 4 in the discussion of vibrational selection rules, the principal factors determining the magnitude of absorption and scattering will be surveyed briefly here.

Infrared absorption spectra (the only type treated in this book) obey the basic laws of electrodynamics, according to which only an oscillating dipole interacts with incident radiation. The absorption magnitude and thus also the absorption band intensity is related to the magnitude of the permanent electric dipole moment of the vibrating molecule, or, more precisely, to periodic changes in this dipole moment during vibration. We shall attempt to explain this problem using an oscillating heteronuclear biatomic molecule.

If μ_0 is the permanent electric dipole moment of a biatomic molecule in an equilibrium configuration, defined by the equilibrium distance of the nuclei r_0, then the value of the instantaneous electric dipole moment, μ, of the vibrating molecule for internuclear distance r follows from the relationship

$$\mu = \mu_0 + (\mathrm{d}\mu/\mathrm{d}r)\,\Delta r. \tag{2.16}$$

The derivative, $(\mathrm{d}\mu/\mathrm{d}r)$, describing the change in the dipole moment on an infinitesimal change in the vibration coordinate, $\mathrm{d}r$, is termed the effective charge and its square is directly proportional to the magnitude of absorption of infrared radiation and hence to the (integrated) intensity of the absorption band in the infrared spectrum of the biatomic molecule. Thus the absorption band intensity for vibration of heteronuclear biatomic molecules is the larger, the larger is their dipole moment; biatomic homonuclear molecules (e.g. H_2, N_2, Cl_2, etc.) without electric dipole moments do not absorb infrared radiation.

With polyatomic molecules, predominantly vibrations localized in the polar parts of the molecules are connected with pronounced absorption of infrared radiation. Less polar parts (e.g. hydrocarbon molecular residues) do not give rise to important absorption. The infrared spectra of molecules of organic compounds containing strongly polar groups (e.g. COOH, NH_3^+) or bonds (e.g. C–F, Si–O) thus yield much more important information about these polar groups than about the hydrocarbon parts of the molecules.

However, a polyatomic molecule need not possess a permanent electric dipole moment in order to absorb infrared radiation. Even a non-polar molecule may undergo vibrational movements, during which the centres of gravity of the positive and negative charge, which overlap in the equilibrium configuration, are separated. This can be demonstrated e.g. on the carbon dioxide molecule, which has no permanent electric dipole moment in the equilibrium configuration. No dipole moment appears during symmetrical stretching vibration; however, antisymmetrical stretching and degenerate deformation vibrations lead to the formation of polar structures and hence both are manifested by absorption of infrared radiation.

Raman spectra are dependent on a change in the dipole moment induced in the molecule by the electric vector, $\boldsymbol{E}$, of the exciting radiation, which is connected with a change in the polarizability of the molecule during vibrations. Similar to the preceding case, this question will be discussed using a biatomic molecule as an example.

If α_0 is the polarizability of a biatomic molecule in the equilibrium configuration, characterized by equilibrium distance of the nuclei r_0, the instaneous value of the polarizability α during vibrations is given by the relationship

$$\alpha = \alpha_0 + (\mathrm{d}\alpha/\mathrm{d}r)\,\Delta r. \tag{2.17}$$

The square of the derivative of the polarizability, $(\mathrm{d}\alpha/\mathrm{d}r)^2$, determines the magnitude of the scattering of the corresponding Raman line and thus its intensity.

The polarity and polarizability of molecules are interconnected; molecules with large permanent dipoles usually exhibit negligibly small or zero polarizability and vice versa. Biatomic homonuclear molecules (e.g. H_2,

N_2, Cl_2) that do not absorb infrared radiation change their polarizability during vibrations and consequently appear in the Raman spectrum.

Changes in the polarizability can be readily explained using biatomic molecules as an example. During vibrational movement, the internuclear distances in the molecule periodically change. As the polarizability depends on these distances, it changes with the same periodicity. A similar situation occurs with polyatomic molecules. For example, symmetrical stretching vibration of carbon dioxide is connected with changes in the "length" of the molecule and in the polarizability and is actively manifested in the Raman spectrum. With antisymmetrical stretching vibration, the "length" of the molecule is the same in both limiting positions; consequently, the polarizability is also constant and thus antisymmetrical stretching vibration does not give rise to Raman scattering.

In polyatomic molecules, less polar parts of molecules, such as hydrocarbon residues, etc., contribute more to Raman scattering. The Raman spectra of organic molecules containing polar groups then provide information predominantly on the non-polar parts, i.e. the opposite type to that derived from infrared spectra.

It is sometimes difficult to predict the activity of individual normal vibrations in infrared and Raman spectra for molecules with more than three atoms. However, since this activity depends on the symmetry of the molecule and on the symmetry of the corresponding normal vibration, a set of rules has been developed for various types of molecules from the point of view of symmetry, using which the activity can be objectively predicted. These are the vibration selection rules and are discussed in detail in Chapter 4.

2.2.1 Infrared and Raman Spectra

Information on vibrations of atomic nuclei in molecules is most frequently obtained from infrared and Raman spectra.

Absorption infrared spectra can be recorded by monitoring the passage of infrared radiation through samples of compounds, as polychromatic infrared radiation incident on a set of molecules is partly absorbed and only part of it passes through the sample. The molecule does not absorb an infrared quantum with arbitrary energy, but only a quantum whose energy, $E = h\nu$, satisfies the Bohr frequency condition, i.e. its magnitude equals the difference in the energies of two vibrational (rotational) levels of the molecule, ΔE. The Bohr frequency condition is, of course, not the only condition limiting the absorption; other conditions follow from the symmetry of molecules and vibrations and are contained in the selection rules (Chapter 4).

If the Bohr frequency condition, also called the resonance condition, is satisfied, then the energy of the photon absorbed during interaction of radiation with the molecule results in transfer of the molecule from a lower vibrational level to a higher level.

If photons of various energies are contained in infrared radiation before interaction, then only those absorbed in the resonance interaction will be missing after interaction. This effect can be put in practice experimentally by monitoring infrared radiation before reaching the sample and after passage through the sample.

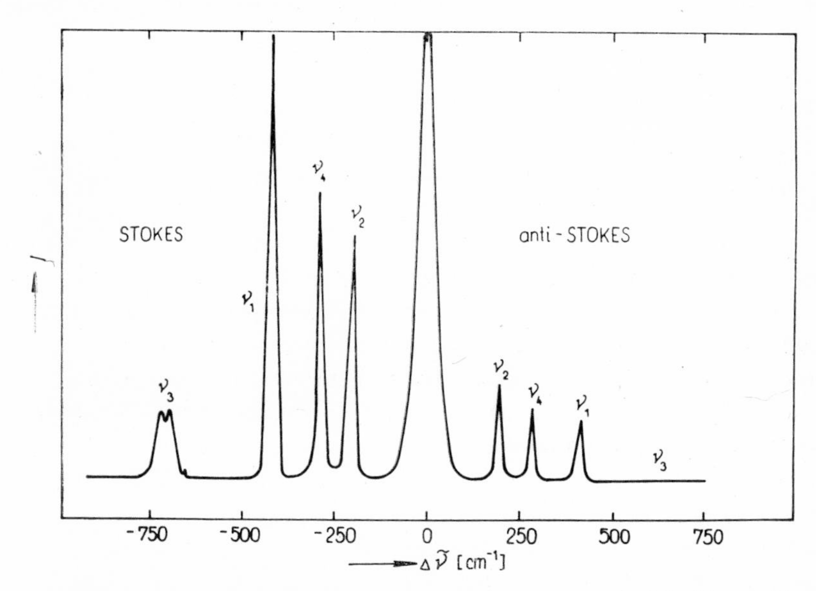

Fig. 7. The anti-Stokes and Stokes regions in the Raman spectrum of liquid tetrachloromethane, CCl_4, with the excitation line in the middle. To facilitate the orientation, the lines in the spectrum are assigned to normal vibrations $\nu_1 - \nu_4$ of the CCl_4 molecule in both regions (the doublet of the ν_3 vibration is not perceptible in the anti-Stokes region).

Similar to infrared spectra, Raman spectra are (rotational-) vibrational spectra of molecules, but differ from the former in the mechanism of their origin; they are spectra of scattered radiation and are measured in the visible spectral range.

The Raman effect results from an energy exchange between a molecule and a photon during elastic collision, during which no absorption of light but its frequency modulation occurs. If the scattered radiation emerging from the irradiated sample, e.g. at right angles to the axis of the exciting radiation, is dispersed, components with smaller and higher energies will appear in addition to the excitation line. If the vibrational state of the molecule changes in the instant of the elastic collision of the photon with the molecule, the energy required is exchanged with the photon. The energy of the photon reflected from the molecule is decreased or increased by the energy exchanged.

If the intensity of the scattered radiation is recorded as a function of its wavenumber, the Raman spectrum is obtained. The excitation line is located in the centre of this spectrum; on both sides of it are Raman lines, ordered symmetrically according to their wavenumbers (Fig. 7). The region of lines with wavenumbers lower than that of the excitation line is called the Stokes region, that with higher wavenumbers is called the anti-Stokes region. The distance of the Raman lines from the excitation line on the energy scale gives the distance between the two vibrational levels between which the transition took place. If a transition from the ground to an excited state occurs, the molecule accepts energy from the exciting radiation and the corresponding Raman lines are in the region of lower wavenumbers; these lines thus form the Stokes region. Transitions from excited states to the ground state are manifested in the anti-Stokes region. As the content of collision-excited molecules in the sample sharply decreases with increasing level separation, the line intensity in the anti-Stokes spectral region also decreases. For this reason, the Stokes region of Raman spectra is more frequently employed in practice.

Regarding the energy of the rotational-vibrational states of molecules, Raman spectra yield information of the same quality as infrared spectra. However, the selection rules for the transition activity in the Raman spectrum are different; additional information concerning the polarity or polarizability of molecular bonds is stored in the intensities of the absorption bands of infrared spectra or in the intensities of Raman lines. Infrared and Raman spectra complement one another partially or completely; as will be shown later, the spectra are perfectly complementary e.g. with molecules having a centre of symmetry. For example, of the twelve kinds of vibration in the ethylene molecule, five are manifested in the infrared spectrum and six in the Raman spectrum; the twelfth vibration is inactive in both types of spectrum.

Monochromatic radiation in the visible region of the electromagnetic spectrum is usually employed for excitation of Raman spectra. Optical lasers are particularly suited for this purpose, as they produce monochromatic radiation of high radiant power.

2.2.2 Properties of Sets of Molecules

In view of the rigorousness of the Bohr frequency condition, the process of absorption should result in line absorption; the same should hold for radiation scattered due to the Raman effect. However, quite broad bands are often found in experimental infrared and Raman spectra.*) The band character of the spectra is a result of the behaviour of large sets of molecules, such as macroscopic samples of compounds, on which the spectra are measured.

In all states of aggregation, except for dilute gases, the statistical nature of molecular orientation during interactions and of the forces thus generated must be taken into consideration. Consequently, the separations of energy levels of individual molecules in the set are not identical and transitions among them cannot be sharp (line).

Not even the rotational-vibrational spectra of gases have line character, although their band widths are relatively small (so that they are often called rotational-vibrational lines). With low-pressure gases, whose spectra exhibit the narrowest bands, the band-width is chiefly determined by the Doppler effect, due to the relative velocities of molecules randomly moving with respect to the stationary spectrometer. On an increase in the pressure, the gas molecules enter into more frequent interactions and band broadening assumes collision character. Additional, often significant broadening occurs in compressed gases and in liquids, where the molecules are located close together and where van der Waals forces and even chemical interactions (complex formation, hydrogen bonding) play a role.

The spectra of crystalline compounds often consist of narrower bands than the spectra of the same compounds in the liquid state. In the ordered crystal structure the mutual interactions of molecules are not as randomly distributed as in chaotically ordered liquids and consequently the bands become narrower. The degree of order in crystals increases with decreasing temperature and the bands simultaneously become narrower. The band

---

*) The term "Raman line" has become deeply rooted in laboratory vernacular (apparently owing to photographic recording of the original spectra); it is therefore also used in the present book. However, the introduction of this term should not be taken as contradictory to the band character of Raman spectra.

width (and shape) thus contains information about the strength and sometimes also the mechanism of interactions among molecules (hydrogen bonding, etc.).

Another aspect of the behaviour of molecules in large sets, the collision population of higher energy states, follows from experimental spectra. Hot bands in infrared spectra and anti-Stokes lines in Raman spectra result from this population. In sets of molecules, some molecules are in collision-populated higher energy states even before interaction with radiation and their presence is manifested in the spectra by transitions starting from higher energy levels.

The energy state distribution in sets of molecules obeys the laws of statistical physics. If a set with only two kinds of molecules, namely, those in the ground state and those in an excited state, could exist, then the distribution of the two kinds would follow from the relationship

$$n_1/n_0 = \exp(-\Delta E/kT). \tag{2.18}$$

The number of excited molecules (n_1) in relation to the number of molecules in the ground state (n_0) would exponentially decrease with increasing separation of the two states, ΔE, and with decreasing temperature T; quantity k is the Boltzmann constant.

In media where population inversion cannot occur, fraction n_1/n_0 cannot attain a value larger than unity. As indicated in Table I, the content of excited molecules in the system would approach one half at very low values of ΔE (the data are related to a temperature of 25 °C). With increasing values of ΔE, fraction $n_1/(n_0 + n_1) = n_1/n$ (where n is the total number of molecules in the system) decreases sharply; for separation $\Delta E = 1000\ \text{cm}^{-1}$ (a separation typical for vibrational states) the population of excited molecules is less than one per cent and for a separation of $2000\ \text{cm}^{-1}$ it is negligible. It thus follows that hot bands and differential frequencies must be sought rather in the long-wave spectral region or in systems where transitions with small ΔE values take place. It follows from the table that, in infrared spectroscopy, there is no danger of saturation effects, common in microwave and radiofrequency spectrometry. In saturation effects, the n_1/n fraction for photons attains the value one half after absorption of the initial photons and additional photons pass through the sample without interaction. The danger of saturation effects lies in the fact that the decrease in the intensity of the incident radiation caused by absorption is not proportional to the number (concentration) of molecules.

Table I

Distribution of two quantum states[a] differing by energy separation, ΔE (at a temperature of 25 °C)

$\Delta E/hc$ (cm^{-1})	$\% n_1$[b]	n_1/n_0[c]	$\Delta E/hc$ (cm^{-1})	$\% n_1$[b]	n_1/n_0[c]
10	48.72	0.95	400	12.66	0.145
20	47.64	0.91	500	8.26	0.090
40	45.06	0.82	600	5.48	0.058
60	42.86	0.75	700	3.29	0.034
80	40.48	0.68	800	2.06	0.021
100	37.89	0.61	900	1.28	0.013
200	27.54	0.38	1000	0.79	0.008
300	18.70	0.23	2000	0.000 06	0.000 06

[a] In this simplified scheme only the molecules in the ground-state (their number is n_0) and in the excited state (n_1) are considered; [b] percentage of molecules in the excited state; [c] the ratio of the numbers of molecules in the excited and ground states.

The presence of collision-excited molecules is manifested in Raman spectra by a series of anti-Stokes lines. These lines appear in the region of higher energies than that of the excitation line, since they are formed as a result of the transition of excited molecules to the ground state. Since, in molecular sets, molecules with higher energies are more scarce, the intensity of the anti-Stokes lines decreases with increasing wavenumber; to a first approximation, the ratio of the intensities in the corresponding pairs of Raman lines (anti-Stokes/Stokes) can be derived from the value of fraction n_1/n_0 for the appropriate separation, ΔE, and temperature T.

In real systems the population of various energy states may be rather large; the population of a certain state must be determined from the expression:

$$n_a/n = g_a \exp(-\Delta E_a/kT) / \sum_{i=0}^{\infty} g_i \exp(-\Delta E_i/kT); \qquad (2.19)$$

n_a is the number of molecules in the given state, n is the total number of molecules and g_a and g_i are the state degeneracies. Expression (2.19) must be applied e.g. in considerations on the collision-distribution of the rotational states of molecules, when it is borne in mind that a typical separation of two rotational states is about 10 cm^{-1}. The distribution of rotation-excited molecules is manifested in the relative intensities of the components of a rotational-vibrational band and even in their occurrence: the spectra of a heated gas exhibit a larger number of rotational transitions (since an increase in the temperature leads to an increase in the population of the higher rotational states of molecules) and corresponding changes in the band intensities.

2.3 Range of Vibrational Spectra

The magnitude of the energy differences in the vibrational levels of molecules corresponds to the energies of quanta of infrared radiation. The infrared region of the electromagnetic spectrum is located between the long-wave (red) region of the visible radiation and the microwave region. The borderlines are not exactly defined; Table II gives wavenumbers (and also wavelengths and frequencies for the sake of comparison), which can be considered as approximate boundaries of these regions. The infrared region is divided into near, medium and far regions. The near infrared region (NIR) borders with the visible range and the far infrared region (FIR) with the microwave region. As can be seen from Table II, the medium (or fundamental) region is located between 4000 and 500 cm^{-1}. Of course, these limits are arbitrary and are also connected with the experimental possibilities of measuring infrared spectra. Therefore it would be suitable to abandon the limit of the far infrared region at 500 cm^{-1}, defined earlier,

Table II

Infrared region in the electromagnetic spectrum

Region	Wavelengths λ (μm)	Frequency ν (THz)	Wavenumber $\tilde{\nu}$ (cm^{-1})
Visible			
}	0.8	375	12 500
Near Infrared			
}	2.5	120	4 000
(Medium) Infrared[a]			
}	20	15	500
Far Infrared			
}	1 000	0.3	10
Microwave			

[a] This is also termed "fundamental region".

and move it to 200 cm^{-1}, which is easily accessible with modern spectrometers.

Chemical practice is very versatile and generally does not permit the neglecting of any region of the infrared spectrum. However, the results of spectra measured in the fundamental region are utilized most frequently.

The medium or fundamental infrared region contains chiefly the bands corresponding to transitions from the ground to the first excited vibrational state, i.e. the fundamental frequencies. The fundamental frequency of the vibration of the hydrogen molecule, the lightest stable molecule, 4160 cm^{-1}, is located at the upper limit of this region. This is a symbolic boundary, since hydrogen does not absorb infrared radiation at low pressures (as its molecule has not a permanent dipole). Above this boundary, only higher harmonics and combination frequencies can appear.

The lower boundary of the fundamental infrared region has no logical relation to the nature of vibrational frequencies. The fundamental frequencies of the diatomic molecules of heavy atoms decrease to a limiting value of 200 cm^{-1} (e.g. 210 cm^{-1} for iodine molecule, I_2); however, under this limit there are a number of fundamental frequencies of the deformation modes of polyatomic molecules (out-of-plane deformations, torsion vibrations), as well as vibrations of the components of crystal cells (translation or libration, etc.).

Commercial infrared spectrometers are at present constructed, with only a few exceptions, with a linear wavenumber scale between 4000 and 200 cm^{-1}, thus ensuring the greatest versatility at the given price, since they can be assembled from standard components, such as sources, detectors, monochromators, etc., without excessive difficulties.

Many analytical and structural diagnosis tasks require the measurement of spectra in the near infrared region. These spectra are measured on ultraviolet spectrometers which are constructed for use in the visible and near infrared regions. The lower boundary of applicability of these instruments is around 3000 cm^{-1}; cells supplied with ultraviolet spectrometers, which are usually made of quartz and the thickness of which is also suitable, are used for the measurement of spectra in the near infrared region. Bands in this region are usually of low intensity and must thus be measured in substantially thicker cells than the bands in the fundamental infrared region.

Spectra below 200 cm^{-1} are measured either on special infrared spectrometers or, more often, on Fourier-transform interferometers. Recently, instruments operating on the basis of microwave techniques have been also used for measurements in the far infrared region.

Considerations on the spectral range in which the Raman effect is monitored are based on quite different conditions. Of principal importance here is the fact that the position of the Raman lines is measured with respect to the position of the excitation line, which is usually located in the visible spectral region.

Detectors sensitive to the whole visible range, up to about 11,500 cm^{-1}, are used in Raman spectrometers. The wavenumber of the detector sensitivity limit thus represents one boundary of experimental Raman spectra; the other boundary is the excitation line wavenumber. If e.g. the green line of the argon ion laser with a wavenumber of 20,490 cm^{-1} (488 nm) is employed for excitation, it is possible to measure the spectrum from this value to 11,500 cm^{-1}, i.e. in a range of about 9,000 cm^{-1}. However, the red line of the helium-neon laser with a wavenumber of 15,800 cm^{-1} (632.8 nm) permits the measurement of the Raman spectrum of the same substance on the same spectrometer only over a range of about 4,300 cm^{-1}.

The measurement of Raman spectra at wavenumbers higher than 3,000 cm^{-1} is rarely required in practice, most often for the study of transitions inactive in the infrared spectrum (e.g. with hydrogen, as was already mentioned). Combination and higher-harmonic frequencies are very weak in Raman spectra and therefore the region above 3,000 cm^{-1} is almost exclusively studied using infrared spectroscopy. On the other hand, Raman spectra are a very valuable source of information on transitions in the far infrared region, as they can readily be measured in the close vicinity of the excitation line, whose width is usually the only limitation determining the lower boundary. Lines separated by 20 to 10 cm^{-1} from the excitation line can be routinely measured on spectrometers with laser excitation sources, due to the very narrow lines produced by lasers.

One more note on Raman spectra. Everything mentioned above is true for the Stokes line region alone, i.e. the region with wavenumbers smaller than that of the excitation line. The lines in the anti-Stokes region have higher wavenumbers than the excitation line and present no difficulties in detection. However, as they correspond to transitions from excited states to the ground state and their intensity sharply decreases with increasing distance from the excitation line, they are, in principle, unsuitable for the purpose of practical spectroscopy.

2.4 Description of Vibrational Spectra

In this section, the actual objects of our interest, actual spectra and their description, will be treated. As has already been mentioned, the vibrational spectrum is a continuous function expressing the dependence of a suitably selected quantity*) on the wavenumber $\tilde{\nu}$. This quantity can generally be the spectral radiant power, $\Phi_{\tilde{\nu}}$, transmittance τ or absorbance D. It has already been pointed out that the value of the absorbance is a measure of the probability of absorption of a photon by a molecule (or, more precisely, by a resonating oscillator) along the whole length of the beam path in the

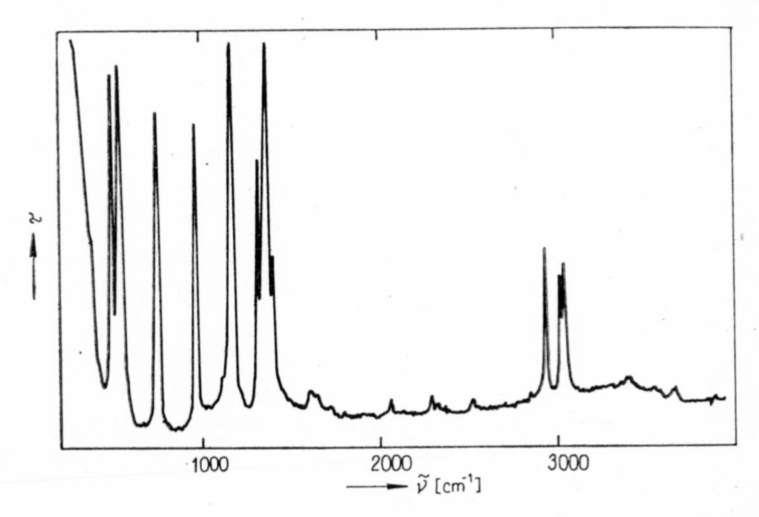

Fig. 8. Infrared spectrum of liquid methanesulphonyl chloride. Measured on the Perkin-Elmer model 621 infrared spectrometer between KBr windows without compensation.

absorbing medium. To a first approximation, the contributions of the absorption probabilities for the individual oscillators are mutually independent and additive; hence the special position of the absorbance among the quantities used for the description of spectra.

In Fig. 8 is given the experimental infrared absorption spectrum of a liquid methanesulphonyl chloride sample. The energy scale of the spectrum (abscissa) is linear in wavenumbers, the degree of absorption of radiation by the substance is expressed in per cent transmittance (ordinate). Most

*) This quantity is colloquially called the intensity (of an absorption band or a Raman line).

commercial infrared spectrometers yield spectra in this form. Figure 9 represents the Raman spectrum of the same sample. The energy scale of the Raman spectrum is again linear in wavenumbers and the intensity of the scattered radiation is expressed on an arbitrary scale.

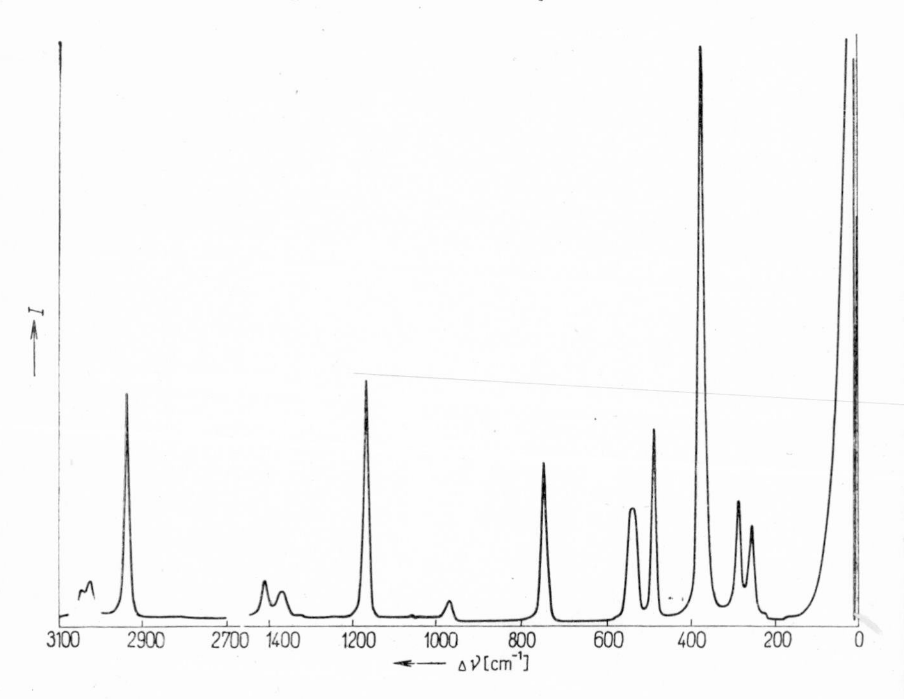

Fig. 9. Raman spectrum of liquid methanesulphonyl chloride. Measured on the Jeol model JRS-SI Raman spectrometer, excited by the blue line of the Ar^+ laser (488 nm).

2.4.1 Spectral Bands and their Description

Generally speaking, infrared and Raman spectra are functions with many extremes. These extremes are called bands in spectroscopy. From the phenomenological point of view the spectra can be considered as a set of spectral lines, which have been broadened due to a number of processes (the Doppler effect, molecular interactions, etc.) to give bands with a finite width. If an individual spectral band can be described by a profile function, P, then the whole spectrum with n bands can be formally transcribed as a superposition of n profile functions P plus the function of the spectral background, P_0. For example, an infrared spectrum expressed by the depend-

Table III

Fractional rational function $P(x) = 1/[1 + (1 - a)\,x^2 + ax^4]$ and the Gauss function $P_g(x)$

a	0.00[a]	0.04	0.09	0.16	0.25	0.36	0.64	1.00	$P_g(x)$
x	$P(x)$								
0.00	1.000 000	1.000 000	1.000 000	1.000 000	1.000 000	1.000 000	1.000 000	1.000 000	1.000 000
0.05	0.997 506	0.997 605	0.997 730	0.997 903	0.998 127	0.998 400	0.999 097	0.999 994	0.998 269
0.10	0.990 099	0.990 487	0.990 973	0.991 654	0.992 531	0.993 605	0.996 349	0.999 900	0.993 092
0.15	0.977 995	0.978 837	0.979 892	0.981 373	0.983 283	0.985 627	0.991 646	0.999 494	0.984 525
0.20	0.961 538	0.962 961	0.964 744	0.967 253	0.970 497	0.974 492	0.984 810	0.998 403	0.972 655
0.25	0.941 176	0.943 257	0.945 871	0.949 555	0.954 334	0.960 240	0.975 610	0.996 109	0.957 603
0.30	0.917 431	0.920 197	0.923 677	0.928 595	0.934 995	0.942 937	0.963 777	0.991 965	0.939 523
0.35	0.890 869	0.894 294	0.898 613	0.904 731	0.912 720	0.922 678	0.949 033	0.985 216	0.918 594
0.40	0.862 069	0.866 083	0.871 153	0.878 352	0.887 784	0.899 591	0.931 113	0.975 039	0.895 025
0.45	0.831 601	0.836 092	0.841 775	0.849 862	0.860 491	0.873 849	0.909 799	0.960 609	0.869 043
0.50	0.800 000	0.804 829	0.810 948	0.819 672	0.831 169	0.845 666	0.884 956	0.941 176	0.840 896
0.55	0.767 754	0.772 762	0.779 113	0.788 183	0.800 159	0.815 300	0.856 557	0.916 165	0.810 846
0.60	0.735 294	0.740 311	0.746 679	0.755 780	0.767 813	0.783 051	0.824 712	0.885 269	0.779 165
0.65	0.702 988	0.707 844	0.714 010	0.722 825	0.734 483	0.749 253	0.789 675	0.848 532	0.746 131
0.70	0.671 141	0.675 674	0.681 427	0.689 648	0.700 513	0.714 267	0.751 844	0.806 387	0.712 025
0.75	0.640 000	0.644 058	0.649 202	0.656 545	0.666 233	0.678 469	0.711 744	0.759 644	0.677 128
0.80	0.609 756	0.613 202	0.617 565	0.623 777	0.631 951	0.642 238	0.669 997	0.709 421	0.641 713
0.85	0.580 552	0.583 267	0.586 698	0.591 569	0.597 951	0.605 942	0.627 280	0.657 028	0.606 046
0.90	0.552 486	0.554 372	0.556 747	0.560 106	0.564 485	0.569 932	0.584 281	0.603 828	0.570 382
0.95	0.525 624	0.526 598	0.527 821	0.529 543	0.531 773	0.534 524	0.541 658	0.551 114	0.534 959

[a] Corresponds to the Cauchy function.

Table III — continued

a	0.00[a]	0.04	0.09	0.16	0.25	0.36	0.64	1.00	$P_g(x)$
x	$P(x)$								
1.00	0.500 000	0.500 000	0.500 000	0.500 000	0.500 000	0.500 000	0.500 000	0.500 000	0.500 000
1.10	0.452 489	0.450 417	0.447 854	0.444 315	0.439 846	0.434 504	0.421 474	0.405 828	0.432 269
1.20	0.409 836	0.405 623	0.400 477	0.393 488	0.384 852	0.374 799	0.351 432	0.325 351	0.368 567
1.30	0.371 747	0.365 411	0.357 788	0.347 636	0.335 399	0.321 565	0.291 010	0.259 329	0.309 927
1.40	0.337 838	0.329 461	0.319 556	0.306 649	0.291 511	0.274 923	0.240 141	0.206 543	0.257 028
1.50	0.307 692	0.297 398	0.285 459	0.270 270	0.252 964	0.234 604	0.198 020	0.164 948	0.210 224
1.60	0.280 899	0.268 836	0.255 140	0.238 153	0.219 375	0.200 092	0.163 508	0.132 387	0.169 576
1.70	0.257 069	0.243 399	0.228 228	0.209 910	0.190 276	0.170 755	0.135 396	0.106 928	0.134 904
1.80	0.235 849	0.220 736	0.204 366	0.185 143	0.165 169	0.145 927	0.112 551	0.086 975	0.105 843
1.90	0.216 920	0.200 526	0.183 218	0.163 465	0.143 564	0.124 969	0.093 984	0.071 265	0.081 900
2.00	0.200 000	0.182 482	0.164 474	0.144 509	0.125 000	0.107 296	0.078 864	0.058 824	0.062 500
2.25	0.164 948	0.145 240	0.126 367	0.106 916	0.089 253	0.074 259	0.052 016	0.037 553	0.029 925
2.50	0.137 931	0.116 788	0.098 009	0.080 000	0.064 712	0.052 459	0.035 398	0.024 961	0.013 139
2.75	0.116 788	0.094 808	0.076 751	0.060 595	0.047 688	0.037 837	0.024 799	0.017 185	0.005 290
3.00	0.100 000	0.077 640	0.060 680	0.046 468	0.035 714	0.027 840	0.017 832	0.012 195	0.001 953
3.25	0.086 486	0.064 092	0.048 419	0.036 071	0.027 164	0.020 866	0.013 122	0.008 884	0.000 661
3.50	0.075 472	0.053 298	0.038 982	0.028 329	0.020 963	0.015 908	0.009 857	0.006 620	0.000 205
3.75	0.066 390	0.044 623	0.031 651	0.022 496	0.016 397	0.012 317	0.007 540	0.005 031	0.000 058
4.00	0.058 824	0.037 594	0.025 907	0.018 051	0.012 987	0.009 671	0.005 862	0.003 891	0.000 015
4.25	0.052 459	0.031 857	0.021 368	0.014 626	0.010 405	0.007 692	0.004 623	0.003 056	0.000 004
4.50	0.047 059	0.027 143	0.017 752	0.011 959	0.008 424	0.006 189	0.003 694	0.002 433	0.000 001
4.75	0.042 440	0.023 244	0.014 848	0.009 862	0.006 888	0.005 033	0.002 986	0.001 961	0.000 000
5.00	0.038 462	0.020 000	0.012 500	0.008 197	0.005 682	0.004 132	0.002 439	0.001 597	0.000 000

Table IV

The sum Cauchy—Gauss function, $P(x) = a/(1 + x^2) + (1 - a) . \exp [-x^2 . \ln (2)]$

a	0.00[a]	0.10	0.20	0.35	0.50	0.65	0.80	0.90	1.00[b]
x	$P(x)$								
0.00	1.00 0000	1.000 000	1.000 000	1.000 000	1.000 000	1.000 000	1.000 000	1.000 000	1.000 000
0.05	0.998 269	0.998 192	0.998 116	0.998 002	0.997 887	0.997 773	0.997 659	0.997 582	0.997 506
0.10	0.993 092	0.992 793	0.992 494	0.992 045	0.991 596	0.991 147	0.990 698	0.990 398	0.990 099
0.15	0.984 525	0.983 872	0.983 219	0.982 240	0.981 260	0.980 281	0.979 301	0.978 648	0.977 995
0.20	0.972 655	0.971 543	0.970 432	0.968 764	0.967 097	0.965 429	0.963 762	0.962 650	0.961 538
0.25	0.957 603	0.955 961	0.954 318	0.951 854	0.949 390	0.946 926	0.944 462	0.942 819	0.941 176
0.30	0.939 523	0 937 314	0.935 104	0.931 791	0.928 477	0.925 163	0.921 850	0.919 640	0.917 431
0.35	0.918 594	0.915 822	0.913 049	0.908 890	0.904 732	0.900 573	0.896 414	0.893 641	0.890 869
0.40	0.895 025	0.891 729	0.888 434	0.883 490	0.878 547	0.873 604	0.868 660	0.865 365	0.862 069
0.45	0.869 043	0.865 299	0.861 555	0.855 938	0.850 322	0.844 706	0.839 089	0.835 345	0.831 601
0.50	0.840 896	0.836 807	0.832 717	0.826 583	0.820 448	0.814 314	0.808 179	0.804 090	0.800 000
0.55	0.810 846	0.806 537	0.802 228	0.759 764	0.789 300	0.782 836	0.776 373	0.772 063	0.767 754
0.60	0.779 165	0.774 778	0.770 390	0.763 810	0.757 229	0.750 649	0.744 068	0.739 681	0.735 294
0.65	0.746 131	0.741 816	0.737 502	0.731 031	0.724 559	0.718 088	0.711 616	0.707 302	0.702 988
0.70	0.712 025	0.707 937	0.703 848	0.697 716	0.691 583	0.685 450	0.679 318	0.675 229	0.671 141
0.75	0.677 128	0.673 415	0.669 702	0.664 133	0.658 564	0.652 995	0.647 426	0.643 713	0.640 000
0.80	0.641 713	0.638 517	0.635 322	0.630 528	0.625 735	0.620 941	0.616 147	0.612 952	0.609 756
0.85	0.606 046	0.603 497	0.600 947	0.597 123	0.593 299	0.598 475	0.585 650	0.583 101	0.580 552
0.90	0.570 382	0.568 592	0.566 803	0.564 118	0.561 434	0.558 750	0.556 065	0.554 276	0.552 486
0.95	0.534 959	0.534 025	0.533 092	0.531 692	0.530 292	0.528 891	0.527 491	0.526 558	0.525 624

[a] Corresponds to the "pure" Gaüss function; [b] corresponds to the "pure" Cauchy function.

Table IV — continued

a	0.00[a]	0.10	0.20	0.35	0 0.50	0.65	0.80	0.90	1.00[a]
x	$P(x)$								
1.00	0.500 000	0.500 000	0.500 000	0.500 000	0.500 000	0.500 000	0.500 000	0.500 000	0.500 000
1.10	0.432 269	0.434 291	0.436 313	0.439 346	0.442 379	0.445 412	0.448 445	0.450 467	0.452 489
1.20	0.368 567	0.372 694	0.376 821	0.383 011	0.389 202	0.395 392	0.401 582	0.405 709	0.409 836
1.30	0.309 927	0.316 109	0.322 291	0.331 564	0.340 837	0.350 110	0.359 383	0.365 565	0.371 747
1.40	0.257 028	0.265 109	0.273 190	0.285 312	0.297 433	0.309 555	0.321 676	0.329 757	0.337 838
1.50	0.210 224	0.219 971	0.229 718	0.244 338	0.258 958	0.273 578	0.288 199	0.297 945	0.307 692
1.60	0.169 576	0.180 708	0.191 840	0.208 539	0.225 237	0.241 936	0.258 634	0.269 767	0.280 899
1.70	0.134 904	0.147 120	0.159 337	0.177 662	0.195 986	0.214 311	0.232 636	0.244 853	0.257 069
1.80	0.105 843	0.118 844	0.131 844	0.151 345	0.170 846	0.190 347	0.209 848	0.222 848	0.235 849
1.90	0.081 900	0.095 402	0.108 904	0.129 157	0.149 410	0.169 663	0.189 916	0.203 418	0.216 920
2.00	0.062 500	0.076 250	0.090 000	0.110 625	0.131 250	0.151 875	0.172 500	0.186 250	0.200 000
2.25	0.029 925	0.043 427	0.056 930	0.077 183	0.097 437	0.117 690	0.137 944	0.151 446	0.164 948
2.50	0.013 139	0.025 618	0.038 097	0.056 816	0.075 535	0.094 254	0.112 973	0.125 452	0.137 931
2.75	0.005 290	0.016 440	0.027 590	0.044 314	0.061 039	0.077 764	0.094 489	0.105 638	0.116 788
3.00	0.001 953	0.011 758	0.021 562	0.036 270	0.050 977	0.065 684	0.080 391	0.090 195	0.100 000
3.25	0.000 661	0.009 244	0.017 826	0.030 700	0.043 574	0.056 448	0.069 321	0.077 904	0.086 486
3.50	0.000 205	0.007 732	0.015 259	0.026 549	0.037 838	0.049 128	0.060 418	0.067 945	0.075 472
3.75	0.000 058	0.006 692	0.013 325	0.023 275	0.033 224	0.043 174	0.053 124	0.059 757	0.066 390
4.00	0.000 015	0.005 896	0.011 777	0.020 598	0.029 419	0.038 241	0.047 062	0.052 943	0.058 824
4.25	0.000 004	0.005 249	0.010 495	0.018 363	0.026 231	0.034 100	0.041 968	0.047 213	0.052 459
4.50	0.000 001	0.004 707	0.009 412	0.016 471	0.023 530	0.030 589	0.037 647	0.042 353	0.047 059
4.75	0.000 000	0.004 244	0.008 488	0.014 854	0.021 220	0.027 586	0.033 952	0.038 196	0.042 440
5.00	0.000 000	0.003 846	0.007 692	0.013 462	0.019 231	0.025 000	0.030 769	0.034 615	0.038 462

ence of absorbance D on wavenumbers $\tilde{\nu}$ can be transcribed as

$$D(\tilde{\nu}) = P_0(\tilde{\nu}) + \sum_{i=1}^{n} P_i(\tilde{\nu}) \tag{2.20}$$

The functions that describe the shape of a band in the spectrum will be called band profile functions, either true, $P(\tilde{\nu})$, corresponding to the actual profile undistorted by instrumental effects, or apparent, $Q(\tilde{\nu})$, distorted by instrumental effects (see 3.3.3.3).

As follows from the simplified theory of collision processes, the profile of an isolated band in the IR spectrum of a substance measured in the condensed phase should be described by the Cauchy (Lorentz) function, namely,

$$P(\tilde{\nu}) = D_{max}/[1 + b_c^2(\tilde{\nu} - \tilde{\nu}_{max})^2], \tag{2.21}$$

where D_{max} is the absorbance at the band maximum, $\tilde{\nu}_{max}$ is the position of the band maximum and b_c is the so-called width parameter of the band, related to the band half-width, $\Delta\tilde{\nu}_{1/2}$, by the relationship

$$b_c = 2/\Delta\tilde{\nu}_{1/2} \tag{2.22}$$

The symmetry of spectral bands is clearly expressed in this function; if the spectral width of the spectrometer slit is substantially smaller (10–20 times) than the half-width of the studied band, function (2.21) can also be used for expressing the apparent profile function, $Q(\tilde{\nu})$.

When collision processes exert negligible influence on the band shape (e.g. with spectra of a gas at an extremely low pressure), the profile function, $P(\tilde{\nu})$, approaches the Gaussian function, expressed e.g. in the form

$$P(\tilde{\nu}) = D_{max} \exp[-b_g^2(\tilde{\nu} - \tilde{\nu}_{max})^2] \tag{2.23}$$

Width parameter b_g is then related to the band half-width by the relationship

$$b_g = 2\sqrt{\ln 2}/\Delta\tilde{\nu}_{1/2} \tag{2.24}$$

This band shape is virtually never encountered in common spectroscopic practice.

The values of the Cauchy (2.21), Gauss (2.23), sum Cauchy–Gauss (2.28) and fractional rational (2.30) functions are tabulated in Tables III and IV, for maximum absorbance values equal to unity. As the independent variable, dimensionless quantity x, defined by

$$x = 2\,|\tilde{\nu} - \tilde{\nu}_{max}|\,/\Delta\tilde{\nu}_{1/2},$$

was used, enabling application of the tables to the description of bands with various half-widths.

In view of the fact that the instrumental function (see 3.3.3.3) can be approximated by the Gaussian function for small spectral slit widths and the band shape by the Cauchy function, the Voigt function was employed for the description of the apparent profile function, defined by

$$Q(\tilde{\nu}) = \int_{-\infty}^{+\infty} \frac{D_{max} \exp\left[-t^2/2\sigma\right]}{\sigma\sqrt{2\pi}\left[1 + b^2(\tilde{\nu} - \tilde{\nu}_{max} - t)^2\right]} \, dt, \tag{2.26}$$

where parameter σ is related to the spectral slit width, s, by the equation

$$s = 2\sigma\sqrt{\ln 4}. \tag{2.27}$$

The Voigt function is identical with the real part of the complex error function. It is relatively difficult to express this function numerically; a generally applicable algorithm can be found elsewhere.*)

This difficulty in numerical computation of the values of the Voigt function prompted search for simpler functions, which would introduce a Gaussian perturbation into the Cauchy function. Jones proposed mixed Gauss–Lorentz curves for the purpose, either of the additive type

$$Q(\tilde{\nu}) = D_{max}\{p \exp\left[-b_g^2(\tilde{\nu} - \tilde{\nu}_{max})^2\right] + (1 - p)/\left[1 + b_c^2(\tilde{\nu} - \tilde{\nu}_{max})^2\right], \tag{2.28}$$

where ($p \in \langle 0, 1\rangle$ gives the degree of Gaussian perturbation of the band), or of the product type

$$Q(\tilde{\nu}) = \frac{D_{max} \exp\left[-b_g^2(\tilde{\nu} - \tilde{\nu}_{max})^2\right]}{1 + b_c^2(\tilde{\nu} - \tilde{\nu}_{max})^2} \tag{2.29}$$

A generalized form of the Cauchy and Gauss function is the profile function defined as a fraction rational function

$$P(\tilde{\nu}) = \frac{D_{max}}{1 + \sum_{i=1}^{n} \frac{1}{i!}\left[b_i^2(\tilde{\nu} - \tilde{\nu}_{max})^2\right]^i}, \tag{2.30}$$

which, in addition to the two mentioned functions, also describes functions whose shapes lie between these two limiting cases (see Fig. 10). At constant $b_i = b_g$, function (2.30) rapidly approaches the Gaussian profile with increasing n; for $n = 5$, deviations of this function from the Gauss function do not exceed 1 % of D_{max}, which is comparable with the experimental error of common measurements.

All the functions discussed above assume axial symmetry of the band profile; experience shows that this condition is never perfectly satisfied by

---

*) Gautschi W.: Algorithm 363. Comm. ACM *12*, 635 (1969).

bands in infrared spectra. Nevertheless, these functions are successfully used for band separation in spectra and for spectra simulation.

The asymmetry of bands is often so great that it must be taken into account. The simplest and most commonly used method is then the use of a profile function composed of two parts; one describes the band shape for $\tilde{\nu} < \tilde{\nu}_{max}$ and the other for $\tilde{\nu} > \tilde{\nu}_{max}$. If functions (2.21) or (2.22) are employed, the band shape is then described by four parameters: the maximum

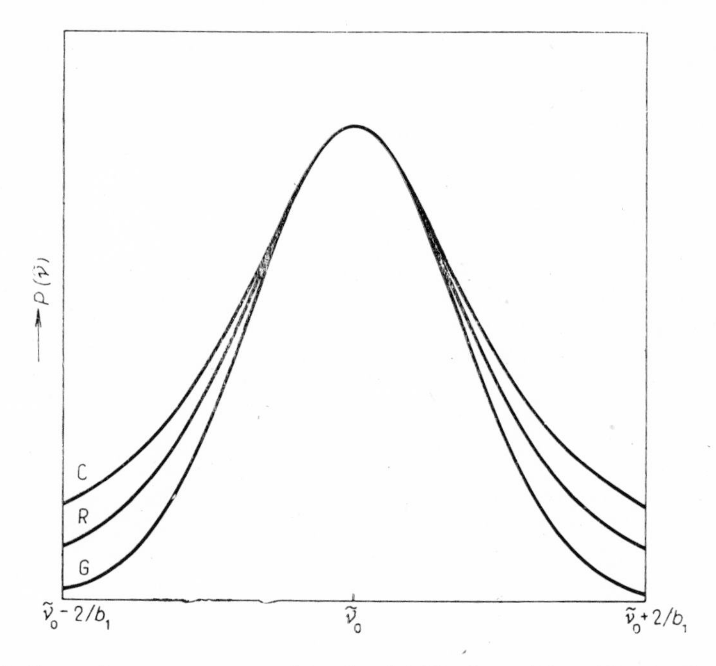

Fig. 10. Comparison of the shapes of the Cauchy (C), Gauss (G) and fractional rational function (R) with coefficients $b_1 = 1$ and $b_2 = 0.5$.

position, $\tilde{\nu}_{max}$, the maximum intensity, D_{max}, and width parameters b_- and b_+, valid for half-profiles towards lower and higher wavenumbers, respectively.

Another method of accounting for the lack of band symmetry is a linear combination of one of the above symmetrical profile functions with its first derivative, P'_{sym}

$$P_{asym}(\tilde{\nu}) = P_{sym}(\tilde{\nu}) + kP'_{sym}(\tilde{\nu}) \tag{2.31}$$

It must be pointed out that the parameter of the symmetrical function,

$\tilde{\nu}_{max}$, no longer signifies the position of the band maximum; D_{max} ceases to represent the maximum band intensity.

Further it should be noted that the lack of band symmetry may also stem from an improperly chosen scale. All bands are strongly non-symmetrical on a linear wavelength scale; symmetrical bands can only be expected on linear wavenumber or frequency scales.

2.4.1.1 Band Parameters

In the previous section, functions suitable for description of band shapes in the whole wavenumber range were discussed. Here attention will be paid to the parameters appearing in these functions, which themselves represent primary spectroscopic quantities.

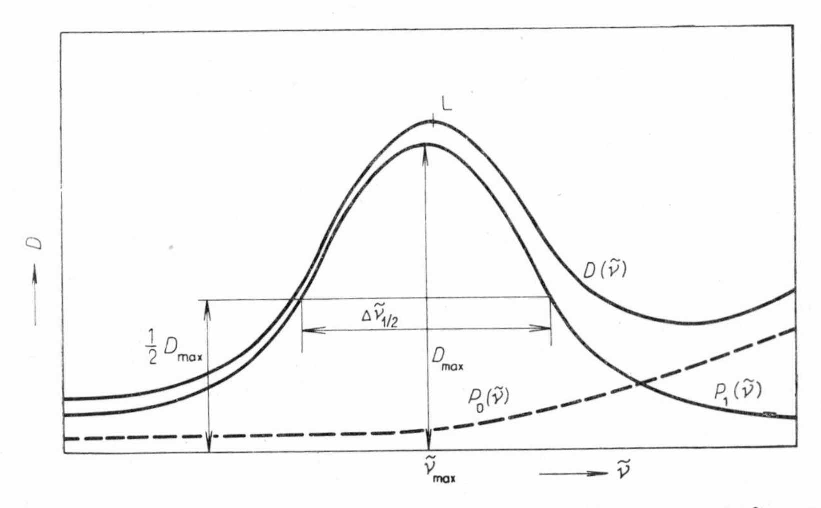

Fig. 11. An isolated spectral band (singlet) and its parameters: $\tilde{\nu}_{max}$, D_{max} and $\Delta\tilde{\nu}_{1/2}$; the non-linear background $P_0(\tilde{\nu})$ of the spectrum is represented by a dashed line. $P_1(\tilde{\nu})$ is the band profile and $D(\tilde{\nu})$ is the shape of the measured spectrum. Point L denotes a local maximum on the spectral curve.

The most important among these quantities is the wavenumber, $\tilde{\nu}$, value, at which the profile function, $P(\tilde{\nu})$, for the given band attains its maximum value, D_{max}. This parameter is called the band (maximum) position and is denoted as $\tilde{\nu}_{max}$ (Fig. 11).

The value of $\tilde{\nu}_{max}$ can be read directly in the spectrum for an isolated band. However, there are relatively few isolated bands in real spectra and increased caution must be exercised with overlapping bands. Overlapping of bands causes shifts of apparent maxima with respect to the true values (see 3.3.4), so that the position of a local maximum in an experimental spectrum need not be identical with the position of the band maximum. With spectra recorded on a linear wavelength scale, the position of the band maximum can be analogously defined by the λ_{max} value, for which profile function $P(\lambda)$ attains a maximum. The maximum position is defined similarly on a frequency scale as ν_{max}.

Another important quantity is the band maximum intensity (band height). The true intensity at the band maximum is the maximum value of the true profile function, $P(\tilde{\nu})$, and is denoted as D_{max}; it clearly follows that

$$D_{max} = P(\tilde{\nu}_{max}) = P(\lambda_{max}) = P(\nu_{max}).$$

The apparent band maximum intensity is analogously defined as the maximum value of the apparent profile function, i.e.

$$D'_{max} = Q(\tilde{\nu}_{max}). \tag{2.33}$$

It must be emphasized that neither of these quantities can be read directly from the spectrum; the absorbance (or intensity) at a local maximum is given by the sum of the contributions from the background and from all the bands in the spectrum (Eq. (2.20)). These effects must be corrected for when the D_{max} or D'_{max} value is determined, as will be described in detail in Section 3.3.4.

Two wavenumber values, $\tilde{\nu}_1$ and $\tilde{\nu}_2$, exist for the true and apparent profile functions, $P(\tilde{\nu})$ and $Q(\tilde{\nu})$, respectively, for which the equation

$$P(\tilde{\nu}_1) = P(\tilde{\nu}_2) = D_{max}/2 \tag{2.34}$$

or

$$Q(\tilde{\nu}'_1) = Q(\tilde{\nu}'_2) = D'_{max}/2 \tag{2.35}$$

is satisfied. In terms of these wavenumbers, quantity

$$\Delta\tilde{\nu}_{1/2} = \tilde{\nu}_2 - \tilde{\nu}_1 \tag{2.36}$$

or

$$\Delta\tilde{\nu}'_{1/2} = \tilde{\nu}'_2 - \tilde{\nu}'_1, \tag{2.37}$$

can be defined; this is called the band half-width. True and apparent band half-widths are differentiated, depending on whether they are derived from the true or apparent profile function.

The wavenumber values, $\tilde{\nu}_1$ and $\tilde{\nu}_2$, are, in addition, also employed for expressing the degree of band assymmetry, q, defined as the ratio,

$$q = (\tilde{\nu}_{max} - \tilde{\nu}_1)/(\tilde{\nu}_2 - \tilde{\nu}_{max}). \tag{2.38}$$

The band half-width is defined on a wavelength scale analogously to the wavenumber scale. It holds that

$$\Delta\lambda_{1/2} = \lambda_1 - \lambda_2 = 1/\tilde{\nu}_1 - 1/\tilde{\nu}_2 \doteq \Delta\tilde{\nu}_{1/2}/\tilde{\nu}^2_{max}. \tag{2.39}$$

A similar relationship can be derived for the band half-width on a frequency scale, where

$$\Delta\nu_{1/2} = c\,\Delta\tilde{\nu}_{1/2}, \tag{2.40}$$

c being the velocity of light.

2.4.1.2 Spectral Background

Function $P_0(\tilde{\nu})$, called the spectral background, appears in Eq. (2.20). In this function are summarized all the contributions to the absorbance which cannot be ascribed to the individual bands, described explicitly by profile functions $P_1(\tilde{\nu})$ to $P_n(\tilde{\nu})$.

The spectral background includes the tails of bands located outside the observed region and very weak bands, especially those of higher harmonics and combination frequencies. Effects arising during the measurement must also be included in the background, e.g. those due to imperfect compensation of the intensities of the reference and measuring beams, the absorbance of the solvent or the cell, scattering or reflection of radiation, etc. In view of the diversity of these contributions, mathematical formulation of function $P_0(\tilde{\nu})$ is very complicated. Some contributions to this function can be determined experimentally, e.g. those arising during measurement. However, we must very often resort to rough approximations using polynomial functions of the type

$$P_0(\nu) = \sum_{i=0}^{m} a_i \,.\, \tilde{\nu}^i. \tag{2.41}$$

With short sections of the spectrum, polynomial (2.41) can usually be limited to the absolute term ($m = 0$).

2.4.1.3 Band Grouping

It has already been pointed out that an isolated band (singlet) is an exception rather than the rule in experimental spectra. Usually bands are grouped in two's (doublets), three's (triplets) or more (generally multiplets). Grouping of bands can be caused by accidental similarity of the wavenumbers of their maxima, but frequently it results from more basic physical laws (splitting of degenerate levels, rotational isomerism, etc.).

From the point of view of formal description, completely and partially resolved and unresolved multiplets can be differentiated. With the former two, local minima can be found on the spectral curve among the individual maxima, separating the multiplet components from one another (Figs. 12A and 12 B). The relative depth of these minima is the criterion of resolution. With unresolved multiplets, their composite character is manifested by ir-

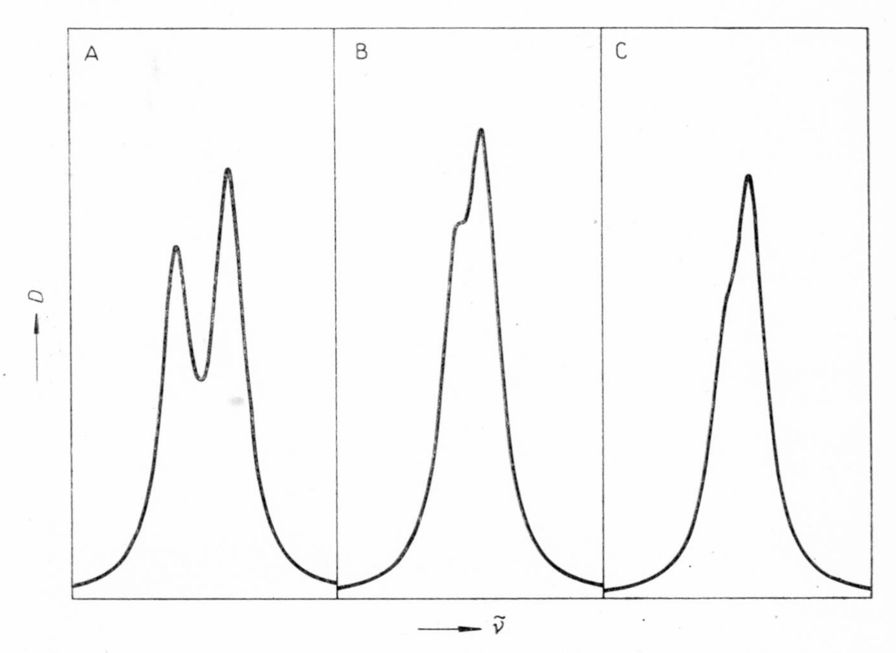

Fig. 12. A pair of bands forming: A — a resolved doublet, B — a partially resolved doublet and C — an unresolved doublet (which we recommend be termed "a band with a satellite".

regularities in the profile (Fig. 12 C) or even by mere asymmetry with respect to the band axis. If one band in a doublet is substantially stronger than the other, they are termed the main band and its shoulder. Band overlapping makes direct reading of the band parameters from spectra difficult. In order to determine true values, the bands must be separated, as described in Section 3.3.4.

2.5 Presentation of Spectra

Spectra are most frequently presented graphically. In these figures the overall character of the spectrum is clearly seen; however, the visual impression is also valuable in evaluation of details of the spectral curve shape. This form of spectrum is mostly obtained directly from the spectrometer and is actually an analog recording.

However, modern spectrometers are sometimes supplemented with an analog/digital converter, which transforms the analog signal (usually voltage), proportional to the transmittance τ or absorbance D of the sample, to digital values for discrete wavenumber values; the digital values are suitably recorded, together with the corresponding wavenumbers. This digitalized spectrum is then represented by a finite number of digital values forming two vectors: the vector of wavenumbers $\tilde{\nu}$ and that of the corresponding transmittances τ or absorbances D. The values can be recorded by an electric typewriter, printer, punched into tapes or cards or be recorded on a magnetic tape for later handling with an off-line computer. Alternatively, a computer can be connected on-line with the analog/digital converter,

Table V

Parameters of the CH_3SO_2Cl bands in the liquid state spectra[a]

Band serial No.	Cauchy function (2.17)				2nd Order polynomial function (2.24)				
	$\tilde{\nu}_{max}$ (cm^{-1})	a_{max}	b_c (cm)	$\Delta\tilde{\nu}_{1/2}$ (cm^{-1})	$\tilde{\nu}_{max}$ (cm^{-1})	a_{max}	b_1 (cm)	$b_2 . 10^3$ (cm^2)	$\Delta\tilde{\nu}_{1/2}$ (cm^{-1})
1	3040.15	0.389	0.105	19.05	3040.14	0.394	0.104	0.0820	19.23
2	3017.63	0.339	0.137	14.60	3017.63	0.333	0.135	5.446	14.51
3	2932.39	0.451	0.154	12.98	2932.39	0.451	0.154	0.027	12.99
4	1410.48	0.064	0.203	9.85	1409.99	0.073	0.146	0.905	13.69
5	1375.76	0.311	0.123	16.26	1375.73	0.283	0.098	7.623	18.24
6	1364.51	0.588	0.097	20.02	1364.37	0.562	0.087	3.550	21.92
7	1321.75	0.165	0.297	6.73	1321.86	0.164	0.249	23.886	7.77
8	1176.64	0.397	0.146	13.70	1176.65	0.402	0.130	0.000	15.38
9	1170.16	0.476	0.182	10.99	1170.17	0.433	0.159	17.946	11.44
10	966.81	0.518	0.173	11.56	966.82	0.488	0.150	15.457	12.18
11	749.50	0.840	0.148	13.51	749.46	0.779	0.120	13.661	14.41

[a] Adapted from the work, Urban Š., Horák M., Vítek A.: *Collection Czechoslov. Chem. Communs. 41*, 3685 (1976).

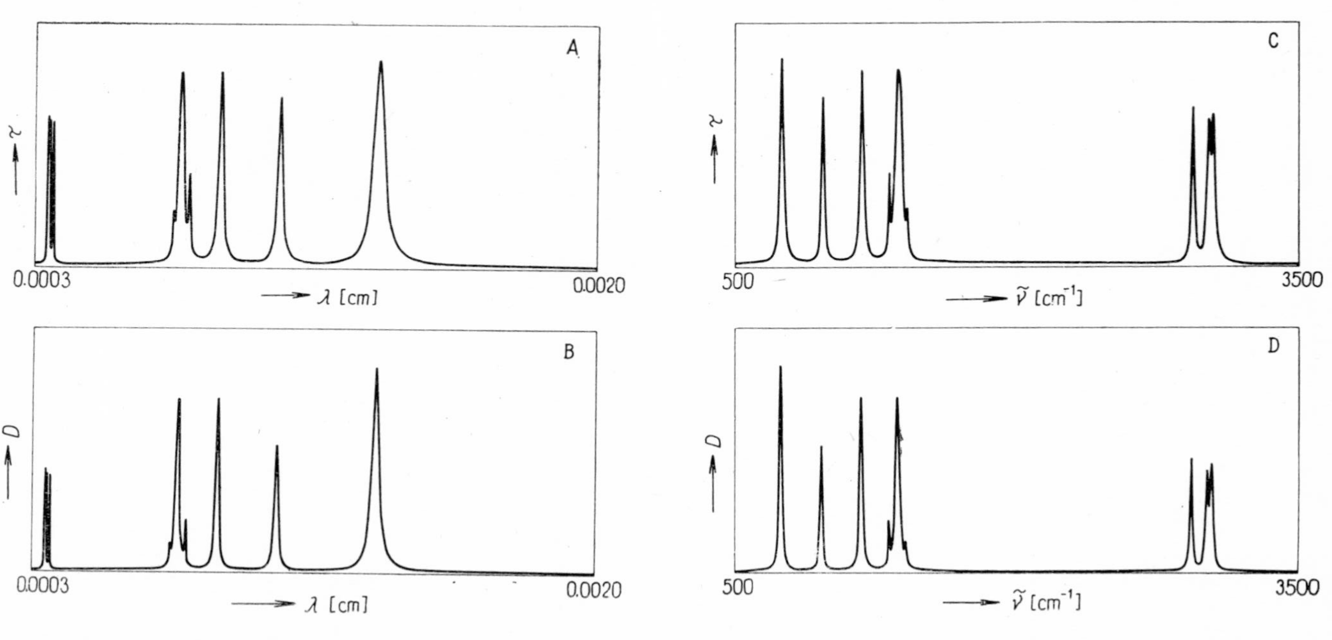

Fig. 13. Methanesulphonyl chloride spectrum simulated by a computer from the data given in the left-hand half of Table V. Simulation was carried out in four variants of linear scales: A — wavelength/transmittance; B — wavelength/absorbance; C — wavenumber/transmittance; D — wavenumber/absorbance.

enabling instantaneous handling of spectra during their measurement. The digital recording is required for operations with profile functions, for band separations, etc.

Methods for band separation in vibrational spectra have recently been developed. If the spectrum is expressed as superposition of a finite number of bands in terms of Eq. (2.20), i.e. of background P_0 and n profile functions P_1 to P_n, information contained in the vibrational spectrum can be transcribed in an exceptionally economical form as a table of the values of these parameters. Table V contains the parameters of bands in the infrared spectrum of methanesulphonyl chloride in a region of 3200 – 400 cm^{-1}. However, the handling of spectra to obtain this form is rather complex and financially demanding.

The spectrum can be reconstructed using a computer and the data in the table of band parameters. The computer-simulated spectrum of methanesulphonyl chloride, on the basis of the parameters from Table V, is given in Fig. 13. Its comparison with the experimental spectrum of methanesulphonyl chloride (Fig. 8) indicates good agreement.

The spectrum was computer-reconstructed in four variants, for combinations of wavenumber and wavelength linear scales and transmittance and absorbance scales. The purpose of this reconstruction is to show the appearance of the spectra of a single substance in all forms which the reader may encounter in practice or which he even can select on the instrument.

Although linear wavenumber scales are now preferred, there are still instruments with linear wavelength scales. On wavelength scales, the short-wave regions in which the characteristic vibrations of functional groups chiefly occur are very dense; especially the region of the stretching vibrations of C–H, N–H and O–H bonds is almost indecipherable. A linear wavenumber scale seems more practically useful in all respects.

Using modern spectrometers, transmittance or absorbance scales can be selected alternatively; the transmittance scale is used for survey spectra, the absorbance scale for quantitative intensity monitoring. The absorbance scale is not suitable for survey spectra chiefly because some bands may be too intense and thus exceed the dimensions of the recording.

Literature

Theory of Vibrational Spectra

1. Herzberg G.: Molecular Spectra and Molecular Structure. I. Diatomic molecules. II. Infrared and Raman Spectra of Polyatomic Molecules. Van Nostrand, Princeton 1945, (new edition 1960).
2. Wilson E. B., Decius J. C., Cross P. C.: Molecular Vibrations. McGraw-Hill, New York 1955.
3. Mayants L. S.: Teoriya i Raschet Kolebanii Molekul. Izdatelstvo Akad. Nauk USSR, Moscow 1960.
4. Malíšek V., Miler M.: Vibrational Spectroscopy. Butterworths, London 1971.
5. Sverdlov L. M., Kovner M. A., Krainov E. P.: Kolebatelnyie Spektry Mnogoatomnykh Molekul, Nauka, Moscow 1970.
6. Steele D.: Theory of Vibrational Spectroscopy. Saunders, Philadelphia 1971.
7. Gans P.: Vibrating Molecules. Chapman and Hall, London 1971.
8. Woodward L. A.: Introduction to the Theory of Molecular Vibrations and Vibrational Spectroscopy. Clarendon Press, Oxford 1972.
9. Volkenshtein M. V., Gribov L. A., Yelyashevich M. A., Stepanov B. I.: Kolebaniya Molekul. Nauka, Moscow 1972.
10. Horák M., Papoušek D. and co-workers: Infrared Spectra of Molecules. ČSAV Publishing House, Prague 1976 (in Czech).

Raman Spectroscopy

1. Kohlrausch K. W. F.: Ramanspektren, in Hand- und Jahrbuch der chemischen Physik, Vol. 9/VI. Becker & Erler, Leipzig 1943.
2. Brandmüller J., Moser H.: Einführung in die Ramanspektroskopie. Steinkopf, Darmstadt 1961.
3. Otting W.: Der Raman-Effekt und seine analytische Anwendung. Springer, Berlin 1962.
4 Szymanski H. A. (ed.): Raman Spectroscopy, Theory and Practice. Plenum Press, New York 1967.
5. Schüler C. J.: Laser Induced Spontaneous and Stimulated Raman Scattering. Pergamon Press, New York 1968.
6. Gilson T. R., Hendra P. J.: Laser Raman Spectroscopy. Wiley, New York 1970.
7. Anderson A. (ed.): The Raman Effect. Vol. 1, Principles. Dekker, New York 1971.
8. Tobin M. C.: Laser Raman Spectroscopy. Wiley, New York 1971.
9. Koningstein J. A.: Introduction to the Theory of the Raman Effect. Reidel, Dordrecht 1972.
10. Schrader B.: Raman und Infrarotspektroskopie. Akad. Verlagsgesellschaft, Frankfurt 1973.

General Treatment

1. Brügel W.: Einführung in die Ultrarotspektroskopie. Steinkopf, Darmstadt 1957.
2. Bauman R. P.: Absorption Spectroscopy, Wiley, New York 1962.

3. Barrow G. M.: Introduction to Molecular Spectroscopy. McGraw-Hill, New York 1962.
4. Waker S., Straw H.: Spectroscopy, Vol. II., Ultraviolet, Visible, Infrared and Raman Spectroscopy. Chapman and Hall, London 1962.
5. Meloan C. E.: Elementary Infrared Spectroscopy. Macmillan, New York 1963.
6. Davies M. (ed.): Infrared Spectroscopy and Molecular Structure. Elsevier, Amsterdam 1963.
7. Kendall D. N. (ed.): Applied Infrared Spectroscopy. Reinhold, New York 1966.
8. Alpert N. L., Keiser W. E., Szymanski H. A.: Theory and Practice of Infrared Spectroscopy. Plenum Press, New York 1970.
9. Brittain E. F. H., George W. O., Wells C. H. J.: Introduction to Molecular Spectroscopy. Academic Press, New York 1970.
10. Chang R.: Basic Principles of Spectroscopy. McGraw-Hill, New York 1971.
11. Banwell, O. N.: Fundamentals of Molecular Spectroscopy. McGraw-Hill, New York 1972.
12. Conley R. T.: Infrared Spectroscopy. Allyn and Bacon, Boston, 1972.
13. Ramsay D. A. (ed.): Spectroscopy. Butterworth, London 1972.
14. Rao K. N., Mathews C. W. (eds): Molecular Spectroscopy: Modern Research. Academic Press, New York 1972.
15. Levine I. N.: Molecular Spectroscopy. Wiley, New York 1974.

Special Aspects of Vibrational Spectroscopy

1. Gribov L. A.: Intensity Theory for Infrared Spectra of Polyatomic Molecules. Plenum Press, New York 1964.
2. Mayants L. S., Averbuch B. S.: Teoriya i Raschot Intensivnostei v Kolebatełnykh Spektrakh Molekul. Nauka, Moscow 1971.
3. Möller K. D., Rothschild W. G.: Far-Infrared Spectroscopy. Wiley, New York 1971.
4. Kimmitt M. F.: Far-Infrared Techniques. Pion Ltd., London 1970.
5. Finch A., Gates P. W., Radcliffe K., Dickson F. N., Bentley F. F.: Chemical Applications of Far-infrared Spectroscopy. Academic Press, New York 1970.
6. Robinson L. C.: Physical Principles for Far-Infrared Radiation. Academic Press, New York 1973.
7. Cyvín S. J.: Molecular Vibrations and Mean Square Amplitudes. Universitetsforlaget, Oslo 1968.
8. Steel W. H.: Interferometry. Cambridge University Press, Cambridge 1967.
9. Cottrell T. L.: Dynamic Aspects of Molecular Energy States. Oliver and Boyd, Edinburgh 1965.
10. Allen H. C. Jr., Cross P. C.: Molecular Vib-Rotors. Wiley, New York 1963.
11. Amat G., Nielsen H. H., Tarrago G.: Rotation-Vibration of Polyatomic Molecules. Dekker, New York 1971.
12. Denney R. C. (ed.): A Dictionary of Spectroscopy. Macmillan, London (1973).

CHAPTER 3

Measurement and Handling of Vibrational Spectra

This chapter surveys the methods for obtaining accurate values of wavenumbers and band-maximum intensities, integrated intensities and of some other principal spectroscopic quantities from infrared and Raman spectra. In order to understand the required operations, some aspects of the measuring technique and sample preparation must be known; these, however, will be treated in this Chapter only from the point of view of distortions they may cause in the experimentaly measured spectra.

3.1 Measurement of Spectra

In the first section of Chapter 3, instruments for the measurement of infrared and Raman spectra, sample pretreatment and errors arising during these operations will be discussed.

3.1.1 Instrumentation

Vibrational spectra are measured on instruments called spectrometers or spectrophotometers. Historically, instruments permitting the determination of wavenumbers (or wavelengths) were termed spectrometers and instruments which could also determine absorption intensities were called spectrophotometers. Practically all contemporary instruments belong in the second category; the terms spectrometer and spectrophotometer are no longer strictly differentiated and consequently the briefer term spectrometer will be used in this book for all instruments.

The world market is saturated with a great variety of commercial infrared and Raman spectrometers from many manufactures. Although commercial prices are relatively high, they are no longer constructed in the laboratory. Even instruments for most specialized operations can be found on the market and adapted for particular tasks if necessary. Only instruments with extremely high resolution or very short measuring periods are exceptions; however, these will not concern us here and consequently all the problems discussed will be related to commercial spectrometers.

The quality of any spectrometer is evaluated according to its maximum efficiency. Correct instrument operation is, of course, assumed; this condition is easy to meet, as all manufacturers inform the customer about correct instrument operation in instrument manuals and at periodical specialized training courses.

Modern infrared spectrometers fall roughly into three classes. The first class involves instruments for routine laboratory work and chiefly for analytical tasks. These are small, portable spectrometers with a limited operational range and are the cheapest among the instruments available; consequently, they are also often employed as university teaching aids. For more involved work, such as the study of molecular structures or other physico-chemical applications, a number of medium-quality instruments are available. Top-quality instruments, suitable for solving tasks of chemical physics, measuring highly resolved spectra of molecules, etc., can also be bought.

The prospective customer must choose carefully and consider the price of the instrument according to the "quality" of the information required and the type of tasks to be solved by infrared spectroscopy. Recently, the willingness and capability of the manufacturer to provide maintenance and repairs for these relatively complicated instruments has become a factor to be considered. So far medium-quality spectrometers have predominated in chemical laboratories; however, it seems that small spectrometers, provided recently with grating monochromators, are spreading rapidly.

The selection of commercial Raman spectrometers is less wide than that of infrared spectrometers, from the point of view both of quality and of price. Medium quality instruments vastly predominate.

3.1.1.1 Infrared Spectrometers

The basic elements of any infrared spectrometer are a source of infrared radiation, a monochromator unit, a detector with an amplifier and a recording device, connected as depicted in the block scheme in Fig. 14. The possibilities of infrared spectroscopy have been limited by difficulties in direct generation of monochromatic infrared radiation. Semiconductor diodes operating as tunable laser sources of infrared radiation have been developed in the last few years, but have found use rather in the construction of high-resolution spectrometers. Therefore, common commercial spectrometers still employ non-specific sources yielding a continuous spectrum with infrared wavelenghts predominating; the individual components are separated in the monochromator.

As sources of infrared radiation, globars (heated silicon carbide rods), Nernst heaters (heated ceramic rods coated with rare earth oxides), tungsten filaments and sometimes carbon are used. The radiant power of

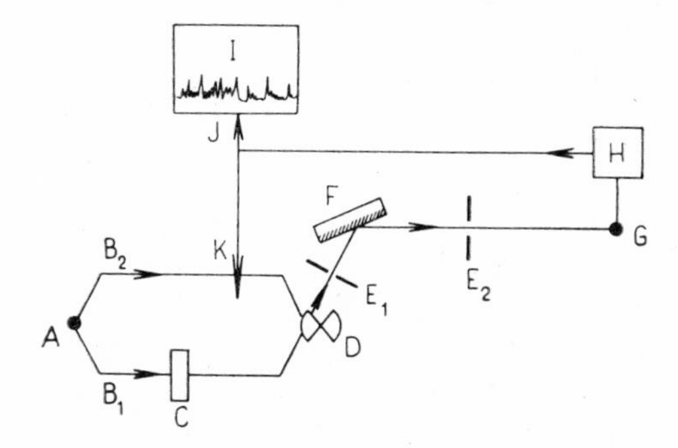

Fig. 14. Block scheme of an infrared spectrometer: A — source, B_1 — measuring beam, B_2 — reference beam, C — cell with the sample, D — chopper (rotating segment), E_1 and E_2 — input and output slit, respectively, F — grating, G — detector, H — amplifier, I — recorder, J — recorder pen, K — attenuating screen.

these sources is distributed similarly to the energy distribution of black-body radiation (Fig. 15); hence, the intensity attains a maximum in the near infrared region and decreases with decreasing wavenumber. With increasing source temperature, the maximum of the curve shifts unsuitably for our purposes, namely, to the visible spectral region.

If the polychromatic source radiation passes through a sample of the compound studied, components with certain wavenumbers are absorbed to a greater or lesser degree. The radiation attenuated by absorption enters the spectrometer monochromator unit, is dispersed into the individual components and these are detected. The optical-mechanical system of spectrometers is usually constructed so that monochromatic radiation with wavenumbers linearly varying in time is incident on the detector. The recording device then records the spectrum in a linear wavenumber scale.

Practically all modern (double-beam) spectrometers employ the principle of so-called optical zero. The source radiation is divided into two equivalent beams, of which one passes through the sample and the other is used as a reference. The two beams enter the monochromator alternately, with a period determined by a rotating mirror chopper. The detector alternately detects signals from one and the other beam, their difference is amplified and controls insertion of an aperture into the reference beam, so that the sample absorption is just compensated. Then identical signals fall on the detector and the instrument does not record any change; only a change in

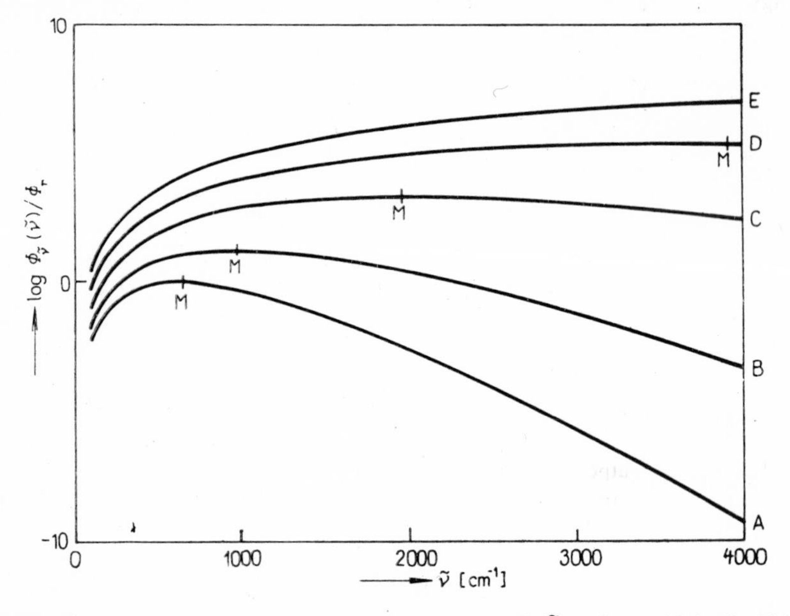

Fig. 15. Distribution of the spectral radiant power, $\Phi_{\tilde{\nu}}(\tilde{\nu})$, of an absolutely black body in dependence on the temperature: A — 330 K, B — 500 K, C — 1000 K D—2000 K, E—5000 K. M—curve maxima, Φ_r — the reference spectral radiant power.

the sample absorption causes a further operation. If the aperture movement is coupled with the movement of the recorder pen, the spectrum is recorded in relative intensities, e.g. in terms of transmittance.

Current development in infrared spectral measuring techniques is characterized by on-line coupling of spectrometers with computers, by means of an analog/digital converter. The spectrometers then operate faster and more accurately. Coupling with a computer leads to suppression of noise, permits work with weaker signals (i.e. with smaller sample amounts) and enables corrections of wavenumber and photometric spectrometer scales. These instruments are particularly well suited for quantitative analytical determinations, as the photometric error is decreased.

3.1.1.2 Raman Spectrometers

In measurement of Raman spectra, the sample is irradiated with a beam of monochromatic radiation in the visible spectral range, which must not be absorbed by the sample. The frequency modulated scattered radiation propagated in all directions is most frequently monitored at right angles to the axis of the incident radiation. The radiation is led into a monochromator unit, dispersed, detected and recorded. A block scheme of a Raman spectrometer is given in Fig. 16.

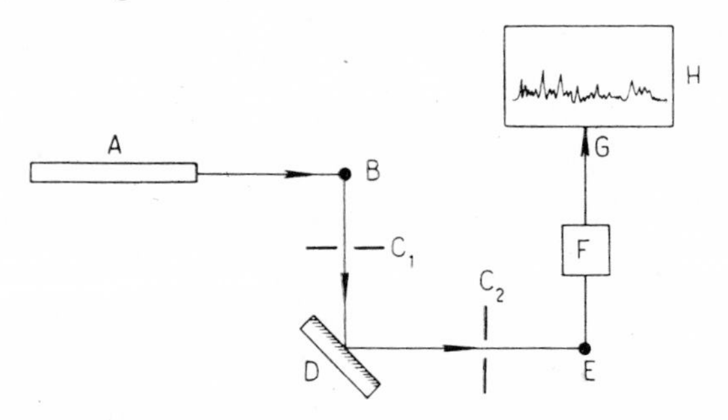

Fig. 16. Block scheme of a Raman spectrometer: A — laser source of excitation radiation, B — cell with the sample, C_1 and C_2 — slits, D — grating, E — detector, F — amplifier, G — recorder pen, H — recorder.

The source of radiation, this time monochromatic in the visible spectral region, is again the limiting component of Raman spectrometers. Raman spectra were originally excited using mercury discharge lamps, which, after filtering, yielded just sufficient radiant power. However, the intensity of the scattered radiation, especially of the Raman lines, was often very low, so that photographic plates were usually employed for the detection; even weak lines could then be discovered using longer exposures.

Soon after the discovery of lasers, the ruby and later the helium-neon laser were used as sources for Raman spectrometers; the latter completely replaced the old excitation sources. Intense laser sources permitted the replacement of the photographic plate by photoelectric recording of Raman spectra. Therefore, modern Raman spectrometers yield spectra virtually identical with infrared spectra in a linear wavenumber scale.

The wavelengths and wavenumbers of lines produced by laser sources most frequently employed for excitation of Raman spectra are summarized in Table VI. Certain parameters of the mercury discharge lamp line used earlier are also given for the sake of comparison. Sources emitting two (or more) lines yield double (multiple) Raman spectra. The radiation from these sources must therefore be filtered in order that a single exciting line be present and errors in spectrum interpretation be avoided. Some multi-line sources (e.g. the argon-ion laser) can be employed for excitation at various wavelengths, selected by filtering off the other lines.

Raman spectroscopy is carried out in the visible spectral region; the choice of transparent materials is thus considerably simplified. While infrared spectrometers require mirror optical components and transparent components made of special materials (see Section 3.1.2.1), quartz and glass can be used in Raman spectrometers.

Table VI

Wavelengths and wavenumbers of the lines of the laser excitation sources employed in Raman spectroscopy

Laser active medium	Line used			
	wavelength λ (nm)	wavenumber $\tilde{\nu}$ (cm^{-1})	output P (W)	colour
Helium-neon	632.8	15 800	0.08	red
Argon	514.5	19 430	1.0	green
	488.0	20 490	1.0	blue
Krypton	647.1	15 450	0.5	red
	568.2	17 590	0.5	green
Cadmium	441.6	22 640	0.2	blue
Mercury[a]	435.8	22 940		blue

[a] The line of the previously used mercury discharge lamp is given for the sake of comparison.

3.1.2 Sample Adjustment for Measurement of Infrared Spectra

Samples of compounds must mostly be adjusted before the measurement of the infrared spectrum. Only exceptionally are the spectra of compounds measured directly; these are always spectra of solids, especially foils of plastics which are fixed in a frame or single-crystals, fixed on a goniometric table. In all other experimental techniques the sample is placed on a support, in a cell, is dissolved or pressed with a medium to form a pellet. The resultant spectrum then contains the absorption bands of the materials of the supports, cell windows, solvents and media, in addition to the absorption of the test compound. All inorganic and organic materials absorb in the infrared spectral region and their use in sample adjustment imposes a greater or smaller limitation on the spectral range of the test compound.

3.1.2.1 Materials Transparent for Infrared Radiation

The most common optical material, glass, has little importance in the infrared region. Various kinds of glass are transparent for near infrared radiation, but strongly absorb from the upper boundary of the fundamental infrared region. Vacuum-fused quartz is somewhat more advantageous, as it is transparent down to the 2300 cm^{-1} region. Hence fused-quartz cells included among common ultraviolet spectrometer accessories are well-suited for the near-infrared and the adjacent part of the fundamental region (their greater thickness is also advantageous).

Among materials best suited for work in the infrared region, sodium chloride should be mentioned first and further the other alkali metal halides and even the halides of the alkaline earths. Windows for cells and other parts through which infrared radiation is to pass are cut from artificially grown single-crystals of these materials; consequently, they are rather expensive.

The parameters of the materials employed most frequently are summarized in Table VII. The basic characteristic of each material is the optimum useful region, within which the materials are most transparent and meet certain other requirements. If necessary, the useful range can be

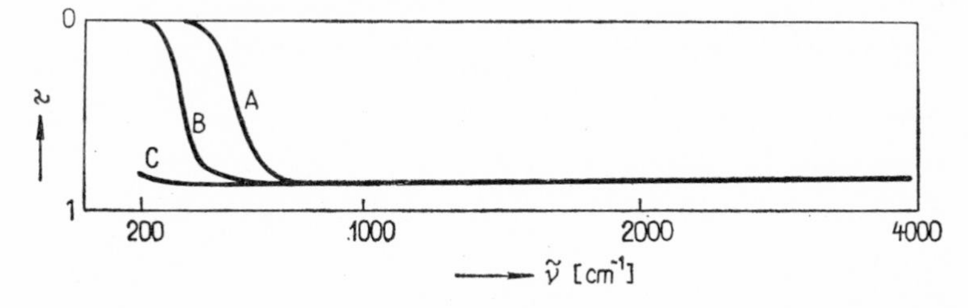

Fig. 17. Transparence of windows made of materials transparent for infrared radiation: A — NaCl (0.4 cm thick); B — KBr (0.4 cm), C — CsI (0.45 cm).

Table VII

Materials transparent to infrared radiation

Material	Optimum useful region (cm^{-1})	Melting point °C	Refractive index[a] n	Loss through reflection %	Solubility in water (g/100 ml, 0 °C)	Solubility in other media
CaF_2[b]	75 000—1000	1360	1.40	5.6	—[d]	in NH_3, in salt solutions
BaF_2[b,c]	75 000—830	1280	1.45	7.7	—[d]	in NH_3, in salt solutions
NaCl[b]	40 000—625	801	1.52	7.5	35.7	sparingly in ethanol
KBr[b]	40 000—330	730	1.53	8.4	53.5	in ethanol, in ether
AgCl	10 000—450	458	2.00	19.5	—[d]	in solution of NH_4Cl and strong bases
KRS-5[e]	10 000—260	414	2.38	28.4	—[d]	in bases, but not in acids
CsBr[f]	10 000—250	636	1.67	24.8		in acids
CsI[f]	500—170		1.74	26.0	44.0	

[a] Calculated for two surfaces; [b] materials unsuitable for cryogenic experiments, as they break when subjected to large temperature changes; [c] material suitable for measuring the spectra of ionic salts in nujol mulls, as it does not exchange the cation or the anion; [d] practically insoluble materials, suitable for measuring the spectra of water and aqueous solutions; [e] composition 42% TlI + 58 % TlBr, poisonous!; [f] materials with a large degree of plasticity, easily deformed mechanically.

extended above the upper limit, but not under the lower limit, which is determined by the rather steep absorption edge. The transparence in the infrared region is given in Fig. 17 for the most commonly used materials, NaCl, KBr and CsI.

Some materials employed more rarely are not listed in Table VII. For example, LiF, which was often used as a material for spectrometer prisms and which is transparent only down to 1450 cm^{-1}, is not mentioned. The so-called irtrans should be briefly mentioned: irtran 1 (MgF_2) and irtran 5 (MgO) are transparent down to 1250 cm^{-1}, irtran 2 (ZnS) down to 700 cm^{-1} and irtran 4 (ZnSe) down to 500 cm^{-1}; all these materials are insoluble in water. Irtran 3 is CaF_2 given in Table VII. Germanium has limited use, chiefly because of its reflectivity, which is 53% on two surfaces, due to the high refractive index of germanium ($n \doteq 4$ for 1000 cm^{-1}). Germanium cells are employed in the study of the refractive indices of liquids and of specific problems of interference of infrared radiation in liquids.

A rather special position is occupied by high-pressure polyethylene without plasticizers, which is employed as foils or plates for the construction of cells for the far-infrared region, between 300 and 10 cm^{-1}.

The data on materials for cell-windows given in Table VII serve for assessment of whether a material is suitable for measurements on aqueous solutions, for cryogenic purposes, etc. The reflectivity of materials is also specified; the reflectivity coefficient R can be calculated from the formula

$$R = [(n - 1)/(n + 1)]^2 \qquad (3.1)$$

where n is the material refractive index.*)

When analyzing infrared spectra, the material of the cell windows and other materials involved in the spectrum measurement must be specified in order to be able to determine the region in which the spectrum can be reliably evaluated.

3.1.2.2 Pretreatment of Samples of Gases and Vapours

Samples of gaseous compounds are readily adjusted for measurement of infrared spectra. They are transferred to an evacuated gas cell and the sample is prepared for the measurement.

---

*) The complex refractive index, expressing the difference in the refractive indices of the two media at the interface at which the reflection occurs, should be used. However, considering the refractive index values for solids, the refractive index of air can be considered to equal unity, resulting in only a small error.

Gas cells are glass or metallic tubes with one or two valves, provided with windows made of materials transparent for infrared radiation on both ends. The windows are either cemented to the cell body (e.g. with picein), or, more frequently, only mechanically pressed onto it, placing silicone rubber O-rings between the window and the body. An advantage of this technique of fixing the windows is easier removal or replacement.

Standard cells are usually constructed so that the optical path of the beam through the gas is 10 cm. However, the spectrum of a gaseous sample can rarely be measured using a single sample in a cell with a constant optical path length. If the pressure of the gas in the cell equals the atmospheric pressure, only very weak bands can be recorded; the absorption of radiation in the other regions is too large. Therefore the pressure in the cell must be decreased somewhat, the suitable bands are measured and the pressure must further be decreased until even the strongest bands in the spectrum are recorded. The spectra of gases are usually presented with the individual regions recorded at various pressures.

For these reasons, the multireflection cell has been developed for measurements on gases; the optical path length for the beam can be varied in this cell in dependence on the number of reflections of the beam in the cell mirror system (called the White mount). If the entrance mirror is removed from the optical system, the cell can be used for direct beam passage, with an optical path length equal to that of a standard cell, i.e. 10 cm. The number of reflections in the mirror system can gradually be varied by lengthening the optical path in steps. The advantage of these cells is evident: the cell is filled with the gas at the lowest possible pressure which permits the measurement of the strongest bands in the spectrum with direct beam passage. The beam path length is gradually increased and less and less intense bands are measured. The whole spectrum of the gaseous sample can usually be measured using a single cell filling.

The spectra of vapours of compounds which are liquid or solid under laboratory temperature and atmospheric pressure must often be measured. Samples of such compounds with low vapour presures are placed in a standard cell before evacuation or are injected into it after evacuation. If the vapour pressure is insufficient for meassurement of the infrared spectrum, the cell with the sample must be heated. This can be done rather simply, but uniform heating of the whole cell including the windows is required. Otherwise, the vapours condense on the colder parts; if condensation occurs on the windows, then gas spectrum recording is prevented by steaming or crystal deposition. The resultant spectrum is the sum of the gas and the liquid or solid spectra and the spectral background increases considerably owing to scattering of the radiation on the deposited sample particles.

The multireflection cell can also be successfully employed for measurement of the spectra of compounds with low vapour pressures. The long optical path enables work at lower pressures. If polar substances are involved, work at lower pressures is an additional advantage, as collision interactions in the gaseous state and complex formation*) are supressed.

Measurement of the spectra of gases is technically relatively simple. Both standard and mutlireflection cells are readily available and cell evacuation and the pressure measurement are operations readily performable in the chemical laboratory. There are only a few aggressive gases attacking the materials of the cell mantle and windows and they are generally well-known. Only measurement of the spectra of compounds of low volatility is a more complicated task; however, if a commercial heated cell or a multireflection cell is available, this can also be carried out routinely.

3.1.2.3 Liquid Sample Pretreatment, Solutions

If the greater variability in the properties of liquids compared with gases is considered (polarity, viscosity, solubility properties, etc.), it can be seen that greater attention must be paid to the adjustment of liquid compound samples.

The treatment of viscous liquids with low vapour pressures is very simple. The infrared spectra can be measured on very thin films between the surfaces of two transparent plates. The drop is placed on one plate and the film is formed by gentle pressing of the other plate onto it. The surfaces of the transparent windows should always be optically planar and well polished, so that the films between them are thin. The two windows with the film between them are placed in a holder and the holder is fixed in the spectrometer cell space.

The cell window material is selected in dependence on the spectral region width required for the measurement. The cheapest and most stable materials should be used, i.e. chiefly sodium chloride; only when the spectrum is to be measured in a region in which NaCl is not transparent is the sample placed between transparent windows of another material. Sodium chloride is most frequently combined with cesium iodide or polyethylene.

---

*) When working with aggressive gases, the possibility of their chemical reaction with mirror surfaces must be borne in mind.

A certain limitation of the measuring technique using liquid films lies in the impossibility of defined variation and determination of the film thickness. If these factors are important, then the measurement must be carried out in cell with distance foils, which determines the cell thickness more accurately.

The measuring technique involving films of liquids is well suited for liquids with low vapour pressures, which do not volatilize from the cell (which is not perfectly closed) during the measurement. The temperature in the cell space of infrared spectrometers with high-power sources of infrared radiation reaches 30 – 50 °C; even liquids of medium volatility volatilize from a cell without a distance foil during the time required for the spectrum measurement at these temperatures. Hence cells with distance foils are also employed because of liquid volatility.

There are two types of cells with distance foils, those which can be demounted and fixed ones. Work with demounted cells is analogous to the preparation of liquid films, but the distance foil is placed on the plate before applying the sample drop. These foils are usually made of lead, as this material adheres well to the window surfaces. However, gold distance foils must be employed for very thin cells, as lead foils thinner than cca 0.04 mm are easily torn. Gold foils are rather rigid and adhere on pressing between the two windows. The advantages of distance foils lie in their maintenance of the cell thickness and at least partial suppression of liquid volatilization from the cell. However, take-apart cells with distance foils are unsuitable for volatile liquids.

Therefore, the most frequently used liquid cells are cemented cells with a fixed width. These are made of two windows with a distance foil cemented on their inner surfaces; lead foils, or gold ones for thin cells, are again used most frequently and are cemented after foil amalgamation with mercury. The foil adheres perfectly to the polished window surfaces, so that even liquids with very low boiling points can be retained in these cells for the time necessary for recording their spectra. One of the windows has two holes drilled in it for filling the cell with the sample; after fixing the windows in the cell holder, the holes are hermetically sealed with lead or teflon stoppers.

Cemented cells with a fixed thickness are employed for measuring the spectra of liquids in the limits determined by the transparence of the window materials. The same rule holds as for primitive cells: windows made of the cheapest and most stable materials, i.e. NaCl and KBr, are employed and rarer materials are selected only when necessary. Similar to gases, only rarely can the whole spectrum of a liquid be measured with a single cell filling; in order to "enhance" weak bands, thicker cells are employed, while thin cells serve to suppress very intense bands (Fig. 18). Hence a series of cemented cells with fixed thicknesses is required. The individual cells are differentiated by the window material and the thickness. The thickest cells employed are of the order of tenths of a centimetre,*) the thinnest about

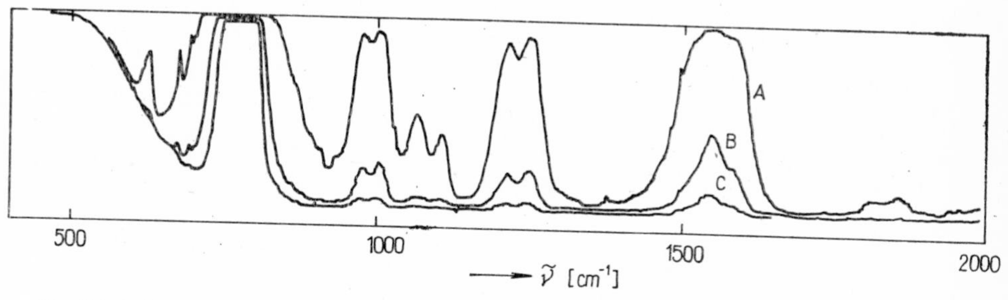

Fig. 18. Infrared spectrum of liquid tetrachloromethane in cells of various thickness: A — 0.1 cm, B — 0.01 cm, C — 0.001 cm.

0.002 cm. Sets of cells with thickness increasing by 0.002 or 0.02 cm are commercially available. The most common thickness is 0.01 cm.

Liquid cells are thus rather thin. The thinnest ones are difficult to fill with viscous liquids and it is even more difficult to completely remove these liquids from fixed cells.

A great danger which must be borne in mind in measuring the spectra of liquids is possible damage to the cell window surfaces. Cells with etched or partially dissolved inner window surfaces exhibit increased scattering and, moreover, the layer thickness between them is not constant. The damaged cells must thus be taken apart (amalgamated foils are loosened by treatment with warm mercury) and the windows are re-ground, repolished and cemented together again. The whole operation is rather tedious and very few experimenters can grind and polish the windows satisfactorily. Therefore the cell window material must be selected carefully before each measurement. Difficulties are encountered in measurement of spectra of damp liquids; dissolved water can destroy the window surfaces completely even if present in a low concentration, although the solvent alone does not attack the

*) For special measurements, especially for the study of associating compounds and for spectral measurements in the near-infrared region, cells with longer optical paths, up to 10 cm, are employed.

windows. The possibility of damage to cell window surfaces made of various materials is pointed in Table VII. Measurements on some reactive liquids or solutions, which may damage the cell window material chemically are also difficult. For example, acid chlorides and peroxides react with CsI, sulphate solutions etch the surface of BaF_2 plates which are otherwise very resistant, etc.

Variable-thickness cells have also been developed for measurements on liquids. Their thickness can be finely adjusted by turning a micrometer screw. Although the construction of these cells is not particularly complicated, still work with them is more difficult than with analogous gas cells. The main difficulty lies in damage to the window internal surfaces and in variations of the actual cell thickness during use. Hence variable-thickness cells are employed only in quantitative analysis and in the measurement of difference spectra; they are most often filled with inert liquids or solvents. Their importance in laboratory practice is limited.

Solutions are treated analogously to liquids; their spectra are measured, with very few exceptions, in cemented cells of fixed thickness. Measurements on solutions require pairs of cells: the solution is placed in the "measuring" cell and the pure solvent in the "reference" cell, in order to compensate for the solvent absorbance.

There are several reasons for studying the spectra of compounds in solutions. First of all this is the easiest method of measuring the spectra of solids. Further it is well suited for studying the spectra of strongly polar and associated compounds (both liquids and solids) which exhibit strong intermolecular interactions in the condensed phase. Quantitative analyses are carried out almost exclusively in solutions.

Any liquid is suitable as a solvent for measurements in the infrared region, provided that it simultaneously satisfies three conditions, namely, is sufficiently transparent for infrared radiation, does not react either with the solute or with the cell window surface and dissolves the chemical compound samples well. These criteria, however, are to a certain degree contradictory and therefore only three solvents satisfy them, carbon disulphide, tetrachloromethane and chloroform. Exceptionally, paraffin oil can be used (it is more frequently used for the preparation of mulls, but is also suitable as a solvent for compounds with low polarity). The infrared spectra of these solvents in the range from 4000 to 200 cm^{-1} are given in Fig. 19; it is evident that none can be employed in the whole spectral region, so that they must be combined. The figure demonstrates the varying solvent usefulness with varying cell thickness; with increasing thickness, the regions of strong absorption broaden, i.e. the regions in which the liquid is applicable as a spectroscopic solvent become narrower.

For specific purposes, when the spectra are measured in narrow spectral regions, liquids which strongly absorb in other regions can be employed as solvents. For example, tetrachloroethylene and n-hexane are used for study-

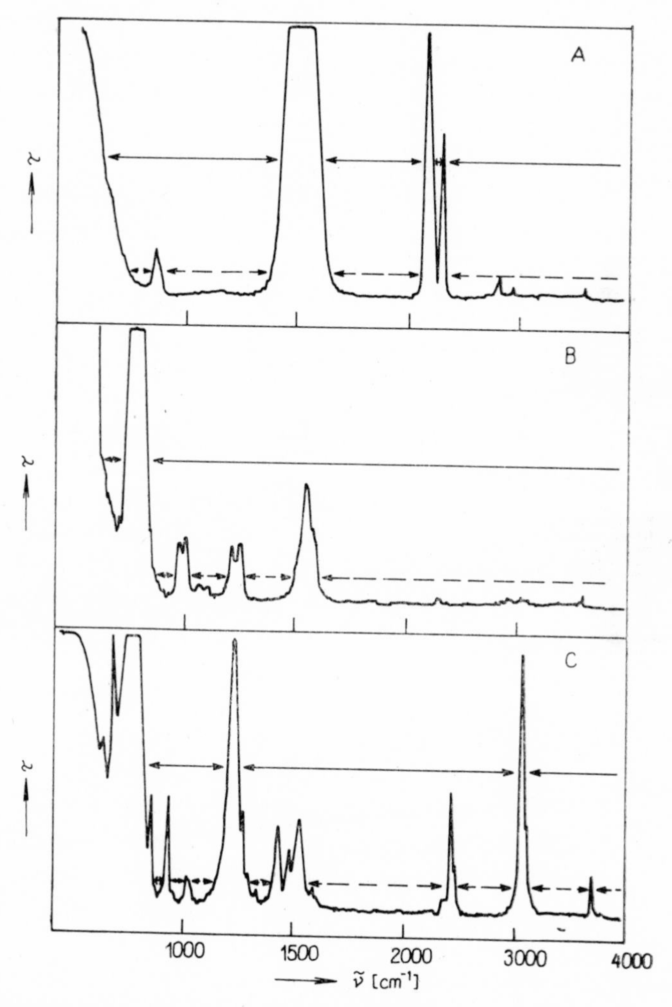

Fig. 19. Infrared spectra of important solvents employed for the measurement of infrared spectra of compound solutions: A — carbon disulphide, B — tetrachloromethane, C — chloroform. The useful range for the solvent in 0.01 cm cells (in which the spectra were measured) is given as a solid line, that in 0.1 cm cells as a dashed line.

ing of the region of the O–H and N–H bond stretching vibrations between 3700–3300 cm^{-1}, or 1,4-dioxane, pyridine, dibutyl ether, etc. for the C=O bond stretching vibration region.

Water and deuterium oxide are generally unsuitable solvents for infrared spectroscopy. Both liquids absorb infrared radiation too strongly and, moreover, dissolve most cell window materials. For measurement of spectra of aqueous solutions very thin cells with windows made of CaF_2, LiF, KRS-5, or other uncommon materials must be used. The spectra of salts or other compounds dissolved in water can be measured between 2700–1800 cm^{-1} and 1400–1000 cm^{-1}, in deuterium oxide between 4000–2900 cm^{-1}, 2000–1300 cm^{-1} and 1100–700 cm^{-1}. For measurement of the spectra of aqueous solutions and suspensions, a method of direct study of a film formed by the aqueous solution between the wettable edges of a frame has been proposed. The unsupported film permits the study of reaction kinetics in aqueous solutions, ageing of precipitates, etc., i.e. atypical applications of vibrational spectroscopy.

The measurement of the spectra of melts can also be mentioned in the section on measurements on liquids. Substances with low melting points can be fused by heating with an infra-lamp or by careful grinding between the transparent material windows. A heated cell, thermostatted during the spectrum recording*) must be employed for compounds with higher melting points. If the melt solidifies and crystallizes during the measurement, the spectrum usually drastically changes. Although the spectra of the liquid and solid samples of the same compound need not be very different, great losses in radiation occur after crystallization, e.g. due to reflection, and the spectral background increases perceptibly.

3.1.2.4 Pretreatment of Samples of Solid Substances

The greatest variability in experimental techniques is required for spectral measurements on compounds that are solid under laboratory conditions and on solid materials in general.

The spectra of unsupported films of polymers can be measured directly; these are prepared from polymer solutions by evaporating the solvent on a smooth surface (e.g. on a mercury surface). The dried films (freed of the solvent) are fixed in a frame and placed in the spectrometer cell space.

---

*) Problems arising in consequence of the emission of the heated cell are discussed in Section 3.3.3.3.5.

Spectra of single-crystals can also sometimes be measured using the direct passage technique. The crystal is fixed in a holder on a goniometric table in the spectrometer cell room. The main difficulty connected with this method is caused by the necessity to optically diminish the cross-sectional area of the beam in the place where the sample is fixed from the usual $80-20$ mm^2 to an area of the order of tenths of a mm^2 (because of the small dimensions of the single-crystals of most compounds). Therefore, a micro-illuminator is placed in the cell space together with the goniometric table; in the microilluminator, the beam is optically narrowed to pass through the sample and is then again broadened to the original dimensions. However, some single-crystals (especially those of ionic compounds) absorb too strongly, so that measurement of their spectra by direct passage is impossible. The spectra of such crystals are then obtained using reflectance techniques (see 5.4.1).

The measurement of spectra of solid samples after conversion into solutions or melts was mentioned in the preceding section. The two most important techniques thus remain to be discussed, the nujol mull and the pellet methods.

In the nujol mull method, the spectra are measured on suspensions of solids in paraffin oil, well-known under the trade-name "nujol". The liquid medium, nujol, occupies the space between the test compound particles, so that the radiation passing through various areas of the sample does not enter regions with two rather different refractive indices as would happen, if e.g. air was present between the particles. Mulls are usually prepared by grinding in an agate dish; however, mulling between the surfaces of two optically planar, matt-polished glass plates is much more suitable. The sample of nujol and the compound are weighed onto one plate and the mull is prepared by brief grinding between the two plates. It is then transferred into a demountable cell.

A good-quality nujol spectrum is obtained when the mean size of the dispersed solid particles is less than the wavelength of the radiation. If the particles are larger than this value, diffraction effects play the greatest role and considerable losses through reflection occur. Mulls prepared by sample grinding between glass plates usually have suitable properties, a relatively low mean particle size and a narrow distribution curve. However, crystals of some compounds, especially with the shape of flat plates or leaves, cannot be ground well under any conditions.

The infrared spectra of nujol suspensions are usually of very good quality (cf. the infrared spectrum of a glycine mull in Fig. 20). A poorly

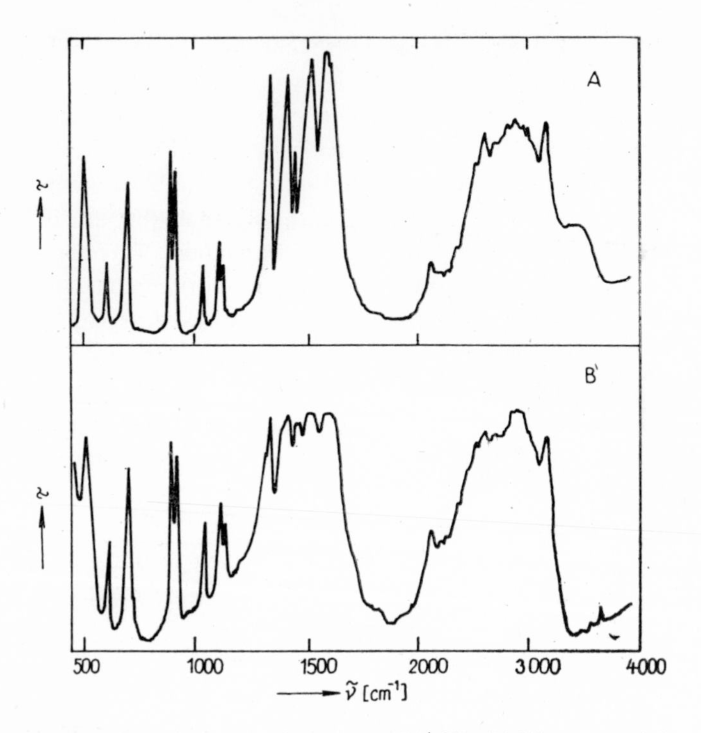

Fig. 20. Infrared spectra of glycine $H_3N^+CH_2COO^-$ measured in: A — KBr pellet, B — nujol mull.

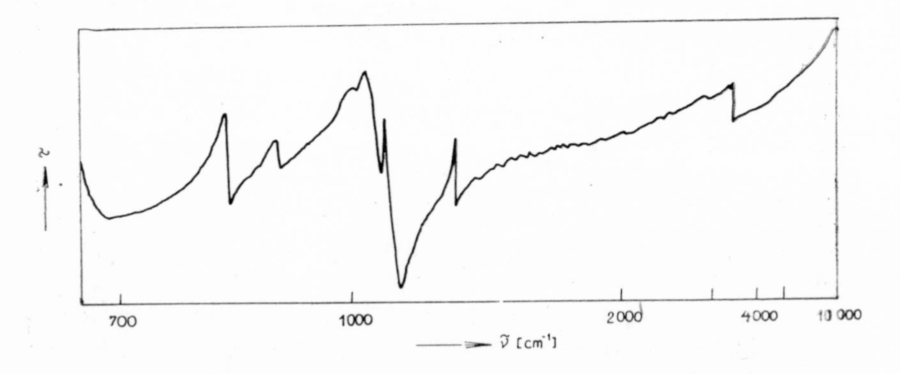

Fig. 21. Infrared spectrum of an ethylene trithiocarbonate sample in a KBr pellet; the recording is an absorption-reflectance spectrum, because of difficulties in preparation of a sample with the required particle size. The bands are distorted by the Christiansen effect and the spectral background increases toward shorter wavelengths (to emphasize the background growth, the spectrum is given on a non-linear wavenumber scale).

prepared mull exhibits background growth toward shorter wavelengths (the wavelength decreases with respect to the constant particle size). The bands in the spectra of poor-quality mulls are distorted; they grow slowly toward smaller wavenumbers and decrease sharply in the opposite direction. This is a manifestation of the Christiansen effect, i.e. the simultaneous effect of absorption and reflection; solids (crystals) exhibit anomalous changes in their refractive indexes in the region of the absorption bands and consequently considerable variations in the reflectivity occur. The effect can be eliminated only by preparing a better-quality sample. A spectrum of a poorly-prepared mull is depicted in Fig. 21.

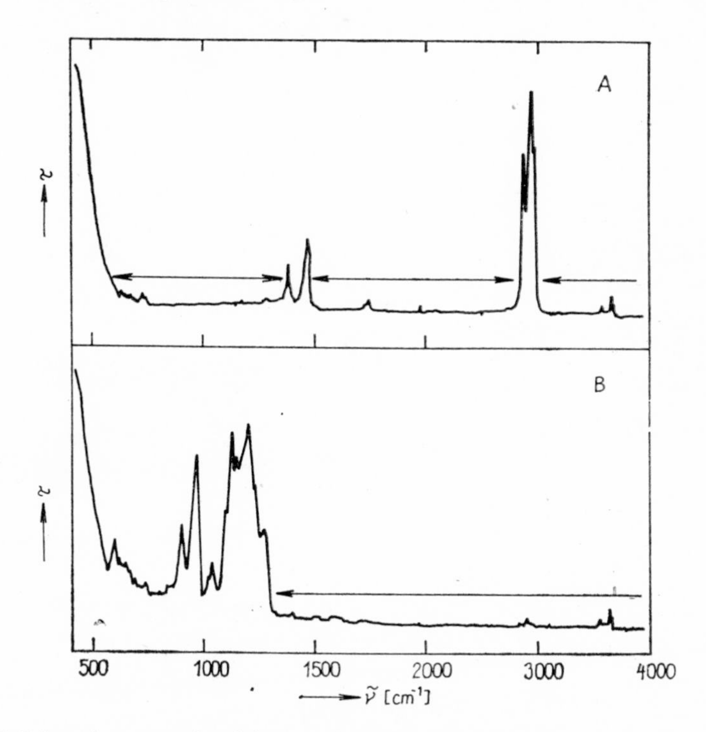

Fig. 22. Infrared spectra of nujol (A) and perfluorokerosene (B); the useful ranges for these liquid media are given as a solid line.

The spectrum of pure paraffin oil (nujol) is shown in Fig. 22. The spectrum is marked by strong stretching vibration bands of C–H bonds around 2900 cm^{-1}, deformation vibration bands around 1400 cm^{-1} and a weak CH_2 groups rocking vibration band at 723 cm^{-1}. Except for the rocking vibration band, the nujol mull spectra cannot be evaluated in the regions of nujol absorption, even if nujol is used for compensation. In order to measure these regions, the spectra of suspensions in another dispersion medium are studied, usually in perfluorokerosene. The spectrum of this liquid is depicted in Fig. 22 together with that of nujol. The spectra of test sample mulls in these two media cover the whole infrared region.

A very important technique for preparing solid samples for infrared spectra measurement is the KBr technique. The weighed and dried sample is mixed with a weighed amount of KBr and mixture is ground well in a vibration mill. The sample must be perfectly dispersed as in the nujol technique. The homogenized sample is pressed at a permanent pressure of up to cca 2 GPa in a steel container (usually in vacuo). A transparent pellet results, which is placed unsupported in a holder in the spectrometer cell room.

The nujol technique was introduced in practice earlier than the KBr-pellet method; the nujol method was then severely critized at the time of introduction of the KBr technique. However, it has been found that this criticism was erroneous, based on the spectra of poorly prepared nujol mulls. At present, the two methods are considered to be of equal quality and selection between them depends on their advantages for the particular task. The nujol technique is easier and faster and the compounds studied are not exposed to extreme conditions during sample preparation. During preparation of KBr pellets compounds and crystals are subjected to large energies in milling and pressing. Some thermolabile compounds may then undergo chemical reactions such as dehydration, dehydrohalogenation, detosylation, etc. Ionic compounds may also exchange ions with the KBr medium; for example, artefacts are found in KBr pellets of carboxylic acid silver salts, potassium salts of carboxylic acids and AgBr. The KBr technique is often employed when chemical mutations of samples are unlikely (and the high energy of milling and pressing does not affect the sample crystal structure). The spectra of glycine in a nujol mull and a KBr pellet are given in Fig. 20.

Materials other than KBr are also employed for preparation of transparent pellets, e.g. KCl, but most frequently "softer" media such as CsBr, polyethylene, adamantane,*) etc. In these media, however, the test compound particles are subject to larger energies during milling and pressing than with KBr, as with the latter, "harder" material the energy was

*) The studied compound is readily regenerated from adamantane pellets; adamantane volatilizes from the pellet even at laboratory temperature.

consumed in destroying the crystal lattice. The effect of larger energies results in higher reactivity, greater tendency to polymorphous conversion, etc.

There are many other, less common, methods for preparing solid samples for measuring infrared spectra. For example, the sample can be sublimed onto the surface of a transparent material, dispersed in a gel support layer, etc.

3.2 Accuracy and Precision of Spectroscopic Data

Three kinds of error can be encountered with vibrational spectra. In first group are gross errors, which can only be avoided. The second group contains systematic errors which should be corrected for and the last group consist of random errors, representing the uncertainty in the measured spectroscopic quantity. In this section, examples of all kinds of error will be given and the methods for their determination and presentation discussed.

Mistakes, imprecise operations and errors occur during sample preparation, cell filling, adjustment of instrumental parameters and during the measurement and evaluation of spectra. Measuring instruments are also imperfect and yield results with a certain degree of accuracy and precision. All this must be borne in mind when determining the wavenumber, intensity and band half-width values. At present, very high demands are placed on the accuracy and precision of spectroscopic values obtainable from experimental spectra and shortcomings stemming from neglected errors are unacceptable.

3.2.1 Errors Affecting the Accuracy of Spectroscopic Data

Gross and systematic errors, affecting the accuracy of the primary spectroscopic quantities, will be dealt with first. Great attention should be paid to the accuracy of the data, as work with inaccurate data leads to generally erroneous and misleading conclusions and hypotheses.

3.2.1.1 Gross Errors

Many circumstances can partially or completely invalidate the information contained in the measured spectra. In order to avoid gross errors, each spectrum should be obtained in multiplicate. In practice this can be done only in exceptional cases (with unstable compounds, in the study of the intensities or shapes of bands, etc.); uncovering of gross errors in a single measured spectrum depends to a great extent on the exterimenter's experience. In any case, the measurement should be repeated immediately even with the slightest suspicion of error, using experimental conditions (sample pretreatment, adjustment and measurement) as different from the previous ones as possible.

Among many errors arising from incorrect spectrometer operation, only one will be pointed out here. This occurs when spectra are recorded at too high a scan-rate compared with the time-constant of the recorder pen. The pen then cannot record the spectrum correctly, the bands become pointed, the pseudomaxima are shifted in the direction of the scan, the intensities are decreased (the more, the higher the band) and the resolution deteriorates. Such a spectrum cannot be either evaluated or corrected; fortunately the error is manifested by characteristically pointed bands.

Another group of gross errors is connected with peculiarities in the chemical structure of the test compound. Various types of spectra can be obtained for associating compounds, in dependence on the selection of the state of aggregation and the experimental conditions; the differences depend on whether the compound exists as a monomeric, oligomeric or polymeric form under the given conditions. Great caution must be exercised in work with reactive compounds. For example, aliphatic amines and especially aminoalcohols react with both CCl_4 and CS_2 used as solvents; insoluble salts or derivatives sometimes separate from the solution.

Great complications can be encountered in the use of the KBr technique. During grinding in a vibration mill and pressing (especially on a rapid increase in the pressure) thermolabile compounds with KBr can undergo elimination or addition reactions. Hydrogen halides can be dissociated from halides, water from alcohols, p-toluenesulphonic acid from tosylates, or, on the other hand, water can be added to double or triple bonds. These reactions need not be quantitative. Problems can also arise during preparation of samples of ionic salts. These salts can exchange ions with alkali metal halides; for example, complete exchange occurs during pressing of silver salts of carboxylic acids with KBr into a pellet, so that a sample containg only silver bromide and the potassium salt of the acid is obtained.

Some ionic compounds exchange ions even in measurements in nujol mulls. Although the sample particles are embedded in non-polar and viscous paraffin oil, they often rapidly exchange ions with the window surface, sometimes even quantitatively. Especially resistant suports, such as BaF_2 plates, must therefore be employed for measuring the spectra of ion-exchanging salts.

In the third category belong the errors arising during manipulation with the sample, after filling the cells, etc. For example, false absorption by impurities in poorly cleaned cells can occur; this error is encountered especially when spectra are measured in thin cells which previously contained viscous liquids. It can happen that a compound crystallizes in the cell and is eluted into new samples and spoils their spectra for a prolonged time.

The spectra of gases or vapours can also be complicated by impurities. The test compound may be absorbed in the O-ring between the window and the cell body and be liberated only during the measurement of the next spectrum. If the O-ring is made of silicone rubber, the compound absorbed can removed by heating the O-ring after removal from the cell.

As has already been pointed out, serious errors may arise during measurement of spectra of compound vapours in a heated cell, in which the sample condenses on insufficiently thermostatted windows. Condensation or even crystallization of vapours lead to considerable losses in the energy through scattering. Similar effects are encountered in spectra when a melted sample crystallizes in the cell.

The sample or the solvent from the reference cell in measurements using compensation techniques may leak or volatilize from loose or improperly closed cells. The empty cell then behaves as a Fabry–Perrot interferometer and interference fringes or their superposition with the sample spectrum result. The cell is usually not emptied instantaneously, so that interference appears gradually during the spectrum measurement. Interference bands also appear in spectra of polymeric material films, mica plates, etc. The interference occurs inside the film or plate and the distances between the interference maxima thus correspond to the film or layer thickness.

Gross errors are also committed when strong absorption in cell window materials, solvents, atmospheric components, etc., is not considered, especially in compensation measurements where this leads to considerable deterioration in the signal-to-noise ratio (see Section 3.3.3.1).

3.2.1.2 Gross Errors in Raman Spectra

In addition to a number of gross errors which may occur in the measurement of both infrared and Raman spectra, there are also errors which are specific for Raman spectroscopy. These errors are connected with differences in sample preparation for the measurement of Raman spectra and sometimes with the diferent physical basis of the two methods.

Cells for sample adjustment for the measurement of Raman spectra are usually glass or quartz capillaries; these are mostly sealed after filling with a liquid, a solution or a solid sample. Undesirable reflections can occur when the capillary is imprecisely adjusted in the laser beam. Sensitive and reactive compounds can decompose, sometimes explosively, due to the high energy of laser beams.

Mechanical contamination of samples for the measurement of Raman spectra must be avoided during their preparation. Dust particles, fibres from filter paper, drops of water, etc., cause parasitic scattering (the Tyndall effect), in which the Raman scattering is obscured. Of course, the sample must not absorb the excitation and the scattered radiation and must not fluoresce.*)

The experimental technique of the measurement of Raman spectra was considerably simplified on introduction of laser sources, especially the helium-neon laser. The application of its red excitation line removed to

Table VIII

Colouration of substances and the properties of the light absorbed

Sample colouration[a]	Properties of the light absorbed		
	colour[b]	wavenumber $\tilde{\nu}$ (cm^{-1})	wavelength λ (nm)
Yellow-green	violet	25 000—23 000	400—435
Yellow	blue	23 000—20 800	435—480
Orange	green-blue	20 800—20 400	480—490
Red	blue-green	20 400—20 000	490—500
Purple	green	20 000—17 900	500—560
Violet	yellow-green	17 900—17 200	560—580
Blue	yellow	17 200—16 800	580—595
Green-blue	orange	16 800—16 500	595—605
Blue-green	red	16 500—13 300	605—750

[a] Sample colouration on irradiation with white light after absorption of the corresponding wavenumber region; [b] the complementary colour.

*) Trouble can also be caused by fluorescence of impurities; for example, contamination with silicone grease, which fluoresces even in trace amounts imperceptible in infrared spectra, is harmful.

a substantial degree the difficulties connected with measurements on coloured compounds, as the red radiation is only absorbed by blue-green compounds, which are not very common among chemical compounds. The difficulties encountered during measurements on coloured substances were substantially more serious when mercury discharge-lamps were used for line excitation, as the blue excitation line is absorbed by yellow compounds and the green line by orange to red compounds. The relationship between the compound colours and the radiation absorbed by them is given in Table VIII.

The red line of the helium-neon laser also excites fluorescence much less frequently than green or blue lines from other excitation sources.

3.2.1.3 Systematic Errors

As in any other measurement, errors stem from constructional limitations of the instruments used in measuring infrared spectra. In contrast to gross errors, which are actually mistakes and should not occur, systematic errors are inherent in all experimental techniques. It is only necessary to learn how to detect them and correct for them.

The first experimental technique limitation is the finite spectrometer resolution. The causes and consequences of differences in the resolution will be dealt with in detail in Section 3.3.3 and 3.3.4 Here it should be pointed out that this fact introduces systematic errors in the determination of the band intensities, half-widths and shapes and sometimes also of band maximum wavenumbers.

Further systematic errors arise owing to incorrect adjustment of the wavenumber and photometric spectrometer scales. Hence the values of wavenumbers, intensities, etc. for a standard sample, obtained on various instruments without correction, will be different. All systematic errors stem from the imperfectness of the construction, manufacture and adjustment of spectral instruments. Calibration methods for correction of systematic errors have been developed; they will be discussed in detail in Sections 3.3.2 and 3.3.3.

3.2.2 Errors Affecting the Precision of Spectroscopic Data

Non-identical experimental results are obtained even if the measurement of spectroscopic quantities is carried out repeatedly on the same instrument under identical conditions. These are random errors caused by instantaneous fluctuations in the temperature, geometric dimensions, line voltage, electric quantities in the electronic components of spectrometers and by free travel in mechanical transmissions or random displacement of the elements, etc. While systematic errors mentioned in the previous section (3.2.1.3) affect predominantly the accuracy of the values obtained, random errors are reflected in their precision.*)

3.2.2.1 Random Errors

In the interpretation of spectra, it is naturally necessary that the initial data should be both accurate and precise. The degree of precision is determined using mathematical statistical methods, the basic laws of which will be treated in Section 3.2.2. However, for detailed information the reader is referred to special statistical literature.

Many experimenters underrate the significance of statistical methods, since they do not realize that the quantitative specification of the precision of the physical quantity determined is of the same importance as the quantity itself, as it enables assessment of the reliability of the conclusions drawn from the measured values. On the other hand, the usefulness of statistics must not be overestimated. Even the most refined statistical methods cannot remove errors committed through negligence during experiments or due to inadequate instrumentation.

3.2.2.1.1 Random Variables, Distribution and Frequency Functions

Let us assume that a very large number of measurements (limiting to infinity) of a quantity, e.g. a single band maximum wavenumber, was carried out. The material obtained (the basic set) can be plotted graphically (Fig. 23a), with wavenumber values on the x-axis and the probability that the measured wavenumber value will be smaller than a given x on the y-axis. The non-descending curve $F(x)$ thus obtained is termed the distribution curve for the statistical distribution of this selected random variable. Its first derivative, $f(x) = F'(x)$ (Fig. 23b) is called the frequency function,

---

*) According to the proposal of the IUPAC Commission for Nomenclature in 1972, the quantitative criterion of the accuracy of a quantity is the minimum value of the ratio, $A = \text{Min}\ x/\Delta x$, where x is the real value of the quantity and Δx is the difference between the largest and the smallest measured value, affected by various systematic errors (e.g. by measurement on various instruments). The quantitative measure of the precision of a quantity is the reciprocal of the relative standard deviation of the set of data, all the elements of which are subject to the same systematic error, i.e. $\Gamma = x/\sigma$.

as it is a measure of the probability that the random variable will lie within the small interval $\langle x, x + dx \rangle$ (the frequency of occurence). Generally, these functions can have different shapes; however, for most physical quantities it can be assumed that the frequency function is Gaussian and can be written as

$$f(x) = \exp\left[-(x - \mu)^2/2 \,.\, \sigma^2\right]/\left[\sigma \,.\, \sqrt{(2\pi)}\right] . \tag{3.2}$$

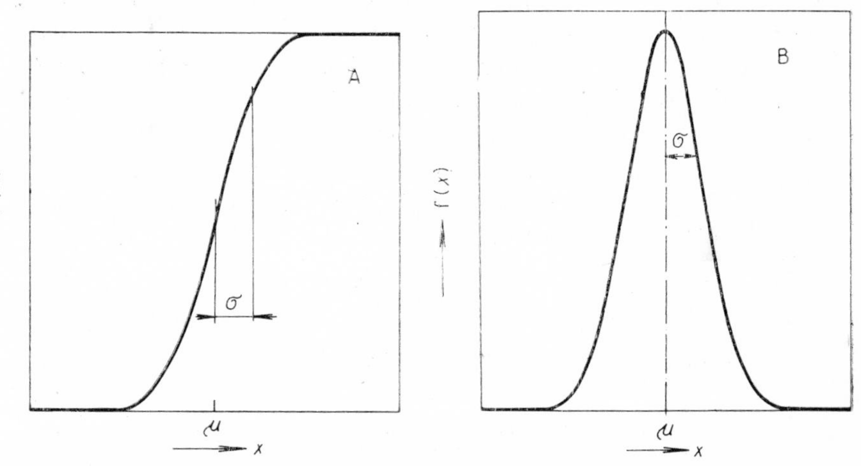

Fig. 23. Functions describing A — normal distribution function $F(x)$, B — frequency function $f(x)$. μ — mean value, σ — standard deviation.

Such a variable is then termed a random variable with normal distribution. Parameters μ and σ^2 are called the mean value and the variance of the random variable, respectively; these two values unambiguously determine the normal distribution. The statistical treatment of experimental data actually serves for the determination of these two parameters.

The mean value of random variable, μ, represents the best possible expression of the real value of the physical quantity studied;*) variance σ^2 or its square root, mean standard deviation σ, is then a quantitative measure of its uncertainty.

3.2.2.1.2 Mean Value, Variance and Standard Deviation

As follows from the exact definitions of the mean value of a random variable

$$\mu = \int_{-\infty}^{+\infty} x \,.\, f(x) \,.\, \mathrm{d}x \tag{3.3}$$

*) Subject to unknown systematic error.

and the variance

$$\sigma^2 = \int_{-\infty}^{+\infty} (x - \mu)^2 \,.\, f(x) \,.\, dx, \tag{3.4}$$

the frequency or the distribution function must necessarily be known for their accurate determination. However, for the construction of these functions, an infinite amount of experimental data is required, which condition obviously cannot be fulfilled. Therefore, a compromise must be resorted to: a smaller number of measurements is carried out, obtaining a random selection of the values of the random variable from the infinite basic set, and the values of parameters μ and σ are estimated on this basis. These assessed values will be denoted as $\hat{\mu}$ and $\hat{\sigma}$.

It can be seen that, with a sufficiently large set of data (cca 20 values or more), the arithmetic mean of all the values, x_i, $i \in \langle 1, n \rangle$, is the best assessment of the mean value, i.e.

$$\hat{\mu} = \bar{x} = \sum_{i=1}^{n} x_i / n. \tag{3.5}$$

The standard deviation for the given selection

$$\sigma_{\text{select.}} = \sqrt{\sum_{i=1}^{n} (x_i - \bar{x})^2 / n} \tag{3.6}$$

and the assessed standard deviation for the infinitely large set (basic set) is then

$$\hat{\sigma} = \sigma_{\text{select.}} \sqrt{[n/(n-1)]} = \sqrt{\left[\sum_{i=1}^{n} (x_i - \bar{x})^2 / (n-1)\right]}. \tag{3.7}$$

However, with a low number of experimental values it may happen that the values of parameters $\hat{\mu}$ and $\hat{\sigma}$ thus determined are subject to too large an error, which is difficult to determine, due to non-uniform distribution of random errors in the set. Then a different assessment is made. For the determination of μ, median $\tilde{x}$ is employed instead of the arithmetic mean; this is the value of the middle element in the set of data arranged according to their magnitude, $(x_1 \leqq x_2 \leqq \ldots \leqq x_n)$, if their number is odd and the average of the middle two, if their number is even:*)

$$\begin{aligned} \hat{\mu} &= \tilde{x} = x_{(n+1)/2} \ (n \text{ is odd}) \\ \hat{\mu} &= \tilde{x} = (x_{n/2} + x_{n/2+1})/2 \ (n \text{ is even}). \end{aligned} \tag{3.8}$$

---

*) With large sets the comparison of the median with the arithmetic mean can be employed to verify whether many gross errors occur in the set or whether the distribution of the random variable is symmetrical.

Variation range R, which equals the difference between the largest and the smallest element of the set,

$$R = | x_n - x_1 |, \tag{3.9}$$

is then employed for assessment of the standard deviation, $\hat{\sigma}$, using the relationship,

$$\hat{\sigma} = k_R \,.\, R; \tag{3.10}$$

coefficients k_R for various n (i.e. for variously large sets) are given in Table IX.

Table IX

The values of coefficient k_R for the calculation of the mean standard deviation from the variation range

n	k_R	n	k_R	n	k_R
2	0.886	5	0.430	8	0.351
3	0.591	6	0.395	9	0.337
4	0.486	7	0.370	10	0.325

From the statistical point of view, the assessed mean value of a random variable, $\hat{\mu}$, is itself a random variable, as infinitely many random selections can be made from the basic (infinite) data set. If they were actually made, a set of assessed values of $\hat{\mu}$ would be obtained, which would again have the normal distribution whose mean value would be the actual mean value, μ, and which would have standard deviation $\sigma_{\hat{\mu}}$. This procedure would be very awkward; fortunately, the assessed standard deviation, $\hat{\sigma}_{\hat{\mu}}$, of the assessed values of $\hat{\mu}$ can be readily calculated

$$\hat{\sigma}_{\hat{\mu}} = \hat{\sigma}/\sqrt{n}, \tag{3.11}$$

where $\hat{\sigma}$ is the assessed standard deviation for the basic set. Then it is possible to determine the confidence limits for the assessed mean value, $\hat{\mu}$.

From the practical point of view, $\hat{\mu}$ is the best possible assessment of the measured physical quantity, e.g. the wavenumber. The assessed standard deviation of the basic set, σ, is a quantitative measure of the precision of each individual measurement. The assessed standard deviation for the mean values, $\hat{\sigma}_\mu$, is a measure of the precision of the mean (3.5) or median (3.8), i.e. of the result obtained by the treatment of a finite set of measurements repeated under completely identical conditions.

3.2.2.1.3 Confidence Limits

It follows from the definition of the frequency function that the integral

$$\int_a^b f(x) \,.\, \mathrm{d}x = P(\xi \in \langle a, b \rangle) \tag{3.12}$$

equals the probability with which a randomly selected value ξ from the basic set falls into interval $\langle a, b \rangle$. With the normal distribution, especially intervals located symmetrically with respect to mean value μ are significant from the point of view of statistical data handling. For example, the probability corresponding to interval $\langle \mu - \sigma, \mu + \sigma \rangle$ equals 0.6827. Because of the symmetry of the frequency function for the normal distribution it holds that

$$P(\xi \in \langle \mu, \mu + a \,.\, \sigma \rangle) = P(\xi \in \langle \mu - a \,.\, \sigma, \mu \rangle) \tag{3.13}$$

and

$$P(\xi \in \langle \mu + a \,.\, \sigma, \infty \rangle) = 0.5 - P(\xi \in \langle \mu, \mu + a \,.\, \sigma \rangle), \tag{3.14}$$

$$P(\xi \in \langle -\infty, +\infty \rangle) = 1, \tag{3.15}$$

$$P(\xi \in \langle \mu, \infty \rangle) = 0.5. \tag{3.16}$$

These relationships will be required for calculating the probability for an arbitrary interval (using Table X).

If the mean value of a certain random variable is known, then the interval symmetrically located around the mean value μ can be found, in which a randomly selected measuring result ξ will fall with a required probability (the two-side estimate). The probabilities are mostly selected rather high, e.g. 99 and 95 %. The intervals

$$P(\xi \in \langle \mu - 1.96\sigma, \mu + 1.96\sigma \rangle) = 0.95 \tag{3.17}$$

and

$$P(\xi \in \langle \mu - 2.58\sigma, \mu + 2.58\sigma \rangle) = 0.99, \tag{3.18}$$

correspond to these probabilities. Values 1.96 and 2.58 are called critical values of the normal distribution for 95% and 99% probability, respectively and are denoted as $u_{0.95}$ and $u_{0.99}$. The u_p values for other values of probability p can be found in Table X; generally,

$$P(\xi \in \langle \mu - u_p \,.\, \sigma, \mu + u_p\sigma \rangle) = p. \tag{3.19}$$

One-side estimates of random variables can be obtained analogously; either the maximum value of the random variable is sought, which can be attained

Table X

The values of $F(a \cdot \sigma) - 0.5$ of the normal distribution function

a	0	1	2	3	4	5	6	7	8	9
0.0	0.000 0	004 0	008 0	012 0	016 0	019 9	023 9	027 9	031 9	035 9
0.1	0.039 8	043 8	047 8	051 7	055 7	059 6	063 6	067 5	071 4	075 3
0.2	0.079 3	083 2	087 1	091 1	094 8	098 7	102 6	106 4	110 3	114 1
0.3	0.117 9	121 7	125 5	129 3	133 1	136 8	140 6	144 3	148 0	151 7
0.4	0.155 4	159 1	162 8	166 4	170 0	173 6	177 2	180 8	184 4	187 9
0.5	0.191 5	195 0	198 5	201 9	205 4	208 8	212 3	215 7	219 0	222 4
0.6	0.225 7	229 1	232 4	235 7	238 9	242 2	245 4	248 6	251 7	254 9
0.7	0.258 0	261 1	264 2	267 3	270 4	273 4	276 4	279 4	282 3	285 2
0.8	0.288 1	291 0	293 9	296 7	299 5	302 3	305 1	307 8	310 6	313 3
0.9	0.315 9	338 6	321 2	323 8	326 4	328 9	331 5	334 0	336 5	338 9
1.0	0.341 3	343 8	346 1	348 5	350 8	353 1	355 4	357 7	359 9	362 1
1.1	0.364 3	366 5	368 6	370 8	372 9	374 9	377 0	379 0	381 0	383 0
1.2	0.384 9	386 9	388 8	390 7	392 5	394 4	396 2	398 0	399 7	401 5
1.3	0.402 2	404 9	406 6	408 2	409 9	411 5	413 1	414 7	416 2	417 7
1.4	0.419 2	420 7	422 2	423 6	425 1	426 5	427 9	429 2	430 6	431 9
1.5	0.433 2	434 5	435 7	437 0	438 2	439 4	440 6	441 8	442 9	444 1
1.6	0.445 2	446 3	447 4	448 4	449 5	450 5	451 5	452 5	453 5	454 5
1.7	0.455 4	456 4	457 3	458 2	459 1	459 9	460 8	461 6	462 5	463 3
1.8	0.464 1	464 9	465 6	466 4	467 1	467 8	468 6	469 3	469 9	470 6
1.9	0.471 3	471 9	472 6	473 2	473 8	474 4	475 0	475 6	476 1	476 7
2.0	0.477 2	477 8	478 3	478 8	479 3	479 8	480 3	480 8	481 2	481 7
2.1	0.482 1	482 6	483 0	483 4	483 8	484 2	484 6	485 0	485 4	485 7
2.2	0.486 1	486 4	486 8	487 1	487 5	487 8	488 1	488 4	488 7	489 0
2.3	0.489 3	489 6	489 8	490 1	490 4	490 7	490 9	491 1	491 3	491 6
2.4	0.491 8	492 0	492 2	492 5	492 7	492 9	493 1	493 2	493 4	493 6
2.5	0.493 8	494 0	494 1	494 3	494 5	494 6	494 8	494 9	495 1	495 2
2.6	0.495 3	495 5	495 6	495 7	495 9	496 0	496 1	496 2	496 3	496 4
2.7	0.496 5	496 6	496 7	496 8	496 9	497 0	497 1	497 2	497 3	497 4
2.8	0.497 4	497 5	497 6	497 7	497 7	497 8	497 9	497 9	498 0	498 1
2.9	0.498 1	498 2	498 2	498 3	498 4	498 4	498 5	498 5	498 6	498 6
3.0	0.498 7	498 7	498 7	498 8	498 8	498 9	498 9	498 9	499 0	499 0
3.1	0.499 0	499 1	499 1	499 1	499 2	499 2	499 2	499 2	499 3	499 3
3.2	0.499 3	499 3	499 4	499 4	499 4	499 4	499 4	499 5	499 5	499 5
3.3	0.499 5	499 5	499 5	499 6	499 6	499 6	499 6	499 6	499 6	499 7
3.4	0.499 7	499 7	499 7	499 7	499 7	499 7	499 7	499 7	499 7	499 8

with probability p (upper estimate), or the minimum value is looked for (lower estimate). Therefore, coefficients a must be found, for which

$$P(\xi \in \langle -\infty, \mu + a \,.\, \sigma\rangle) = p, \tag{3.20}$$

or

$$P(\xi \in \langle \mu - a \,.\, \sigma, +\infty\rangle) = p. \tag{3.21}$$

Using relationships (3.14)–(3.19), it is readily derived that in both cases

$$a = u_{2p-1}. \tag{3.22}$$

If the estimate of $\hat{\sigma}$ is sufficiently reliable, i.e. is determined from a sufficiently large number of values, the intervals appearing in Eqs. (3.17)–(3.19) can be applied; otherwise critical values u_p must be replaced by critical values $t_{p,v}$ of the Student distribution given in Table XI, the number of degrees of freedom, v, being put equal to $v = n - 1$. Consequently,

$$P(\xi \in \langle \mu - t_{p,v} \,.\, \sigma, \mu + t_{p,v} \,.\, \sigma\rangle) = p. \tag{3.23}$$

Table XI

The critical values, $t_{p,v}$, of the Student distribution

v	$p = 0.05$	$p = 0.01$	v	$p = 0.05$	$p = 0.01$
1	12.706	63.657	18	2.101	2.878
2	4.303	9.925	19	2.093	2.861
3	3.182	5.841	20	2.086	2.845
4	2.776	4.604	21	2.080	2.831
5	2.571	4.032	22	2.074	2.819
6	2.447	3.707	23	2.069	2.807
7	2.365	3.499	24	2.064	2.797
8	2.306	3.355	25	2.060	2.787
9	2.262	3.250	26	2.056	2.779
10	2.228	3.169	27	2.052	2.771
11	2.201	3.106	28	2.048	2.763
12	2.179	3.055	29	2.045	2.756
13	2.160	3.012	30	2.042	2.750
14	2.145	2.977	40	2.021	2.704
15	2.131	2.947	60	2.000	2.660
16	2.120	2.921	120	1.980	2.617
17	2.110	2.898	∞	1.960	2.576

The standard deviation or the confidence limits can also be utilized in discovering randomly occuring gross errors. It sometimes happens that some experimental value perceptibly deviates from the set; the question is whether this value should remain in the set or not. The Grubbs test is most frequently employed for the solution of this question.

Let us assume that the measured values have been ordered according to their magnitude ($x_1 \leqq x_2 \leqq \dots \leqq x_n$). From the assessed mean value, μ, and the standard deviation of the selection, $\sigma_{\text{select.}}$, either

$$G = (\hat{\mu} - x_1)/\sigma_{\text{select.}} \tag{3.24}$$

or

$$G = (x_n - \hat{\mu})/\sigma_{\text{select.}}, \tag{3.25}$$

is calculated, according to whether the suspected value is the smallest or the largest, respectively. The calculated values are then compared with the data in Table XII, containing the critical values of G for 95 and 99% probability. If the calculated value of G is larger than the tabulated value, the experimental value is considered extreme and as a gross error is excluded from further treatment. In the opposite case the deviation is considered to be random and the value is left in the set.

Table XII

Critical values G for the Grubbs test

n	$p = 0.05$	$p = 0.01$	n	$p = 0.05$	$p = 0.01$
3	1.412	1.414	15	2.493	2.800
4	1.689	1.723	16	2.523	2.837
5	1.869	1.955	17	2.551	2.871
6	1.996	2.130	18	2.577	2.903
7	2.093	2.265	19	2.600	2.932
8	2.172	2.374	20	2.623	2.959
9	2.237	2.464	21	2.644	2.984
10	2.294	2.540	22	2.664	3.008
11	2.343	2.606	23	2.683	3.030
12	2.387	2.663	24	2.701	3.051
13	2.426	2.714	25	2.717	3.071
14	2.461	2.759			

For a low number of measured values, the non-parametric Dixon test is employed. Quantity Q is calculated from the ordered set of experimental values using the relationships

$$Q = (x_2 - x_1)/(x_n - x_1) \tag{3.26}$$

or

$$Q = (x_n - x_{n-1})/(x_n - x_1); \tag{3.27}$$

this quantity is then compared with the values given in Table XIII. If the calculated value is larger than the tabulated one, the extreme experimental value is discarded.

Table XIII

Critical values Q for the Dixon test

n	$p = 0.05$	$p = 0.01$	n	$p = 0.05$	$p = 0.01$
3	0.941	0.988	18	0.313	0.407
4	0.765	0.889	19	0.306	0.398
5	0.642	0.780	20	0.300	0.391
6	0.560	0.698	21	0.295	0.384
7	0.507	0.637	22	0.290	0.378
8	0.468	0.590	23	0.285	0.372
9	0.437	0.555	24	0.281	0.367
10	0.412	0.527	25	0.277	0.362
11	0.392	0.502	26	0.273	0.357
12	0.376	0.482	27	0.269	0.353
13	0.361	0.465	28	0.266	0.349
14	0.349	0.450	29	0.263	0.345
15	0.338	0.438	30	0.260	0.341
16	0.329	0.426			
17	0.320	0.416			

3.2.2.1.4 Law of Error Propagation

In mathematical operations with quantities which are subject to random errors, the errors are naturally transferred to the results. If the quantities handled are random variables with normal distribution, it can be assumed in the first approximation that the resultant quantity also exhibits normal distribution and can be characterized by the estimates of the mean value and of the standard deviation (or variance).

Generally the estimate of the mean value of function $\varphi(x_1, x_2, \ldots, x_n)$ of n random variables, $\hat{\mu}_\varphi$, equals the value obtained on substituting the estimates of the mean values, $\hat{\mu}_{x_i}$, for the random variables; hence

$$\hat{\mu}_\varphi = \varphi(\hat{\mu}_{x_1}, \hat{\mu}_{x_2}, \ldots, \hat{\mu}_{x_n}). \tag{3.28}$$

The general relationship

$$\hat{\sigma}_\varphi = \sqrt{\sum_{i=1}^{n} \sum_{j=1}^{n} \left[\frac{\partial \varphi}{\partial x_i} \frac{\partial \varphi}{\partial x_j} \cdot \varrho_{i,j} \cdot \sigma_{x_i} \cdot \sigma_{x_j} \right]}, \tag{3.29}$$

holds for the estimate of standard deviation $\hat{\sigma}_\varphi$, where $\varrho_{i,j}$ is the correlation coefficient, which will also be mentioned in Section 3.2.2.2. The $\varrho_{i,j}$ value always equals unity for $i = j$. When $i \neq j$ and the correlation coefficient cannot be determined, its value is set equal to zero;*) then Eq. (3.29) simplifies to give

$$\hat{\sigma}_\varphi = \sqrt{\sum_{i=1}^{n} \left[\left(\frac{\partial \varphi}{\partial x_i} \right)^2 \hat{\sigma}_{x_i}^2 \right]}. \tag{3.30}$$

If the above general formulae are applied to the basic arithmetical operations, the following relationships are obtained:
for multiplication of quantity x by constant a

$$\hat{\mu}_\varphi = a \,.\, \hat{\mu}_x, \tag{3.31}$$

$$\sigma_\varphi^2 = a^2 \,.\, \sigma_x^2; \tag{3.32}$$

for the sum of quantities x and y

$$\hat{\mu}_\varphi = \hat{\mu}_x + \hat{\mu}_y, \tag{3.33}$$

$$\sigma_\varphi^2 = \hat{\sigma}_x^2 + \hat{\sigma}_y^2 + 2\hat{\varrho}_{x,y}\hat{\sigma}_x\hat{\sigma}_y; \tag{3.34}$$

for the product of quantities x and y

$$\hat{\mu}_\varphi = \hat{\mu}_x \cdot \hat{\mu}_y + \hat{\varrho}_{x,y}\hat{\sigma}_x\hat{\sigma}_y, \tag{3.35}$$

$$\hat{\sigma}_\varphi^2 = \hat{\mu}_y^2 \cdot \hat{\sigma}_x^2 + \hat{\mu}_x^2 \cdot \hat{\sigma}_y^2 + 2\hat{\mu}_x \cdot \hat{\mu}_y \cdot \hat{\varrho}_{x,y} \cdot \hat{\sigma}_x \cdot \hat{\sigma}_y; \tag{3.36}$$

for division of quantity x by quantity y

$$\hat{\mu}_\varphi = \hat{\mu}_x / \hat{\mu}_y, \tag{3.37}$$

$$\hat{\sigma}_\varphi^2 = \hat{\sigma}_x^2/\hat{\mu}_y^2 + \hat{\sigma}_y^2\hat{\mu}_x^2/\hat{\mu}_y^4 - 2\hat{\varrho}_{x,y}\hat{\sigma}_x\hat{\sigma}_y\hat{\mu}_x/\hat{\mu}_y^3. \tag{3.38}$$

---

*) This approximation corresponds to the assumption that variables x_i and x_j are mutually independent.

3.2.2.2 The Least Squares Method

While the previous section dealt with the precision of individual random variables, here this problem connected with sets of such quantities will be discussed. In the analysis of these sets secondary quantities are usually obtained from the primary spectroscopic data (e.g. the concentrations of the components of a polycomponent mixture). The least squares method is advantageous for this treatment of sets of data.

3.2.2.2.1 Linear Regression

So far, a single random variable was considered. However, several quantities are often determined simultaneously in a single experiment. Ordered groups of n random variables are thus obtained (the n-dimensional random vector); in the simplest case these are pairs of values x_i, y_i (e.g. the maximum band intensity and the temperature of measurement). If the points with coordinates (x_i, y_i) are plotted, it is frequently found that their distribution is not random; the random variables, x and y, are then correlated and the degree of correlation can be determined by statistical methods.

A curve can be constructed in the field of points (x_i, y_i), which is called the regression line; the simplest case is the regression straight line.*) Linear dependence between random variables x and y is then assumed

$$y = a + b \,.\, x \tag{3.39}$$

and it is required that regression coefficients a, b be found such that the sum of the squares of the residual deviations of the calculated y values from the experimental values,

$$S = \sum_{i=1}^{n} (y_i^{\text{exp}} - y_i^{\text{calc}})^2 = \sum_{i=1}^{n} (y_i^{\text{exp}} - a - b \,.\, x_i)^2 \tag{3.40}$$

is a minimum. The value of S can be considered as a function of two unknown variables, a and b; the condition that S should be minimum is equivalent to the requirements,

---

*) Rigorously, regression is a functional relationship between the independent variable, whose value is given unambiguously (i.e. is not subject to an error), and a dependent variable subject to random errors. If this condition is met (e.g. with the relationship between the number of CH_2 groups in a molecule and the $\nu(CH_2)$ band intensity), then the two variables cannot be interchanged. In "regressions" between two random variables which are subject to errors, the variables can be interchanged provided that this can be logically justified (cause-consequence). The independent variable is usually plotted on the x-axis and the dependent variable on the y-axis.

$$\partial S/\partial a = 0 \quad \text{and} \quad \partial S/\partial b = 0, \tag{3.41}$$

i.e. to the system of equations

$$\begin{aligned} a \,.\, \Sigma x_i + b \,.\, \Sigma x_i^2 &= \Sigma (x_i \,.\, y_i), \\ a \,.\, n + b\, \Sigma x_i &= \Sigma y_i \,. \end{aligned} \tag{3.42}$$

The solution of the system of two equations (3.42) with two unknowns, a and b, yields

$$a = \frac{\Sigma x_i\, \Sigma (x_i y_i) - \Sigma x_i^2\, \Sigma y_i}{(\Sigma x_i)^2 - n\, \Sigma x_i^2} \tag{3.43}$$

and

$$b = \frac{\Sigma x_i\, \Sigma y_i - n\, \Sigma (x_i y_i)}{(\Sigma x_i)^2 - n\, \Sigma x_i^2} \,. \tag{3.44}$$

As the selection of the set of pairs x_i and y_i was random, the values of a and b are also random variables; the solutions of Eqs. (3.43) and (3.44) are, strictly speaking, estimates of the mean values, $\hat{\mu}_a$ and $\hat{\mu}_b$. Hence they must be handled as random variables and must not be attributed the properties of constants. From the value of the sum of the squares of the deviations, S, the variance for the y_i values with respect to the regression straight line can be calculated

$$\sigma_{y,x}^2 = S/(n - 2), \tag{3.45}$$

as can the estimates of standard deviations $\hat{\sigma}_a$ and $\hat{\sigma}_b$ of assessed values $\hat{\mu}_a$ and $\hat{\mu}_b$, respectively,

$$\hat{\sigma}_a = \sigma_{y,x} \cdot \left[\frac{\Sigma x_i^2}{n\, \Sigma x_i^2 - (\Sigma x_i)^2} \right] \tag{3.46}$$

and

$$\hat{\sigma}_b = \hat{\sigma}_{y,x} \cdot \left[\Sigma x_i^2 - (\Sigma x_i)^2/n \right]. \tag{3.47}$$

The correlation coefficient, $\varrho \in \langle -1, +1 \rangle$, is a quantitative measure of the "perfectness" of the fit of the straight line to the field of points. If the value of ϱ is zero, there is no functional relationship between variables x and y; it is said that they are stochastically independent. On the other hand, the extreme values, ± 1, correspond to strictly linear dependence.*) Rigorous definition of the correlation coefficient requires knowledge of the frequency function, $f(x, y)$, for the given two-dimensional distribution,

---

*) The necessary condition is that x and y have two-dimensional normal distribution, i.e. that y have normal distribution for each value of x and vice versa.

$$\varrho_{x,y} = \text{cov}(x, y)/(\sigma_x \sigma_y) = \frac{\int_{-\infty}^{+\infty} \int_{-\infty}^{+\infty} (x - \mu_x)(y - \mu_y) \, . \, f(x, y) \, dx \, dy}{\sigma_x \sigma_y}, \quad (3.48)$$

where cov (x, y) is the covariance of random variables x and y and is equal to the integral in the numerator of the fraction on the right-hand side of Eq. (3.48).

For the given selection of pairs of x_i and y_i the selection coefficient of the correlation is calculated,

$$r = \frac{n \Sigma x_i y_i - \Sigma x_i \Sigma y_i}{\sqrt{[n \Sigma x_i^2 - (\Sigma x_i)^2]} \, . \, \sqrt{[n \Sigma y_i^2 - (\Sigma y_i)^2]}}, \quad (3.49)$$

which may, to a first approximation, be used as an estimate of the correlation coefficient, $\hat{\varrho}_{x,y}$.

The quantity,

$$t_r = r \sqrt{(n - 2)}/\sqrt{(1 - r^2)}, \quad (3.50)$$

can be employed for testing the degree of linear dependence between variables x and y, using critical values $t_{p,v}$ of the Student distribution (Table XI) with the number of degrees of freedom $v = n - 2$. If $t_r > t_{p,v}$, then a functional dependence between x and y exists with probability p.

3.2.2.2.2 General Polydimensional Linear Regression

In the same way as the functional dependence between two random variables was followed in the previous section, the correlations of larger numbers of variables or their functions can be followed generally. The ordered set of random variables, $[x_1, x_2, \ldots, x_k] = \boldsymbol{x}$, will be called the k-dimensional random vector; then a set of functions of this vector can be defined, $a_1(\boldsymbol{x}), a_2(\boldsymbol{x}), \ldots, a_m(\boldsymbol{x})$, and the linear relationship,

$$p_1 a_1(\boldsymbol{x}) + p_2 a_2(\boldsymbol{x}) + \ldots + p_m a_m(\boldsymbol{x}) = y, \quad (3.51)$$

can be assumed. If a set of n equations (3.51) ($n > m$) is available, it is possible to solve this (overdetermined) system with respect to unknown parameters p_1 to p_m (i.e. vector $\boldsymbol{p}$), again on the condition that the sum of the squares of the deviations,

$$S = \sum_{j=1}^{n} [y_j - \sum_{i=1}^{m} p_i a_i(\boldsymbol{x}_j)]^2 \quad (3.52)$$

is a minimum, i.e. that m relationships

$$\partial S/\partial p_i = 0, \qquad i \in \langle 1, m \rangle, \quad (3.53)$$

are valid. If the indicated differentiation is carried out, a set of m linear equations for m unknown parameters p_1 to p_m, termed the set of normal equations, is obtained

$$\sum_{i=1}^{m} p_i \sum_{j=1}^{n} [a_i(x_j) \, . \, a_k(x_j)] = \sum_{j=1}^{n} y_j a_k(x_j), \qquad k \in \langle 1, m \rangle. \tag{3.54}$$

Eqs. (3.51) – (3.54) can be advantageously transcribed in the economical matrix form. Let the matrix of values $a_i(x_j)$ be denoted as **A**, the vector of the unknown parameters $\boldsymbol{p}$ and the vector of the right-hand sides of the equations $\boldsymbol{y}$. It then holds that*)

$$\mathbf{A}p = y, \tag{3.55}$$

$$S = (y - \mathbf{A}p)^{\mathrm{T}} (y - \mathbf{A}p), \tag{3.56}$$

$$\mathbf{A}^{\mathrm{T}}\mathbf{A}p = \mathbf{A}^{\mathrm{T}}y. \tag{3.57}$$

If the product, $\mathbf{A}^{\mathrm{T}}\mathbf{A}$ is then denoted as **B**, the solution can be simply written in the form,

$$p = \mathbf{B}^{-1}\mathbf{A}^{\mathrm{T}}y; \tag{3.58}$$

$\mathbf{B}^{-1}$ is called the inversion matrix with respect to matrix **B**. In substitution of the vector of parameters $\boldsymbol{p}$ back into Eq. (3.52) or (3.56), the sum of the squares of deviations S is calculated, which enables an estimation of the (unknown) variance of the y_j values using the relationship*)

$$\hat{\sigma}_y^2 = S/(m - n). \tag{3.59}$$

Matrix

$$\mathbf{C} = \hat{\sigma}_y^2 \, . \, \mathbf{B}^{-1}. \tag{3.60}$$

is called the covariance matrix of the given system, as its elements equal

$$c_{i,j} = \mathrm{cov}\,(p_i, p_j) = \varrho_{i,j}\sigma_{p_i}\sigma_{p_j}, \qquad i \neq j \tag{3.61}$$

and

$$c_{i,i} = \hat{\sigma}_{p_i}^2. \tag{3.62}$$

The estimates of the parameter standard deviations obtained can be employed for the calculation of the confidence limits according to Eqs. (3.17) – (3.23), substituting $v = n - m$ for the number of degrees of freedom.

---

*) Superscript ^{T} denotes a transposed matrix, i.e. a matrix in which the columns were interchanged with the rows.

*) If variance σ_y^2 (or standard deviation σ_y) can be assessed by an independent method, then conclusions can be drawn on the basis of comparison of the two values concerning the acceptability of function (3.51) or (3.55) for the description of the experimental y_j values.

In calculation of derived quantities, dependent on two or more parameters p_i, the values of the correlation coefficients from Eq. (3.61) can be used for a more rigorous assessment of the standard deviations of the derived quantities according to Eq. (3.29) and the following relationships.

So far it was assumed that all elements of vector y have the same variance, σ_y^2. However, it may happen that the individual y_j values are subject to variously large errors, expressed by standard deviations σ_j or their estimates $\hat{\sigma}_j$. Then generally different weight, $w_{j,j}$, equal to

$$w_{j,j} = 1/\sigma_j^2, \qquad j \in \langle 1, n \rangle, \tag{3.63}$$

must be attributed to each square of the deviation between the calculated and experimental y_j. Eq. (3.52) thus changes into

$$S = \sum_{j=1}^{n} w_{j,j}[y_j - \sum_{i=1}^{m} p_i a_i(x_j)]^2 \tag{3.64}$$

or, written in the matrix form, into

$$S = (y - \mathbf{A}p)^T \mathbf{W}(y - \mathbf{A}p), \tag{3.65}$$

where $\mathbf{W}$ is the diagonal matrix of the weights, whose elements equal

$$\left.\begin{aligned} w_{k,j} &= 0, \ k \neq j \\ w_{j,j} &= 1/\sigma_j^2 \end{aligned}\right\} k, j \in \langle 1, n \rangle. \tag{3.66}$$

The further relationships analogously change into

$$\mathbf{A}^T\mathbf{W}\mathbf{A}p = \mathbf{A}^T\mathbf{W}y \tag{3.67}$$

and

$$p = \mathbf{B}^{-1}\mathbf{A}^T\mathbf{W}y, \tag{3.68}$$

where $\mathbf{B} = \mathbf{A}^T\mathbf{W}\mathbf{A}$. If the vector of the parameters, p, is substituted into Eq. (3.64) or (3.65), the sum of the weighed squares of the residual deviations is obtained; if this is employed to calculate the variance of the experimental y_j values using Eq. (3.59), it must be borne in mind that actually the ratio of the y_j value variance with respect to the regression function to the a priori estimate of the variance used in Eq. (3.63) is calculated.*)

Inversion matrix $\mathbf{B}^{-1}$ can be employed without alterations for the calculation of the covariance matrix using Eqs. (3.60) – (3.62).

A typical example of the application of polydimensional linear regression is the quantitative analysis of polycomponent mixtures (Section 5.2.4). The linear regression discussed in the previous Section is a special case of

---

*) This ratio must be close to unity, if the assumption concerning the applicability of function (3.51) or (3.55) to the description of the experimental y_j values is fulfilled.

this general regression ($p_1 = a$, $p_2 = b$, $a_1(\boldsymbol{x}_j) = 1$, $a_2(\boldsymbol{x}_j) = x_j$). Another example is the construction of a function for a polynomial of the $(m - 1)$th degree ($a_i(\boldsymbol{x}_j) = x_j^{i-1}$).

3.2.2.2.3 General Non-Linear Regression

The values of random variable y cannot always be expressed as a linear function of parameters p_1 to p_m (3.51). A typical example of a non-linear dependence of the parameters is the expression of the absorbance in terms of the superposition of the individual profile functions. The solution of such tasks is generally called non-linear regression

$$a(\boldsymbol{p}, \boldsymbol{x}) = y. \tag{3.69}$$

Again, a vector of parameters $\boldsymbol{p}$ is required such that the sum of the squares of the deviations

$$S = \sum_{j=1}^{n} [y_j - a(\boldsymbol{p}, \boldsymbol{x}_j)]^2 \tag{3.70}$$

is a minimum. There are many methods for the solution of this problem. Their common feature is the iteration procedure, i.e. the final solution is approached by a series of successive approximations. Here only the modified Newton–Raphson method will be discussed, which is, in our opinion, most important practically. However, the solution is not as straight forward as with linear regression. First, an initial estimate of the parameters, $\boldsymbol{p}^{\text{est.}}$ must be known. If this estimate is close to the correct solution, the function on the left-hand side of Eq. (3.69) can be expanded in a Taylor series, taking only the first term into consideration,

$$a(\boldsymbol{p}^{\text{est}}, \boldsymbol{x}) + \sum_{i=1}^{m} \partial a(\boldsymbol{p}, \boldsymbol{x})/\partial p_i \cdot (p_i - p_i^{\text{est}}) = y, \text{ for } \boldsymbol{p} = \boldsymbol{p}^{\text{est}}, \tag{3.71}$$

i.e.,

$$\sum_{i=1}^{m} \partial a(\boldsymbol{p}, \boldsymbol{x})/\partial p_i \,.\, \Delta p_i = y - a(\boldsymbol{p}^{\text{est}}, \boldsymbol{x}), \text{ for } \boldsymbol{p} = \boldsymbol{p}^{\text{est}}. \tag{3.72}$$

If the matrix of partial derivatives

$$J_{i,j} = \partial a(\boldsymbol{p}, \boldsymbol{x}_j)/\partial p_i, \text{ for } \boldsymbol{p} = \boldsymbol{p}^{\text{est}}, \quad i \in \langle 1, m \rangle, \quad j \in \langle 1, n \rangle, \quad m \geqq n \tag{3.73}$$

is denoted as $\mathbf{J}$ (the Jacobian of system (3.69)), it is possible to write in the matrix form,

$$\mathbf{J}\, \Delta \boldsymbol{p} = \boldsymbol{y} - \boldsymbol{a}(\boldsymbol{p}_{\text{est}}, \boldsymbol{x}). \tag{3.74}$$

In view of the unknown values of parameter corrections Δp_i, Eq. (3.74) represents an overdefined system of linear equations, which can be directly solved after conversion into a system of normal equations,

$$\mathbf{J}^T\mathbf{J}\,\Delta \boldsymbol{p} = \mathbf{J}^T[\boldsymbol{y} - \boldsymbol{a}(\boldsymbol{p}^{est}, \boldsymbol{x})] \tag{3.75}$$

and

$$\Delta \boldsymbol{p} = \mathbf{B}^{-1}\mathbf{J}^T[\boldsymbol{y} - \boldsymbol{a}(\boldsymbol{p}^{est}, \boldsymbol{x})], \tag{3.76}$$

where $\mathbf{B} = \mathbf{J}^T\mathbf{J}$. The values of $\Delta \boldsymbol{p}$ obtained are employed for the calculation of a new, better estimate of the parameter vector,

$$\boldsymbol{p}^{est'} = \boldsymbol{p}^{est} + \Delta \boldsymbol{p}. \tag{3.77}$$

In a favourable case the sum of the squares of the deviations becomes smaller; the iteration process is then said to converge. The whole calculation is then repeated with the new estimate, $\boldsymbol{p}^{est'}$, starting from Eq. (3.71). The procedure is continued, until the best, or required, agreement between the calculated and the experimental values of y is attained. However, if function (3.69) is considerably non-linear and approximation by a Taylor expansion, (3.71), is too rough, the sum of the squares of the deviations does not decrease and, on the contrary, the process is often divergent. There are various methods for dealing with the problem of convergence. The simplest approach consist of "shortening" the correction vector, $\Delta \boldsymbol{p}$; Eq. (3.77) then becomes

$$\boldsymbol{p}^{est'} = \boldsymbol{p}^{est} + k\,\Delta \boldsymbol{p}, \tag{3.78}$$

where $k \in \langle 0, 1 \rangle$.

A more complicated, but better procedure is the Levenbergh method of the dampened least squares of the deviations, which has two variants. In the first one, matrix $\mathbf{B}$ in Eq. (3.76) is modified by adding positive number d to each diagonal term, i.e.

$$\mathbf{B} = \mathbf{J}^T\mathbf{J} + d\mathbf{E}, \tag{3.79}$$

where $\mathbf{E}$ is the unit matrix. In the second variant, each diagonal element of the matrix of the normal equations is multiplied by number $(1 + d)$; the non-diagonal elements are not changed.

$$\left.\begin{aligned} B_{i,j} &= [\mathbf{J}^T\mathbf{J}]_{i,j}, \quad i \neq j \\ B_{i,i} &= (1 + d)\,[\mathbf{J}^T\mathbf{J}]_{i,i} \end{aligned}\right\} i, j \in \langle 1, n \rangle. \tag{3.80}$$

Positive coefficient d is chosen empirically for each concrete case and its value generally varies during the iteration process. However, it can be seen that the optimum value of d is approximately proportional to the sum of the squares of the deviations, S, i.e.,

$$d = \text{const. } S, \tag{3.81}$$

the value of the constant usually lies in the interval, const. $\in \langle 0.1, 10 \rangle$.

The iteration process ends when the value of S decreases below a chosen limit (assessed e.g. on the basis of standard deviation σ_y, see Eq. (3.59)).

Similar to the linear regression, matrix

$$\mathbf{C} = \sigma_y^2(\mathbf{J}^T\mathbf{J})^{-1} = S/(n - m) \,.\, (\mathbf{J}^T\mathbf{J})^{-1} \tag{3.82}$$

can be employed to estimate the standard deviations and covariances for the parameters sought, using Eqs. (3.60) – (3.62).*)

In the same way as with linear regression, the weighing method can be applied here, with all the consequences discussed in Section 3.2.2.2.2, on introducing the matrix of the weights $\mathbf{W}$. Eq. (3.75) then becomes

$$\mathbf{J}^T\mathbf{W}\mathbf{J} \,.\, \Delta p = \mathbf{J}^T\mathbf{W} \,.\, [y - a(p^{est}, x)]; \tag{3.83}$$

and the following relationship, (3.76), changes analogously.

A typical example of application of non-linear regression is numerical separation of bands in spectra (Section 3.3.4.3.3).

3.2.2.3 Statistical Noise and its Suppression

Spectroscopic data, either in graphical or digital form, are always distorted to a certain degree by statistical noise. This noise must be removed for many applications.

The best procedure for noise suppression would be multiple repetition of the experiment under identical conditions and averaging of all the spectra obtained. On the condition that the noise is a random variable with zero mean value, the noise is removed after carrying out a sufficient number of observations (measurements). This procedure is possible only exceptionally with common commercial spectrometers because of its tediousness. Moreover, the requirement that the experimental conditions remain constant over the time necessary to obtain many (e.g. several dozen) spectra may be difficult to meet. Only the most modern infrared spectrometers directly coupled with single-purpose computers (e.g. Perkin–Elmer model 580) permit use of this procedure for noise elimination.

For a great majority of common spectral operations, in which extreme precision is not required, graphical smoothing of the noise by constructing a smooth curve through the noise oscillations is sufficient. This procedure

---

*) It should be pointed out that matrix $\mathbf{B}$ of the normal equations, modified according to Eq. (3.79) or (3.80), cannot be used for this calculation.

is usually satisfactory for the determination of band maxima (except for very weak or very strong bands).

If the spectrum is in digital form, it can be assumed, with sufficiently frequent sampling, that the spectrum can be replaced by a smooth curve over several neighbouring points, which is approximated by a polynomial of a sufficiently high order, m, in order to record the spectrum non-linearity. If the spectrum is represented by pairs of coordinates $(\tilde{\nu}_i, y_i)$ (y_i can be the transmittance τ_i or the absorbance D_i of the given point in the spectrum or another quantity), then the calculation of the corrected (smoothed) values, $y_{i,corr}$, is carried out as follows:

1. An odd number of points, $2n + 1 (2n > m)$, is selected within the interval to be smoothed;

2. the values of gradually decreasing statistical weights, w_0 to w_n, are chosen, by means of which the effect of the neighbouring points of the spectrum on the value to be smoothed is governed;

3. smoothing is started with the i-th point using the following procedure;

4. the overdetermined set of equations

$$\sum_{k=0}^{m} p_k (\tilde{\nu}_j - \tilde{\nu}_i)^k = y_j, \tag{3.84}$$

where $j \in \langle i - n, i + n \rangle$, is solved with respect to the unknown coefficients of the interpolation polynomial, p_k, $k \in \langle 0, m \rangle$, with statistical weights $w_{|i-j|}$, (for the weighed least squares method see Section 3.2.2.2.2);

5.
$$y_{i,corr} = p_0; \tag{3.85}$$

is substituted as the corrected (smoothed) value;

6. subscript i is increased by unity and the procedure is repeated starting from point 4.

In construction of the polynomial by the least squares method, the largest statistical weight is thus given to the studied point; the weight decreases with increasing distance from this point. The interpolated value of the polynomial function for the coordinate of the studied point is then an estimate of the real functional value with the noise eliminated.

The method is discussed in detail in refs.[1,2] We would like to point out that spectrum convolution actually occurs in this procedure (for explana-

---

1 Savitzky A., Golay M. J. E.: Smoothing and Differentiation of Data by Simplified Least Squares Procedures. *Anal. Chem. 36*, 1627 (1964).

2 Steiner J., Termonia Y., Deltour J.: Comments on Smoothing and Differentiation of Data by Simplified Least Squares Procedure. *Anal. Chem. 44*, 1906 (1972).

tion see Section 3.3.3.3). Therefore the sampling must be so frequent that the interval in which the selected polynomial is constructed is substantially narrower than the spectral width of the spectrometer slit with which the studied spectrum was measured; otherwise serious distortion of the shape of the spectral curve cannot be avoided.

If the spectrum is sampled at equidistant points, it is advantageous to employ coefficients k_j given in Tables XIV and XV for the calculation. The interpolated corrected functional value for the i-th point, $y_{i,corr}$, then equals

$$y_{i,corr} = [k_0 \cdot y_i + \sum_{j=1}^{n} k_j \cdot (y_{i-j} + y_{i+j})]/N, \tag{3.86}$$

where N is the normalization factor also given in Tables XIV and XV.

A better numerical method involves the use of spline functions. These are again polynomial functions of the n-th order; the difference lies in the fact that a different polynomial function is defined for each interval defined by two successive experimental points (x_{i-1}, x_i). An additional condition of the continuity of the spline function and of all its derivatives up to the $n - 1$th degree must also be satisfied. For most applications in spectroscopy, cubic spline functions S_3, defined by the relationship,

$$y = S_3(x) = P_i(x) = a_i + b_i x + c_i x^2 + d_i x^3, \tag{3.87}$$
$$x \in \langle x_{i-1}, x_i \rangle, \qquad i \in \langle 1, n \rangle, \qquad (x_0 = -\infty)$$

are sufficient.

Further conditions can be imposed on spline functions. For our purposes the condition that the value of the mean square of the second derivative (a quantitative measure of "smoothness"),

$$Q = \int_{x_{min}}^{x_{max}} S_3''(x)^2 \, . \, dx, \tag{3.88}$$

be minimum is very suitable, fulfilling simultaneously another condition, namely, that the sum of the weighed squares of the deviations be equal*) to a non-negative value of S,

$$S = \sum_{i=1}^{n} [(S_3(x_i) - y_i)]^2 \, . \, w_i, \qquad w_i = 1/\sigma(y_i)^2; \tag{3.89}$$

(a quantitative measure of the "precision" of the fit).

---

*) If a straight line is the smoothest possible curve, the sum of the squares may be smaller.

Table XIV

Coefficients K_j for the calculation of smoothed function values using 2nd and 3rd order polynomials

n	12	11	10	9	8	7	6	5	4	3	2
j	K_j										
0	467	79	329	269	43	167	25	89	59	7	17
1	462	78	324	264	42	162	24	84	54	6	12
2	447	75	309	249	39	147	21	69	39	3	—3
3	422	70	284	224	34	122	16	44	14	—2	
4	387	63	249	189	27	87	9	9	—21		
5	348	54	204	144	18	42	0	—36			
6	287	43	149	89	7	—13	—11				
7	222	30	84	24	—6	—78					
8	147	15	9	—51	—21						
9	62	—2	—76	—136							
10	—33	—21	—171								
11	—138	—42									
12	—253										
Norm factor *N*	5175	805	3059	2261	323	1105	143	429	231	21	35

Table XV

Coefficients K_j for the calculation of smoothed function values using 4th and 5th order polynomials

n	12	11	10	9	8	7	6	5	4	3
j					K_j					
0	4253	1011	44003	1393	883	11063	677	143	179	131
1	4125	975	42120	1320	825	10125	600	120	135	75
2	3750	870	36660	1110	660	7500	390	60	30	—30
3	3155	705	28190	790	415	3755	110	—10	—55	5
4	2385	495	17655	405	135	—165	—135	—45	15	
5	1503	261	6378	18	—117	—2937	—198	18		
6	590	30	—3940	—290	—260	—2860	110			
7	—255	—165	—11220	—420	—195	2145				
8	—915	—285	—13005	—255	195					
9	—1255	—285	—6460	340						
10	—1122	—114	11628							
11	—345	285								
12	1265									
Norm factor N	30015	6555	260015	7429	4199	46189	2431	429	429	231

Application of spline functions necessitates the use of computers. For this reason the procedure for the determination of coefficients a_i, b_i, c_i, d_i is not discussed here; it can be found e.g. in ref[1]. The computer procedure is given in Appendix A.

3.3 Determination of the Parameters of Bands in Vibrational Spectra

Without respect to the purpose that a vibrational spectrum is to serve, the primary spectroscopic quantities must first be determined, e.g. the wavenumbers of the band maxima and their intensities or parameters describing their shapes, i.e. the quantities collectively termed the spectral band parameters. For the sake of simplicity, the principal rules of their determination will first be explained on an ideal isolated band; only in Section 3.3.3 will real, i.e. more or less overlapping bands, be treated.

There are no basic differences in the determination of the wavenumbers for bands in infrared and Raman spectra; the rules discussed in Section 3.3.1 hold for both. On the other hand, the determination of the intensities of Raman lines and of their depolarization factors is a quite specific problem and hence a separate section, 3.3.4, is devoted to it.

In all the following sections, the methods with which the highest precision can be attained in the determination of the band parameters will be given. Of course, it is not always necessary to resort to the most complicated procedures. For example, with survey spectra the demands placed on the precision of the determination of the band wavenumbers and intensities will not be as stringent as e.g. in the detailed study of conformational equilibria in a molecule on the basis of splitting of certain characteristic bands. Each spectrum interpreter must estimate the degree of precision that is required and, on the other hand, determine the degree of reliability that can then be expected from the conclusions based on the spectroscopic data obtained.

In most cases an infrared or a Raman spectrum is recorded graphically, plotting the transmittance or the absorbance (the apparent relative intensity with Raman spectra) against the wavenumber or, less frequently, against the wavelength. Hence a graphical recording is handled with the initial data in the form of the geometric coordinates of points on the spectral curve,

---

1 Reinsch C.: Smoothing by Spline Functions. *Num. Math. 10*, 177 (1967).

whose relationship with the spectral quantities is mediated. The task of calibration methods is to make this relationship as precise as possible. The precision of the determination of spectroscopic quantities is determined primarily by two factors; the precision of the determination of geometric coordinates in the recording (the precision of reading) and the precision of the instrument calibration.

Furthermore, distortion of the spectrum recording to a greater or lesser degree by ever present noise (Section 3.2.2.3) must be considered. In handling the spectrum an attempt to eliminate the noise must first be made.

3.3.1 Determination of the Background

The determination of all the basic parameters depends on the assessment of the spectral background (zero base-line).

The simplest procedure involves construction of a tangent common to the two lowest points in the vicinity of the studied band (Fig. 24). However,

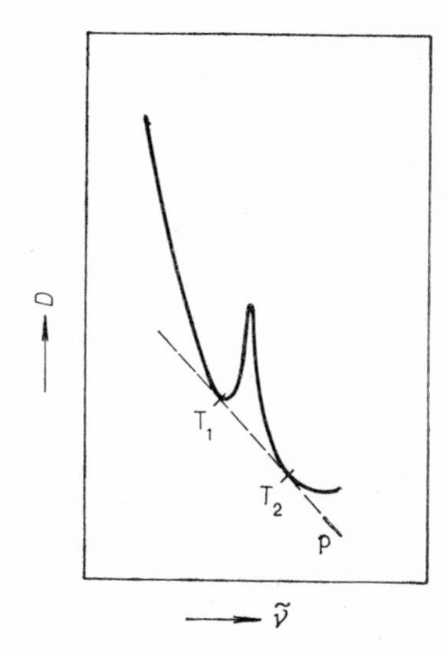

Fig. 24. Construction of baseline p as a linear approximation of the background of a spectral band with a lower intensity. T_1 and T_2 — tangent points.

it should be pointed out that this frequently employed method overestimates the background contribution and the results must be corrected in more precise determinations, especially the values of the unit and integrated intensities. The method represents the background slope reasonably well and can thus be employed without greater difficulties for the determination of the positions of isolated band maxima (see Section 3.3.2).

More objective data can be obtained by measuring the spectra of the solvent or the blank solution against each other in the cells. The background caused by imprecise cell compensation (solvent absorption, non-uniform optical properties of the cell surfaces, etc.) can then be determined. However, the summary effect of the contributions from the distant spectral bands of the compound studied cannot be detected in this way, nor can changes in the solvent absorption due to interaction with the

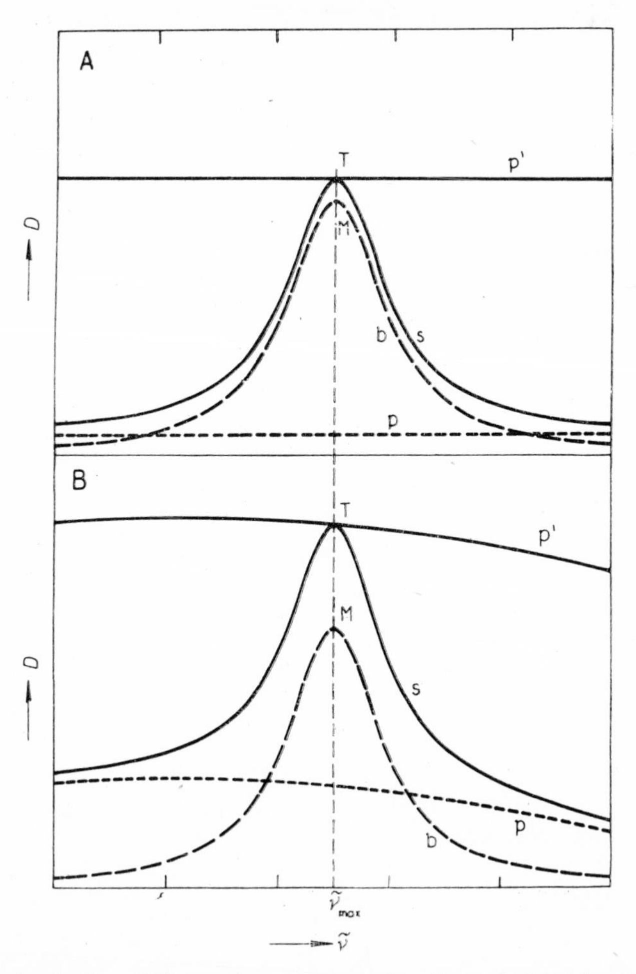

Fig. 25. Determination of the maximum wavenumbers for an isolated band with background: A — parallel with the wavenumber axis, B — nonparallel with the wavenumber axis. M — band maximum, T — tangent point, s — spectrum shape, b — band profile, p — background, p′ — parallel shifted background, $\tilde{\nu}_{max}$ — maximum wavenumber.

studied substance. However, these effects are usually of little importance; they are more marked only at higher concentrations of the studied substance. Some authors then recommend recording two spectra at somewhat different concentrations and employing the calculated difference spectrum.

The most effective, but simultaneously the most involved, method for background determination is numerical band separation (Section 3.3.4.3.3).

3.3.2 Isolated Band Maximum Wavenumbers

Let us assume that we deal with an ideal spectrum after elimination of statistical noise. It has been pointed out in Section 2.4.1.1 that the position of local maxima on the spectral curve need not coincide with the position of the maxima (profile curves) of the individual bands. This unfortunately is also true for an isolated band; the only exception is the limiting case of a single band superimposed on a linear background parallel with the wavenumber axis (Fig. 25a). Otherwise, point T must be found on the spectral curve plotted on an absorbance scale, at which this curve has a common tangent with the vertically shifted background (Fig. 25b). The horizontal (wavenumber) coordinate of this tangential point T then corresponds to the horizontal coordinate of the maximum of the band profile function.

Direct location of the tangential point is difficult even on a smooth ideal curve and is almost impossible on a spectral curve with noise. Therefore, it is mostly located using the band median. At various heights on the band, curves (mostly straight lines) parallel with the background curve are constructed. The distance between points A_i and B_i (see Fig. 26), where the curves intersect the spectral curve, are halved; a curve is then constructed through points S_i and is extrapolated to the band maximum point, M, whose geometric coordinates (e.g. with respect to the chart coordinates) can then be read directly.

If the background, linear but not parallel with the wavenumber axis on a transmittance scale, is shifted, it must be borne in mind that it will be somewhat distorted, due to logarithmic transformation from the absorbance to the transmittance scale. This effect can be seen in Fig. 27. However, if the background is not too sloped, this effect can be neglected.

If the background is parallel with the wavenumber axis, the position of an isolated band maximum can practically be read directly on the spectrometer wavenumber scale. However, the detection of the instant of maximum recorder-pen deflection is rather difficult even with very slow scan rates. It is more advantageous to record the wavenumber values at a number of various

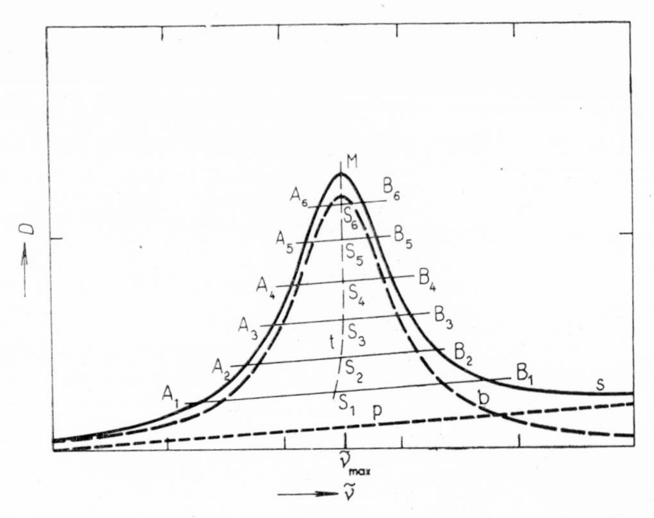

Fig. 26. Construction of band median t. s — spectrum shape, b — band profile, p — background, $\overline{A_iB_i}$ — band chords, S_i — chord centres, M — extrapolated band maximum, $\tilde{\nu}_{max}$ — the maximum wavenumber.

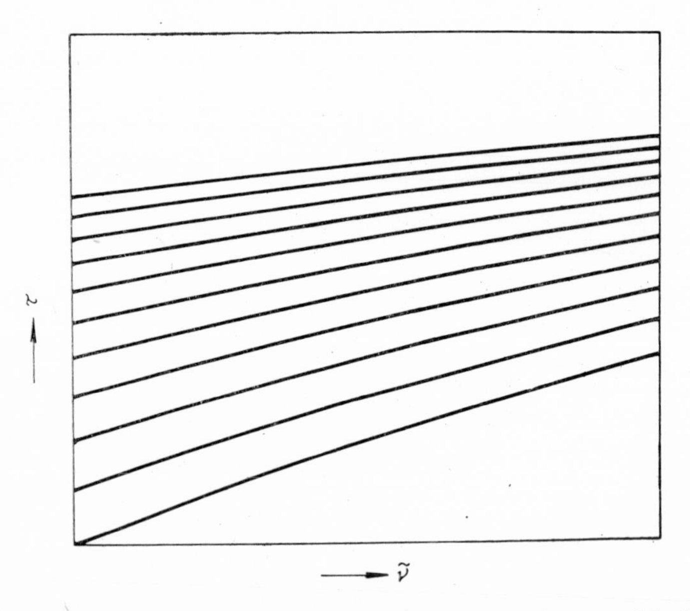

Fig. 27. A set of originally equidistant parallels, approximating the spectral background on the absorbance sacle, becomes a set of non-uniformly distributed curves on non-linear transformation from the absorbance into the transmittance scale. The non-linear transformation exerts the largest effect in the region of the smallest transmittance values.

transmittance or absorbance values on both sides of the band and to calculate the maximum position as the extrapolated value of the averages of these pairs for the transmittance (or absorbance) value at the local maximum of the spectral curve. This is hence a variant of the median method.

With spectra in digital form the local maximum (which can be considered to a first approximation as an estimate of the maximum band position with an isolated band) can be found by constructing a polynomial of a sufficiently high degree through the experimental points in the vicinity of the maximum, using the least squares method (Section 3.2.2.2.2), or by construction of a spline function in connection with a smoothing process (Section 3.2.2.3). In both cases, the maximum coordinate is then found analytically from the condition*) $d\tau/d\tilde{\nu} = 0$ or $dD/d\tilde{\nu} = 0$. The maximum coordinate can then be refined using the separating programs described in Section 3.3.4.3.

The maximum wavenumbers obtained are subject to both random and systematic errors. First methods for estimation of the magnitude of random errors and methods for their suppression will be discussed; then our attention will turn to a discussion of wavenumber scale calibration, i.e. elimination of systematic errors.

3.3.2.1 Precision of the Determination of the Maximum Wavenumber

Random errors stem from instrumental imperfectness (free travel, imprecise recorder chart adjustment, electric fluctuations) and from shortcomings in the work of the experimenter and in his tools for spectrum handling. All these factors are mutually combined and affect the overall precision of the wavenumber determination.

This precision naturally cannot be determined for each individual measurement because of time and operational factors. It is usually determined by tests on a suitably selected compound, whose spectrum exhibits bands with different intensities and half-widths. By repeated measurements and complete spectrum evaluation under precisely defined conditions identical with those employed commonly in experiments, within the shortest possible time interval, the short-time instrument precision is determined. On the basis of similarities in the shapes of bands of the studied compound and those in the reference spectrum, the variance (or standard deviation) values

---

*) If the background in the band region is not parallel with the wavenumber axis, the background slope is substituted in the right-hand side of the condition, i.e. the derivative of the background function on the wavenumber, $dP_0(\tilde{\nu})/d\tilde{\nu}$.

obtained can be transferred to the wavenumbers from currently measured spectra.

While instrumental effects strongly vary with the spectrometer type, the errors in spectrum evaluation depend predominantly on two circumstances: on the resolution with which the coordinates can be read in the chart scale and which varies roughly around ± 0.2 mm and on the reproducibility of the determination of the band maximum. If operating in the region of suitable intensities, i.e. at band maximum absorbance from 0.15 to 0.85, a reproducibility of cca $\pm 5\,\%$ of the band half-width can be guaranteed. If the absorbance lies outside these limits, the measurement uncertainty increases sharply. In order to attain the optimum precision of the wavenumber values, a sufficiently expanded wavenumber scale is required and the concentrations or cell thickness must be selected so that the band maximum absorbances fall in the above interval.

Instrumental effects must be minimized by meticulous maintenance of the experimental conditions specified for the given type of measurement in spectrometer manufacturers' manuals.

If the variance of the determined standard compound band wavenumber is followed over a long time, data on the long-term precision are obtained, which, however, also include some effects that can be considered systematic in short-time observation (e.g. the effect of different laboratory temperatures in various periods of the year). It is in any case advantageous to know these data, since they determine the limit of the precision of determination of band wavenumbers more objectively.

Manufacturers specify the precision of the band wavenumber determination for most commercial spectrometers. As this is usually not determined for a particular instrument but is based on the manufacturer's experience with the given instrument type, it should at least be verified by a test. This value mostly includes only instrumental effects and not the additional errors arising during the handling of spectra.

3.3.2.2 Accuracy of Wavenumber Values; Calibration of the Wavenumber Scale

The accuracy with which the spectral band maximum wavenumbers are determined depends first of all on correct adjustment of the spectrometer wavenumber scale. Using an instrument with an incorrectly adjusted scale, statistical analysis of the variance enables the determination of the maximum position with a standard deviation of e.g. $\pm 1\ \mathrm{cm}^{-1}$, but with a wavenumber error of e.g. $-10\ \mathrm{cm}^{-1}$. The difference between the actual and the experimental wavenumber can be caused by incorrect adjustment of the monochromator optical elements and represents a systematic error of the spectrometer wavenumber scale.

The first large group of sources of systematic errors arises from the fact that modern spectrometers yield spectra recorded on chart paper with a wavenumber scale that serves as a coordinate system in determining the band maximum positions. With some instruments, e.g. models UR-10 and UR-20 from Carl Zeiss (Jena), the coordinate system is drawn simultaneously with the spectrum recording; in some other instruments, an auxiliary pen marks certain wavenumbers at the edge of the chart.

Spectra could also, in principle, be recorded on blank paper, on which the initial and terminal wavenumbers are marked. Then the wavenumber accuracy would virtually be determined only by the accuracy of the spectrometer wavenumber scale. However, when using a pre-printed coordinate system, errors stemming from inaccuracy of its scale must first be eliminated.

The scale of the chart paper may be shifted with respect to the spectrometer wavenumber scale e.g. due to free travel of the recorder. All wavenumbers read from chart scale are then systematically shifted. Moreover, not even the correct chart length need be maintained; the paper usually shortens due to drying during prolonged storage in a dry room. If uniform shortening (or lengthening) of paper over its whole length is assumed, then both these effects can be described by a linear dependence of the spectrometer wavenumber scale, $\tilde{\nu}_{\mathrm{exp}}$, on the wavenumber read from the chart, $\tilde{\nu}_{\mathrm{chart}}$,

$$\tilde{\nu}_{\mathrm{exp}} = a + b \, . \, \tilde{\nu}_{\mathrm{chart}} . \tag{3.90}$$

Coefficients a and b are determined from at least two points on the spectrum, usually at the begining and the end of the measurement. If the chart length is correct, coefficient b equals unity and a represents the systematic shift in the scale owing to free travel.

However, not even the spectrometer wavenumber scale is absolutely correct, chiefly due to the limited precision of the mechanical finish of the optical-mechanical parts. When the inaccuracy of the scale exceeds the error in the wavenumber determination, the deviations, $\Delta\tilde{\nu}_{\mathrm{cal}}$, between the theoretical wavenumbers values, $\tilde{\nu}_{\mathrm{theor}}$, and the experimental values, $\tilde{\nu}_{\mathrm{exp}}$, must be determined,

$$\Delta\tilde{\nu}_{\mathrm{cal}} = \tilde{\nu}_{\mathrm{theor}} - \tilde{\nu}_{\mathrm{exp}} , \tag{3.91}$$

i.e. the wavenumber scale must be calibrated. It should be borne in mind that the value of $\Delta\tilde{\nu}_{cal}$ is different for various wavenumbers. It must be determined experimentally over the whole scale and expressed as a function of the wavenumber.

The spectrometer wavenumber scale is carefully calibrated especially after instrument instalation. The calibration is rather tedious and thus the calibration graph is usually used for a longer time, until e.g. a test reveals that changes in the scale have occured (after new adjustment, moving the instrument, etc.). Random tests are, however, made even when nothing happens to the instrument. If a deviation from the previously determined $\Delta\tilde{\nu}_{cal}$ value exceeding the precision of the wavenumber determination is found, the instrument is completely recalibrated. The time-changes in the wavenumber scale are rather complex; it is almost never possible to "shift" the calibration curve over the whole range on the basis of the change in $\Delta\tilde{\nu}_{cal}$ for a single wavenumber.

Details concerning calibration will not be discussed here, as they are thoroughly treated in the publication of the Commission on Molecular Structure and Spectroscopy IUPAC and other works.[1-6] Principally, the calibration involves the measurement, under rigorously prescribed conditions, of the spectra of the calibration standards on the spectrometer tested. At present, a number of gaseous compounds, e.g. carbon monoxide and dioxide, nitrous oxide, water vapour, acetylene, hydrogen cyanide, methane, hydrogen chloride and bromide, ammonia, etc., are employed for calibration of the wavenumber scale from 4300 to 600 cm^{-1}; water vapour is mostly used in the wavenumber region below 600 cm^{-1}. The accurate values of the band wavenumbers for the rotational-vibrational spectra of these gases

---

[1] IUPAC — Commission on Molecular Structure and Spectroscopy: Tables of Wavenumbers for Calibration of Infrared Spectrometers. Butterworth, London 1961. Also: *Pure Appl. Chem. 1*, 537 (1961).

[2] Rao K. N., Humhreys C. J., Rank D. H.: Wavelength Standards in the Infrared. Academic Press, New York 1966.

[3] Blaine L. R., Plyler E. K., Benedict W. S.: Wavelength Calibration of Small Grating Spectrometers in the Far Infrared (600—166 cm^{-1}). *J. Res. Nat. Bur. Stand.*, Sect. A *66*, 223 (1962).

[4] Silvera I. F., Hardy W. N., McTague J. P.: Precision Frequency Calibration for Raman Spectrometers. *Rev. Sci. Instrum. 43*, 58 (1972).

[5] Jones R. N., Nadeau A.: Calibration of Low Resolution Spectrometers in the Range 4000—600 cm^{-1}. *Pure Appl. Chem. 37*, 649 (1974).

[6] Jones R. N., Nadeau A.: The Stability and Uniformity of the Indene Wavenumber Standard for the Infrared and Comparative Resolution Studies on Dispersive and Interferometric Spectrometers. *Can. J. Spectrosc. 20*, 33 (1975).

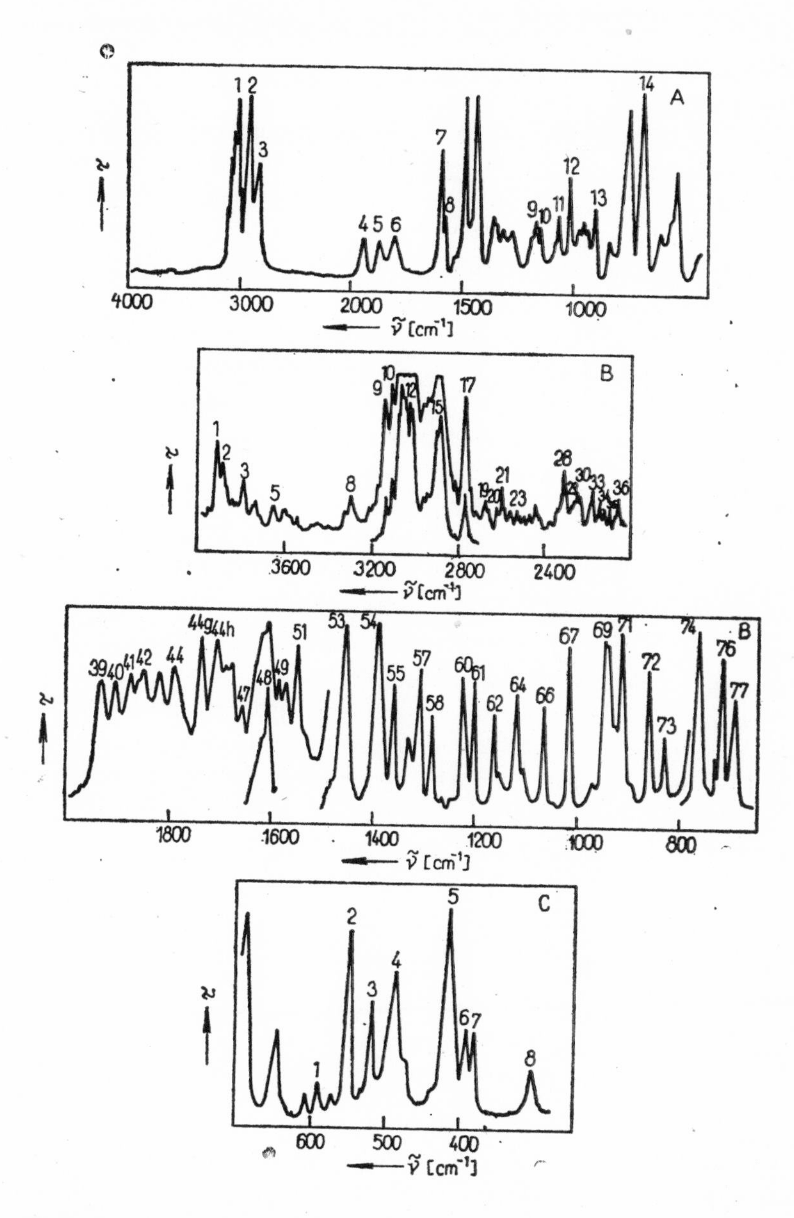

Fig. 28. Orientative infrared spectra of calibration standards suitable for calibration of spectrometers with medium and low resolution: A — polystyrene, B — indene to which 0.8 % camphor and 0.8 % cyclohexanone have been added, C — 1 : 1 : 1 mixture of indene camphor and cyclohexane.

have been determined in relation to the emission lines of certain rare gases and are known with relatively good precision. The spectra of two calibration standards employed for calibration of spectrometers with a lower resolution, polystyrene and indene, are given in Fig. 28; the wavenumbers of bands suitable for calibration purposes and especially for control of the applicability

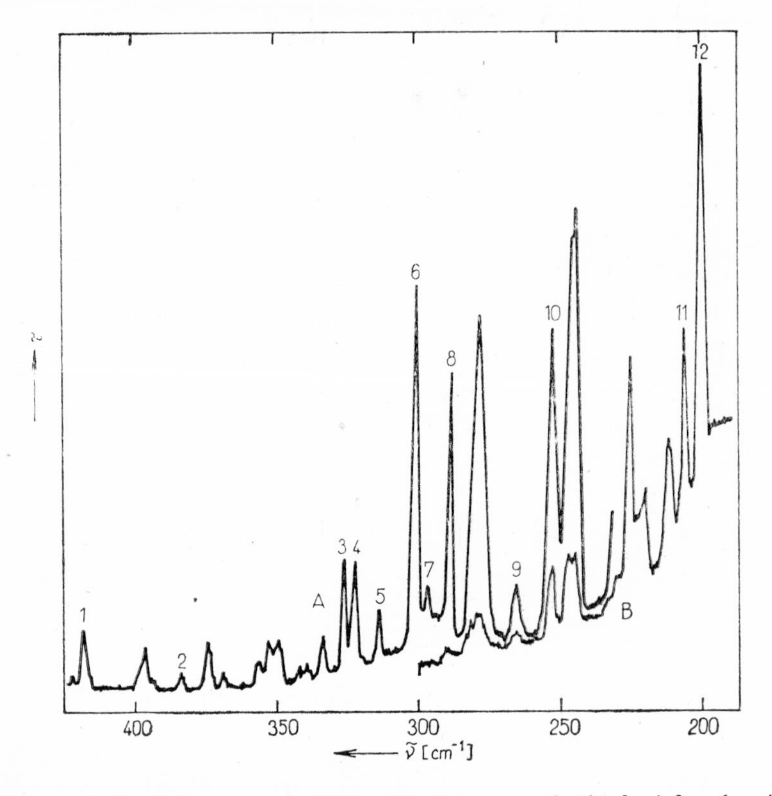

Fig. 29. The infrared absorption spectrum of water vapour in the far infrared region, recommended for calibration of spectrometers with medium or low resolution: recording B corresponds to a lower water vapour partial pressure than recording A. The band numbers correspond to the data in Table XVII.

of calibration graphs are given in Table XVI. In our laboratory, the test of applicability of a completed calibration curve involves recording of two intense bands of a polystyrene foil (bands Nos. 7 and 8 in Table XVI) for each measured spectrum; if the wavenumbers of both bands were located within the limits specified by the calibration graph, the graph is considered applicable (valid).

Table XVI

The wavenumbers of bands of calibration standards usable for the infrared spectral region[a]

Polystyrene[b]

Band No.	Wavenumber $\tilde{\nu}^{c}$ (cm^{-1})	Band No.	Wavenumber $\tilde{\nu}^{c}$ (cm^{-1})	Band No.	Wavenumber $\tilde{\nu}^{c}$ (cm^{-1})
1	3027.1	6	1801.6	11	1069.1
2	2924 ±2	7	1601.4	12	1028.0
3	2840.7	8	1583.1	13	906.7
4	1944.0 ±1	9	1181.4	14	698.9 ±0.5
5	1871.0	10	1154.3		

Indene + 0.8 % wt. camphor + 0.8 % wt. cyclohexanone[d]

Band No.	Wavenumber $\tilde{\nu}^{e}$ (cm^{-1})	Band No.	Wavenumber $\tilde{\nu}^{e}$ (cm^{-1})	Band No.	Wavenumber $\tilde{\nu}^{e}$ (cm^{-1})	Band No.	Wavenumber $\tilde{\nu}^{e}$ (cm^{-1})
1	3927.2[f]	23	2525.5	44^{α}	1741.9[g]	62	1166.1
2	3901.6	28	2305.1	44^{β}	1713.4[h]	64	1122.4
3	3798.9	29	2271.4	47	1661.8	66	1067.7[f]
5	3660.6[f]	30	2258.7	48	1609.8	67	1018.5
8	3297.8[f]	33	2172.8	49	1587.5	69	947.2
9	3139.5	34	2135.8[f]	51	1553.2	70	942.4
10	3110.2	35	2113.2	53	1457.3[f]	71	914.7
12	3025.4	36	2090.2	54	1393.5	72	861.3
15	2887.6	39	1943.1	55	1361.1	73	830.5
17	2770.9	40	1915.3	57	1312.4	74	765.3
19	2673.3	41	1885.1	58	1288.0	76	718.1
20	2622.3	42	1856.9	60	1226.2	77	692.6[f]
21	2598.4[f]	44	1797.7[f]	61	1205.1		

Indene —camphor — cyclohexanone 1 : 1 : 1[d]

Band No.	Wavenumber	Band No.	Wavenumber	Band No.	Wavenumber	Band No.	Wavenumber
1	592.1	3	521.4	5	420.5	7	381.6
2	551.7	4	490.2[b]	6	393.1	8	301.4

[a] Only certain bands in the spectra of the calibration standards can be used for calibration purposes; [b] polystyrene band numbering corresponds to the data from spectrum 28A, the wavenumber values are taken from the literature[1], [c] if not stated otherwise, the precision of the wavenumbers given is ±0.3 cm^{-1}; [d] the band numbering for the indene-camphor-cyclohexanone mixture corresponds to the data obtained from the spectra in Fig. 28B, C and the wavenumber values were taken from the literature[6], [e] if not stated otherwise, the precision of the wavenumbers given is ±0.5 cm^{-1}; [f] the wavenumbers are given with a precision of ±1 cm^{-1}; [g] camphor band; [h] cyclohexanone band.

The spectrum of water vapour in a region from 500 to 200 cm^{-1} is depicted in Fig. 29; the wavenumbers of the water bands (Table XVII) are very often applied for spectrometer calibration in the long wavelength infrared region. However, the water vapour spectrum in the long wavelength region consists of a large number of close rotational lines (these are purely rotational transitions) and their resolution depends on the possibilities and parameters of the instrument tested. The published wavenumber values can be used for calibration only when the calibrated spectrometer resolution corresponds to that with which the published spectra were recorded. Otherwise, gross errors may be committed due to various degrees of resolution of close bands observed under different experimental conditions.

Table XVII

Wavenumbers of bands from the water vapour rotational spectrum in the region between 450 and 200 cm^{-1}

Band No.[a]	Wavenumber $\tilde{\nu}$[b] (cm^{-1})	Band No.[a]	Wavenumber $\tilde{\nu}$[b] (cm^{-1})	Band No.[a]	Wavenumber $\tilde{\nu}$[b] (cm^{-1})
1	419	5	315	9	266
2	385	6	303	10	254
3	328	7	298.5	11	208.5
4	324	8	289.5	12	203

[a] Band numbering corresponds to the data from the spectrum in Fig. 29; [b] the wavenumber values were compiled and given with precision corresponding to the resolution with which the spectrum in Fig. 29 was obtained.

The calibration graph is obtained by plotting the $\Delta\tilde{\nu}_{cal}$ values against the values of $\tilde{\nu}_{exp}$; the $\Delta\tilde{\nu}_{cal}$ values are determined from Eq. (3.91); they are plotted against the experimental wavenumbers in order to be able to use them immediately for correction of the experimental wavenumbers. For computer handling it is advantageous to express the calibration graph in terms of spline functions (Section 3.2.2.3). The use of simple polynomial functions that is rather frequent in practice cannot be recommended, not even for short spectral regions.

If a calibration graph is constructed from $\Delta\tilde{\nu}_{cal}$ values for measured standards, it is found that the error in the wavenumber scale can be determined only by interpolation in certain spectral regions, owing to non-uniform band distribution in the calibration standards. In order to remove this drawback, interference can be employed for the calibration. If an empty liquid cell is placed in the spectrophotometer cell space, it behaves as a Fabry–Perrot interferometer. Reflections occur inside the cell and give rise to interference; a recording of the interference fringes after passage of radiation through an empty cell is depicted in Fig. 30. The wavenumber

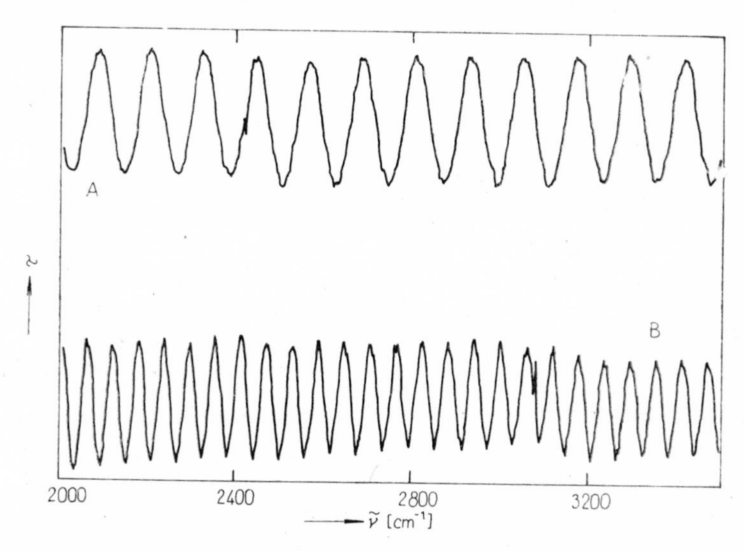

Fig. 30. Recording of interference bands on an infrared spectrometer: A — NaCl cell 0.004 cm thick, B — 0.01 cm NaCl cell.

of the interference maxima, $\tau_{k,\max}$, is a function of the cell thickness, d, or, more precisely, of the distance between the internal surfaces of the windows at which reflection occurs, according to the relationship,

$$d = k/(2 \, . \, n_{air} \, . \, \tilde{\nu}_{k,\max}); \qquad k \in \langle 1, \infty \rangle, \tag{3.92}$$

where k is the order of the interference maximum and n_{air} is the refractive index of air.*) If standard compound spectra are recorded for calibration purposes and the interference recording of an empty cell of a suitable thickness is made immediately afterwards (a greater number of maxima in the interference recording is obtained using a thicker cell), then the region lacking calibration bands can be covered using the equidistant "measure" represented by the interference maxima.[1] The wavenumber of some inter-

---

*) This value is usually set equal to unity (n_{air} = 1.000 29).

[1] Tabačík V., Horák M.: Die Wellenzahleichung von Infrarot-Spektrometern mittlerer Dispersion. *Jenaer Rdsch. 9*, 202 (1964).

ference maxima located close to calibration bands are corrected by means of the $\Delta\tilde{\nu}_{cal}$ values at the ends of this region and are handled by the least squares method. In the calibration graph are then plotted not only the $\Delta\tilde{\nu}_{cal}$ values obtained directly from the calibration standards, but also the $\Delta\tilde{\nu}_{int}$ values determined as the differences of the experimental interference maximum wavenumbers from the values calculated from Eq. (3.92).

It should be borne in mind that the calibration precision is affected by the precision of the determination of the calibration line wavenumbers. These are always subject to a certain error, which is then reflected in the overall precision of the wavenumber determination according to the law of error propagation, discussed in Section 3.2.2.1.4.

For certain special tasks, special calibration standards are sometimes useful. For example, a solution of 4-chloroaniline in tetrachlorethylene has proven advantageous as a standard for wavenumber determination in a region around 3400 cm^{-1}. We determined the wavenumbers of the two relatively narrow bands of NH_2 group stretching vibrations (3481 and 3397 cm^{-1}) by referring them to the ammonia bands. Work with the 4-chloroaniline solution is much faster and simpler than direct calibration with ammonia. The compounds for control of the wavenumber scale must satisfy the principal conditions of standard selection (e.g. the substance must not exhibit rotational isomerism or tautomerism) and must have sufficiently narrow bands. The wavenumbers of these bands are determined, if they are not already known, by reference to the wavenumbers of some recommended test compound, whose absorption bands are close to the bands of the selected standard. The best procedure is repeated measurement of the spectra of the basic and of the derived standard.

3.3.3 Intensities of Isolated Infrared Bands

While the wavenumbers of band maxima can be determined relatively easily and precisely, the determination of band intensities is more tedious and yields less precise values. If the causes of the poorer results of this operation are to be found, the factors affecting the spectrometer photometric precision and some other circumstances must be analyzed in greater detail.

3.3.3.1 Noise

In the analysis of the factors affecting the spectrometer photometric precision, the noise in the electronic components should be given the greatest attention.

The noise consists of random current fluctuations that appear spontaneously in the detector and the connected leads. The level of noise fluctuations is constant for a given set of experimental conditions. Methods have been developed for suppressing the noise (e.g. synchronous detection, a decrease in the detector temperature, cumulation of repeatedly recorded spectra in the memory of a computer and isolation of the signal from the noise), but it is impossible to completely remove noise from spectra.

The signal-to-noise ratio is a spectroscopically important quantity. In infrared spectroscopy this is the ratio of the radiant power of the measuring beam to the noise. It is evident that, with large values of this ratio, when the signal is orders of magnitude higher than the noise, no difficulties are encountered in measuring spectra; problems arise when the signal intensity decreases. In the extreme case, when the signal intensity decreases below the noise level, the spectrometer recorder registers only the noise.

The noise magnitude is principally determined by the detector quality. At present, infrared spectrometers mostly contain thermocouples, which exhibit good sensitivity in a wide spectral region (the whole range, 4000 to 200 cm^{-1}, can be spanned using a single thermocouple), low response-time and low noise.

The signal-to-noise ratio, S/N, is the basic quantity to which the spectrometer parameters are related during adjustment of experimental conditions. It generally holds that the S/N ratio is directly proportional to the square of the slit geometric width,*) w_s, to the square root of the recorder time-constant, t_{rec},**) and to the spectral radiant power of the reference beam, $\Phi_{\tilde{\nu}}^{ref}$; i.e. $S/N \sim w_s^2 . \Phi_{\tilde{\nu}}^{ref} . t_{rec}^{1/2}$. A value of at least 100 : 1 is usually selected for this ratio, i.e. conditions are sought under which the noise fluctuations do not exceed 1 % on the transmittance scale. To attain this value, the source intensity and consequently $\Phi_{\tilde{\nu}}^{ref}$ is varied in the region close to 4000 cm^{-1} (or in other regions where interfering absorption by atmospheric components is absent) with a constant monochromator adjustment, or, on the other hand, the adjustment of the slits in the monochromator unit, w_s, is varied with a constant source output, until the required S/N value, 100, is attained. If the monochromator slits must be widened in order to obtain the required signal-to-noise ratio, the spectrometer resolution deteriorates. Hence correct experimental conditions for the measurement of spectra represent a com-

*) Provided that the slit geometric width, w_s, is not too low, it is proportional to the slit spectral width, s, see Section 3.3.3.3.2.

**) The time required for the pen to move from 100% to 0% on the transmittance scale on sudden interruption of the measuring beam.

promise between instrument resolution and reliable recording of the intensity of the radiation passed through the sample. Spectrometers are constructed so that attaining of these conditions is possible using standard programs and amplifier gain.

If the noise level is regulated by increasing the recorder time-constant, t_{rec}, it must be kept in mind that the scan-rate must be decreased accordingly; otherwise the spectral bands are seriously distorted (see Section 3.2.1.1).

The adjusted signal-to-noise ratio is, however, obtained only in regions where the sample does not absorb light; when the test compound starts to absorb light, the signal-to-noise ratio decreases owing to the decrease in the

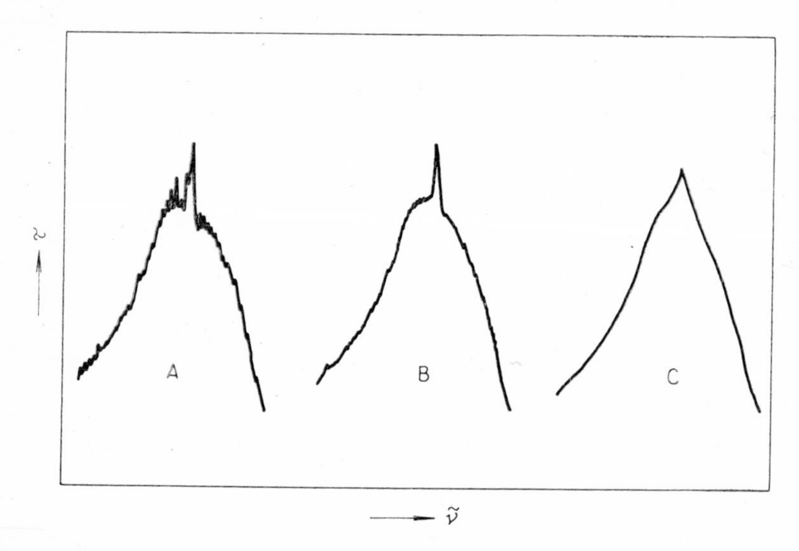

Fig. 31. Spectral recording: A — with excessive noise, B — with an optimum signal-to-noise ratio and C — without noise (in this case too large a recorder pen time constant was selected).

signal intensity and reaches a minimum value at the absorption band peak. Hence the absorption band peak is most distorted by noise, even if this need not always be immediately perceptible (due to distortion in the transmittance scale). Using instruments with expandable intensity scales for measurement of weak bands, an increase in the noise fluctuations is observed in the spectra, proportional to the degree of scale expansion.

Correctly measured spectra should always exhibit spectral curves affected somewhat by noise. The extent of this effect must be selected so that the spectrum handling is not adversely influenced. If noise fluctuations are sufficiently dense (slow scan), the spectral curve is constructed through the centre of the fluctuations (Fig. 31). Absence of noise in the spectrum indicates that instrumental parameters have been adjusted incorrectly; either the measurement was carried out too rapidly, or too large a recorder time-constant has been selected. Large noise on the recording also indicates inadequancies in the measurement. This can be caused e.g. by a discrepancy between the adjustment of the monochromator and the amplifier, especially by too narrow a monochromator slit and too large an amplifier gain.

Considerable problems arise in connection with parasitic absorption by the cell material, the solvent, etc., leading to a decrease in the signal-to-noise ratio. In preparation of the experiment, all possible precautions must be taken to eliminate the causes of parasitic absorption. The optimum signal-to-noise ratio can be attained only by working carefully with a correctly adjusted spectrometer; hence, any further unnecessary deterioration in the ratio would damage the photometric precision.

For example, when measuring a spectrum over a range of 4000 to 200 cm^{-1}, the region from 400 to 200 cm^{-1} is measured using cells with CsI windows and the range from 4000 to 400 cm^{-1} in cells with KBr windows. The loss in the radiation intensity on passage through a cell with CsI windows is cca 60 % due to the high reflectivity of CsI ,while this value only slightly exceeds 10 % with KBr windows. Hence CsI is employed only in the region of its specific transparence, as KBr is virtually not transparent for infrared radiation below 400 cm^{-1}; otherwise it would be advantageous to measure the whole spectrum in cells with CsI windows when they have already been filled for the measurement from 400 to 200 cm^{-1}.

Similar problems are encountered in work with solutions. No solvent is fully transparent over the range of infrared spectra; in the regions of strong solvent absorption the signal-to-noise ratio may deteriorate to such an extent that the spectrum of the dissolved compound cannot be recorded all. The same holds for absorption by nujol and perfluorokerosene, which are employed for the preparation of suspensions of insoluble compounds or salts.

Therefore, the spectra of solutions are measured only in regions where the solvent absorption intensity is not too large. Spectra can be recorded if the solvent passes at least 10 % of the incident energy; however, the photometric precision is very poor. Reliable data can be obtained in regions where the solvent passes at least 50 % of the incident energy. The limits of the photometric precision are, of course, different for each spectrometer type.

Absorption by the solvent increases with increasing cell thickness (cf. Fig. 19). If the solvent absorbs completely in a thin cell, then widening of the layer leads to widening of the region of complete absorption on both sides of the spectral band axis.

As has been pointed out, the recorder records only the noise current when the signal-to-noise ratio is very unfavourable. This phenomenon may sometimes be overlooked when spectra are measured on double-beam spectrometers with optical zero; even if the signal completely disappears, the pen may continue to move in the direction in which it moved in the preceeding region with a favourable signal-to-noise ratio (pen drift). However, the affected part of the spectrum is usually marked by chaotic

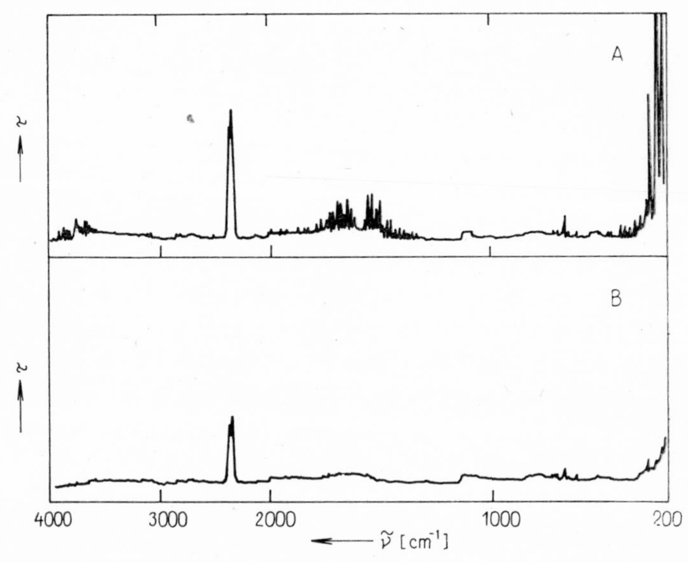

Fig. 32. Infrared absorption spectrum of water vapour and carbon dioxide present in the infrared spectrometer atmosphere (a single-beam recording): A — direct measurement, B — recording after five-minute flushing of the cell room with dried air.

pen movement. If data from an affected region are required, it is necessary to obtain them e.g. using another solvent that is sufficiently transparent in this region.

Considerable difficulties are caused by absorption of infrared radiation by some atmospheric components. The main components, nitrogen, oxygen and the rare gases, do not absorb infrared radiation, but carbon dioxide and water exert an effect. The carbon dioxide concentration in the air is almost constant (cca 0.05 % vol.), but the water vapour concentration varies very widely, depending on the time of the year, temperature, climatic conditions, etc.

The spectrum of atmospheric components (Fig. 32), carbon dioxide and water vapour, obtained on a single-beam spectrometer, exhibits relatively intense bands. Carbon dioxide absorbs especially strongly between 2400 and 2300 cm^{-1} (antisymmetrical stretching vibration ν_3) and weakly around 670 cm^{-1} (degenerate deformation vibration ν_2). Symmetrical stretching vibration ν_1 does not give rise to absorption in the infrared region at all. The non-linear water molecule absorbs between 4000 and 2500 cm^{-1} (antisymetrical and symmetrical vibrations, ν_3 and ν_1, respectively) and between 2000 and 1400 cm^{-1} (deformation vibration ν_2) and is characterized by strong absorption in the long-wave spectral region (around 300 cm^{-1}, purely rotational bands).

Problems with carbon dioxide and water vapour stem chiefly from the fact that, as atmospheric components, they easily penetrate into the internal parts of spectrometers, which then actually become cells with a very long path-length. In commercial instruments the beam usually travels a distance greater than 1 metre from the source to the detector; with this long optical path even substantially lower interferant concentrations than those commonly encountered in the atmosphere have an adverse effect. Absorption by CO_2 and H_2O is eliminated with all infrared spectrometers; some manufacturers use monochromators closed in an evacuated housing, but passage of air freed of H_2O and CO_2 through the monochromator in a closed circuit has proven preferable. A spectrum of the atmosphere after 5-minute drying in the apparatus supplied by Perkin – Elmer model 621 spectrometer is given in Fig. 32.

The interference from carbon dioxide and water vapour causes deteriortion of the signal-to-noise in the regions of their absorption. The rotational-vibrational lines of atmospheric components are, however, substantialy narrower than the half-width of bands of liquids and solids, so that the spectra are unfavourably affected only in very narrow regions. Pseudomaxima or pseudominima then occur on the broad bands in the spectra of the test compounds and consequently incorrect interpretation may lead to the assumption that the bands are doublets, triplets, etc. The appearance of a pseudominimum or a pseudomaximum in the region of CO_2 or H_2O absorption depends on whether the reference or the measuring beam path is longer in the given experimental arrangement. Even if drying devices are employed, compensation is imperfect; for complete compensation, special devices are used, which either moisten or dry the air in the reference beam space.

3.3.3.2 Photometric Scale and its Calibration

The spectrum recording in the chart coordination system also involves an intensity scale. This scale must correspond to the spectrometer photometric scale.

If the spectra are measured in the transmittance scale, a chart with a linear scale in per cent transmittance is employed; charts with non-linear absorbance scales are also available permitting conversion of data from the transmittance scale. Before the measurement the chart scale must be matched with the spectrometer photometric scale, so that their extreme points (i.e. 100 % and 0 % transmittance) are identical. The adjustment is carried out e.g. at 4000 cm^{-1}, i.e. in a region where absorption by atmospheric components does not interfere and the signal-to-noise ratio is favourable. It is incorrect to adjust the instrument e.g. in the region of strong absorption of water vapour; complete transparence is adjusted with free passage for both beams (measuring and the reference) and complete absorption after interruption of the passage of the measuring beam.

Unfortunately, comparison of the terminal points does not suffice; it is necessary to know the scale deviations at any point. The spectrometer photometric scale is not and cannot be without error (although deviations from the accurate values do not usually exceed 1 – 2 % of the transmittance scale with contemporary spectrometers); systematic scale errors must be considered in precise measurements and be determined by calibration. The transmittance scale calibration graph obtained must be used in evaluation of the experimental band intensities if the same chart scale is used for the calibration and the measurement.

The spectrometer photometric scale is most often calibrated*) using a chopper, consisting of circular mutually insertable segments. If the chopper revolves very rapidly, the average decrease in the radiant power is proportional to the ratio of the time intervals in which the radiation is passed and stopped and the latter are proportional to the ratio of the areas of the open and full parts of the segments. In the test the area ratio is set at a certain value and the corresponding transparence value is read on the spectrometer scale. The scale is tested in steps of 1 – 2 % over the whole scale range. In order to be able to calibrate over this wide range, two choppers must be

*) Spectrometer photometric scales can also be calibrated using the absolute intensity values of calibration standards, according to the procedure specified in the paper by K. Doerfel and R. Wegwart: Einsatz absoluter Intensitäten für die Kalibrierung der quantitativen IR-Spektroskopie. *Jenaer Rundschau 19*, 287 (1974).

employed with double-beam spectrometers, placed in the measuring and the reference beam.

The photometric transmittance scale is corrected in terms of values $\Delta\tau_{cal}$, determined from the relationship

$$\tau_{exp} + \Delta\tau_{cal} = \tau_{th}, \tag{3.93}$$

where τ_{th} is the value read on the rotating chopper, τ_{exp} is the value read on the spectrometer photometric scale and $\Delta\tau_{cal}$ is the correction factor sought, which is generally a function of the transmittance τ_{exp}.

The rate of chopper rotation cannot be chosen arbitrarily. Each double-beam spectrometer is provided with a device for synchronous signal detection, part of which is the rotating chopper. When the movement of the two choppers is in phase, a stroboscopic effect appears.

The calibration, similar to adjustment of the chart scale, is carried out in regions where atmospheric components do not absorb, e.g. at 4000 cm^{-1}.

For more detailed information on errors of the photometric scale and its calibration using a rotating chopper see ref.[1])

3.3.3.3 True and Apparent Intensity

If a spectrum is measured on a spectrometer with a correctly calibrated photometric scale, at a favourable signal-to-noise ratio and with correctly adjusted instrumental controls, accurate values of the band intensity still need not necessarily be obtained. Apparent intensity values are obtained in the measurement and the true, absolute values must then be calculated from the apparent ones. Distortion occuring during experiments is connected with the spectrometer resolution and with the effect of the so-called instrument function.

3.3.3.3.1 Instrumental Function, Spectral Slit Width

A perfect monochromator that would separate photons of equal energy $h\nu$ from the whole spectrum is fictitious. Diffraction effects on the slits and on the grating, which is the basic dispersion element of the monochromator unit in modern spectrometers, lead to photons of the same energy being focused into slightly divergent beams. If a slit is placed across the path of radiation dispersed on a grating, photons of various energies will pass through it (see Fig. 33) even if it is infinitely narrow.

---

[1] Jones R. N., Excolar D., Hawranek J. P., Neelakantan P., Young R. P.: Some Problems in Infrared Spectrophotometry. *J. Mol. Struct.* *19*, 21 (1973).

However, an infinitely narrow slit cannnot be employed, as the radiant power passed would be infinitely small. A finite slit opening is thus used in practice, its lower limit being given by the acceptable signal-to-noise ratio (see Section 3.3.3.1). Consequently, the range of photon energies which pass through the exit slit at the given monochromator setting and fall on the detector is even wider.

For a given instrument and a given adjustment of the actual (geometric) slit width, w_s, a function of two variables, $g(\tilde{\nu}_m, \tilde{\nu})$ can be found, which gives the probability with which a photon of radiation with wavenumber $\tilde{\nu}$ passes through the monochromator adjusted to wavenumber $\tilde{\nu}_m$; this function is

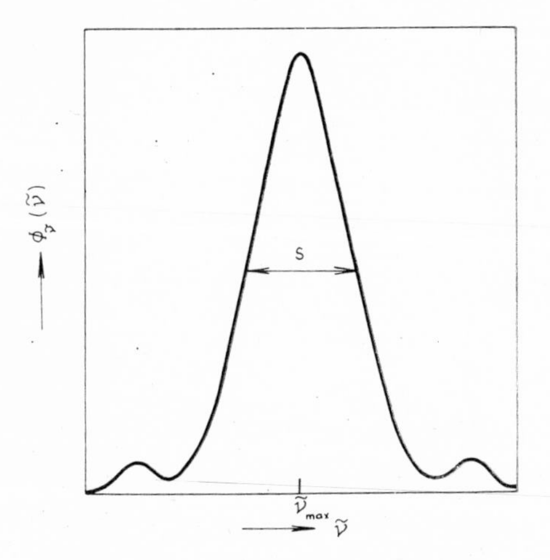

Fig. 33. The recording of the measured values of the spectral radiant power for monochromatic radiation with wavenumber $\tilde{\nu}_m$, after passage through a very narrow slit. Owing to diffraction phenomena, non-zero radiant power $\Phi_{\tilde{\nu}}(\tilde{\nu})$ is recorded even at wavenumbers other than $\tilde{\nu}_m$.

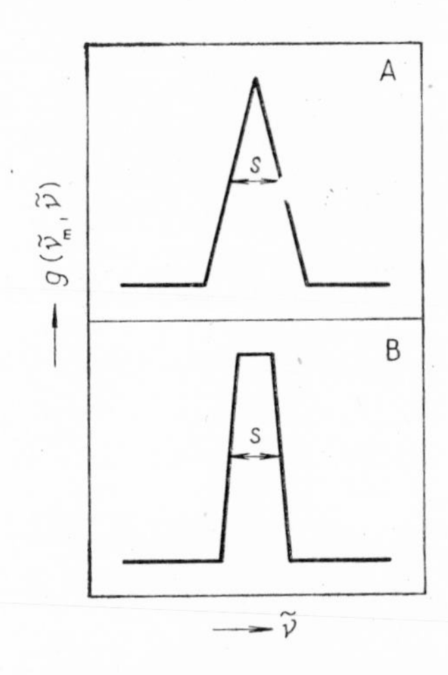

Fig. 34. Approximations of the instrument function $g(\tilde{\nu}_m, \tilde{\nu})$: A—triangular, B—trapezoidal. s — spectral slit width.

termed the instrument function. Mathematically, this is analogous to the statistical frequency function (Section 3.2.2.1.1). From the probability character of function $g(\tilde{\nu}_m, \tilde{\nu})$ follows the normalization condition,

$$\int_0^\infty g(\tilde{\nu}_m, \tilde{\nu}) \, . \, d\tilde{\nu} = 1 \tag{3.94}$$

and non-negative values of the instrument function over the whole wavenumber interval, $\tilde{\nu} \in \langle 0; \infty \rangle$; for a given $\tilde{\nu}_m$, function $g(\tilde{\nu}_m, \tilde{\nu})$ attains a maximum*) at $\tilde{\nu} = \tilde{\nu}_m$.

If polychromatic radiation enters the monochromator, which can be characterized as dependence of the radiant power on the wavenumber, $\Phi_{\tilde{\nu}}(\tilde{\nu})$, then on monochromator adjustment to wavenumber $\tilde{\nu}_m$ the radiant power passing through the exit slit and falling on the detector is

$$\Phi(\tilde{\nu}_m) = \int_0^\infty g(\tilde{\nu}_m, \tilde{\nu}) \,.\, \Phi_{\tilde{\nu}}(\tilde{\nu}) \,.\, d\tilde{\nu}. \tag{3.95}$$

In order to calculate integral (3.95), both function $\Phi_{\tilde{\nu}}(\tilde{\nu})$ and function $g(\tilde{\nu}_m, \tilde{\nu})$ must be known. The latter (instrument) function has, however, a complicated shape; to a first approximation it can be represented by the Gaussian curve for extremely narrow slits. If the geometric slit width of the monochromator is larger than the wavelength of the studied radiation, the shape of the instrument function can be approximated by a triangle (for entrance and exit slits of the same width) or by a trapezium (Fig. 34). A principal feature of the instrument function is a rapid decrease to zero when $\tilde{\nu}$ departs from $\tilde{\nu}_m$.

In the same manner as the band half-width is defined for profile functions, an analogous quantity, termed the spectral slit width, s, is defined for the instrument function; consequently,

$$g(\tilde{\nu}_m, \tilde{\nu}_m - s/2) = g(\tilde{\nu}_m, \tilde{\nu}_m + s/2) = g(\tilde{\nu}_m, \tilde{\nu}_m)/2. \tag{3.96}$$

The value of the spectral slit width corresponds to the wavenumber difference between the points on the slopes of a band with infinitely small half-width, at which the apparent intensity equals one half of the maximum apparent intensity.

If the instrument function is expressed in terms of parameters $\tilde{\nu}_m$ and s, the following relationship is obtained on approximation by the Gaussian curve,

$$g(\tilde{\nu}_m, \tilde{\nu}) = \exp\left[-(\tilde{\nu} - \tilde{\nu}_m)^2 \,.\, 4 \,.\, \ln 2/s^2\right]/\sqrt{(2\pi)}, \tag{3.97}$$

while, for approximation by a triangle,

$$g(\tilde{\nu}_m \tilde{\nu}) = \begin{cases} 0; & \text{for } |\tilde{\nu} - \tilde{\nu}_m| \geqq s \\ (1 - |\tilde{\nu} - \tilde{\nu}_m|/s); & \text{for } |\tilde{\nu} - \tilde{\nu}_m| < s. \end{cases} \tag{3.98}$$

*) This statement holds exactly only for an ideal spectrometer with a quite accurate wavenumber scale; with real spectrometers, the maximum is located at a value of $\tilde{\nu} = \tilde{\nu}_m + \Delta\tilde{\nu}_{cal}(\tilde{\nu}_m)$, where $\Delta\tilde{\nu}_{cal}(\tilde{\nu}_m)$ is the calibration correction of the wavenumber scale (see Section 3.3.2.2). For the sake of simplicity, this fact is neglected and it is assumed that $\Delta\tilde{\nu}_{cal}(\tilde{\nu}_m) = 0$.

As has been mentioned in Section 2.1, the ratio of the radiant powers of the reference and the measuring beam is measured on double-beam spectrometers, i.e.

$$\tau'(\tilde{\nu}_m) = \Phi^{\text{measur.}}(\tilde{\nu}_m)/\Phi^{\text{comp.}}(\tilde{\nu}_m). \tag{3.99}$$

If it can be assumed that in the studied spectral region the spectral radiant power does not change in the vicinity of $\tilde{\nu}_m$ with non-negligible values of $g(\tilde{\nu}_m \tilde{\nu})$,*) it can be written for the reference beam that

$$\Phi^{\text{comp.}}(\tilde{\nu}_m) = \int_0^\infty g(\tilde{\nu}_m, \tilde{\nu}) \,.\, \Phi_{\tilde{\nu}}^{\text{comp.}}(\tilde{\nu}) \,.\, \mathrm{d}\tilde{\nu} \doteq \Phi_{\tilde{\nu}}^{\text{comp.}}(\tilde{\nu}_m) \,. \tag{3.100}$$

On substituting into Eq. 3.99,

$$\begin{aligned}\tau'(\tilde{\nu}_m) &\doteq [\int_0^\infty g(\tilde{\nu}_m, \nu) \,.\, \Phi_{\tilde{\nu}}^{(\text{measur.})}(\tilde{\nu}) \,.\, \mathrm{d}\tilde{\nu}]/\Phi_{\tilde{\nu}}^{\text{comp.}}(\tilde{\nu}_m) \doteq \\ &\doteq \int_0^\infty g(\tilde{\nu}_m, \tilde{\nu})\, \Phi_{\tilde{\nu}}^{(\text{measur.})}(\tilde{\nu})/\Phi_{\tilde{\nu}}^{\text{comp.}}(\tilde{\nu}) \,.\, \mathrm{d}\tilde{\nu} = \\ &= \int_0^\infty g(\tilde{\nu}_m, \tilde{\nu})\, \tau(\tilde{\nu}) \,.\, \mathrm{d}\tilde{\nu}. \end{aligned} \tag{3.101}$$

Hence, true transmittance $\tau(\tilde{\nu})$, defined by Eq. (2.9), is not measured in practice, but apparent transmittance $\tau'(\tilde{\nu})$, reflecting the finite spectral width of the monochromator slit, is recorded.

In discrete representation of the spectrum as a vector of transmittance values, $\boldsymbol{\tau}$, the exact integral, (3.101), must be replaced by an approximate summation formula,

$$\tau'(\tilde{\nu}) = \sum_{\tilde{\nu}} k(\tilde{\nu}_m, \tilde{\nu}).\, g(\tilde{\nu}_m, \tilde{\nu}).\, \tau(\tilde{\nu}), \tag{3.102}$$

where coefficients $k(\tilde{\nu}_m, \tilde{\nu})$ depend on the type of integration formula used and on the step between the individual points. If products $k(\tilde{\nu}_m, \tilde{\nu}) \,.\, (\tilde{\nu}_m, \tilde{\nu})$ are transcribed as elements of matrix $\mathbf{G}$, relationship (3.102) can be simply written as

$$\boldsymbol{\tau}' = \mathbf{G}\boldsymbol{\tau}. \tag{3.103}$$

Matrix $\mathbf{G}$ has a band structure and its non-zero elements are concentrated around the main diagonal.**)

---

*) This assumption is fulfilled, provided that no absorption occurs in the reference beam by a substance exhibiting intense bands in the given region with a half-width smaller than or comparable with the spectral slit width. Hence it is not fulfilled in the regions of absorption by water vapour and carbon dioxide.

**) The statement holds only when homothetic elements of vectors $\boldsymbol{\tau}$ and $\boldsymbol{\tau}'$ correspond to the same wavenumber values.

Table XVIII

Cotes coefficients $H_{n,i}$ for numerical integration

i	0	1	2	3	4	5	6	7
n	$H_{n,i}$							
1	0.50000000	0.50000000						
2	0.16666667	0.66666667	0.16666667					
3	0.12500000	0.37500000	0.37500000	0.12500000				
4	0.07777778	0.35555556	0.13333333	0.35555556	0.07777778			
5	0.06597222	0.26041667	0.17361111	0.17361111	0.26041667	0.06597222		
6	0.04880952	0.25714286	0.03214286	0.32380952	0.03214286	0.25714286	0.04880952	
7	0.04346065	0.20700231	0.07656250	0.17297454	0.17297454	0.07676250	0.20700231	0.04346065
8	0.03488536	0.20768959	—0.03273369	0.37022928	—0.16014109	0.37022928	—0.03722928	0.20768959

If the points in the original and resultant spectrum are distributed equidistantly, with step h in the wavenumber, coefficients $k(\tilde{\nu}_m, \tilde{\nu})$ can be calculated using the Newton–Cotes formulae for even orders n; consequently,

$$k[\tilde{\nu}_m, \tilde{\nu}_m - (n/2 - i) . h] = \frac{h(-1)^{n-1}}{n . i!(n-i)!} \int_0^n \frac{\prod_{j=0}^{n} q - j}{q - i} . dq = h . H_{n,i}; \quad i \in \langle 0, n \rangle. \tag{3.104}$$

Numbers $H_{n,i}$ are termed n-th order Cotes coefficients; their values for $n = 2$ to $n = 8$ are given in Table XVIII. In view of the symmetry of formula (3.104), coefficients $H_{n,i}$ and $H_{n,n-i}$ are equal.

The process leading by the mechanism described above to spectral distortion is called convolution of spectra by the instrument function[1].

Table XIX

Distortion of absorbance values D in the band maxima and of band half-widths $\Delta\tilde{\nu}_{1/2}$ due to instrument function[a]

Acetonitrile band at 750 cm^{-1}, real half-width 15.6 cm^{-1}		
Spectral slit width s (cm^{-1})	D (%[b])	$\Delta\tilde{\nu}_{1/2}$ (%[b])
0.32	100.0	100.0
0.94	97.4	99.7
1.56	97.4	99.4
Naphthalene band at 1009 cm^{-1}, real half-width 1.8 cm^{-1}		
Spectral slit width s (cm^{-1})	D (%[b])	$\Delta\tilde{\nu}_{1/2}$ (%[b])
0.43	100.0	100.0
1.12	78.8	136.5
1.68	63.5	170.2

[a] Two bands were selected, the real width of the first being large and that of the other very small; only the narrow band responds to changes in the spectral slit width; [b] the percent change is given for the sake of lucidity.

[1] Seshadri K. S., Jones R. N.: The Shapes and Intensities of Infrared Absorption Bands — A Review. *Spetrochim. Acta 19*, 1013 (1963).

The distortion of the bands is the more pronounced, the larger is the spectral slit width; The apparent band half-width increases, the apparent band maximum intensity decreases and the spectrometer resolution deteriorates. Examples of the effect of the instrument function on the numerical values obtained experimentally are given in Table XIX.

If the spectral slit width is at least five-times smaller than the band half-width, the experimental band shape approaches the true shape. With contemporary grating spectrometers, spectral slit width values of cca $1-2\ \mathrm{cm}^{-1}$ are commonly attained, so that the effect of the instrument function will be perceptible chiefly with very narrow components of rotational vibrational bands of gases (and sometimes also crystals); on the contrary, broad bands in the spectra of liquids and solutions will be virtually unaffected by the instrument function (see Table XIX).

3.3.3.3.2 Relationship Between Geometric and Spectral Slit Width

If a grating monochromator is considered, in which the entrance and exit slits have the same geometric width w_s, then the relationship between the spectral slit width, s, and the effective slit width, w_{ef}, can be written as

$$s = \frac{2d}{Nnf}\left[1 - \left(\frac{n}{2\tilde{\nu}d}\right)^2\right]^{\frac{1}{2}} . \, w_{ef}, \qquad (3.105)$$

where N is the number of reflections on the grating, n is the interference order, d is the distance between grooves on the grating and f is the focus length of the spectrometer optics.

The effective geometric slit width, w_{ef}, consists of the contributions from the actual geometric slit width, w_s, and from error phenomena

$$w_d = f/(B\tilde{\nu}), \qquad (3.106)$$

(where B is the monochromator aperture magnitude) and aberration effects w_a also include the effects of imperfect optical adjustment of the instrument. The overall effective geometric width then equals

$$w_{ef} = (w_s^2 + w_d^2 + w_a^2)^{1/2}. \qquad (3.107)$$

As the calculation of the spectral slit width is rather difficult, instrument manufacturers supply nomograms for approximate estimation. With many instruments it is even impossible to read the geometric slit width, w_s, directly, since the slit controls have an arbitrary scale, denoted as "slit program".

However, it follows generally from formula (3.105) that the spectral slit width is proportional to the geometric slit width, provided that diffraction effects are not marked in the region where the geometric slit width approaches wavelength λ.

3.3.3.3.3 Corrections for Errors Caused by Spectrum Convolution by the Instrument Function

While the maximum position for an isolated spectral band is unaffected by spectrum convolution, even if the ratio of the spectral slit width to the actual band-width is unfavourable, the other two parameters of the profile function, the band half-width and the band maximum intensity are strongly distorted by convolution. By improving the s ratio to the half-width up to a value of 5 (the spectral width is 20 % of the band half-width), gross distortion is eliminated, but the effect of the instrument function is still not negligible. Thus it can happen with modern spectrometers that these parameters need to be corrected.

A tedious, but the most succesful correction method is extrapolation of the parameter values to zero slit width. This operation involves a certain danger that the instrument function shape changes at small apertures; if the geometric slit width approaches the wavelength, diffraction occurs on the slit and the relationship between the geometric and the spectral slit width becomes more complicated. On closing the slit, the spectrum quality also deteriorates, owing to deterioration in the signal-to-noise ratio. Extrapolation to zero cell thickness or to zero concentration is also employed in correcting the integrated intensities (the Wilson–Wells method, see Section 3.3.3.4.2).

Ramsay derived simple relationships for correction of the apparent intensity of the band maximum

$$a_{max} = K_a \cdot Q'_{max} \tag{3.108}$$

and for the half-width

$$\Delta\tilde{\nu}_{1/2} = K_b \cdot \Delta\tilde{\nu}'_{1/2}, \tag{3.109}$$

for isolated symmetrical bands which can be approximated by the Cauchy curve. Coefficients K_a and K_b depend both on the maximum absorbance and on the ratio of the spectral slit width to the apparent band half-width; their values are given in Tables XX and XXI.

Assuming the applicability of the Cauchy profile, correction coefficient

Table XX

Coefficients K_a for the band maximum intensity correction, using formula (3.108)

$s/\Delta\nu_{1/2}$	0.10	0.15	0.20	0.25	0.30	0.35	0.40	0.45	0.50	0.55	0.60	0.65
a_{max}	K_a											
0.2	1.01	1.02	1.03	1.04	1.06	1.09	1.13	1.17	1.23	1.30	1.40	1.52
0.3	1.01	1.02	1.03	1.04	1.06	1.09	1.13	1.17	1.23	1.30	1.40	1.52
0.4	1.01	1.02	1.03	1.04	1.06	1.09	1.13	1.17	1.23	1.30	1.40	1.53
0.5	1.01	1.02	1.03	1.04	1.06	1.09	1.13	1.17	1.23	1.30	1.41	1.53
0.6	1.01	1.02	1.03	1.04	1.06	1.09	1.13	1.17	1.24	1.31	1.41	1.53
0.7	1.01	1.02	1.03	1.04	1.06	1.09	1.13	1.17	1.24	1.31	1.41	1.54
0.8	1.01	1.02	1.03	1.04	1.06	1.09	1.13	1.17	1.24	1.31	1.41	1.54
0.9	1.01	1.02	1.03	1.04	1.06	1.09	1.13	1.17	1.24	1.31	1.42	1.55
1.0	1.01	1.02	1.03	1.04	1.06	1.09	1.13	1.17	1.24	1.31	1.42	1.55
1.2	1.01	1.02	1.03	1.04	1.06	1.09	1.13	1.17	1.24	1.31	1.42	1.56
1.4	1.01	1.02	1.03	1.04	1.06	1.09	1.13	1.18	1.24	1.32	1.43	1.57
1.6	1.01	1.02	1.03	1.04	1.06	1.09	1.13	1.18	1.24	1.32	1.43	1.58
1.8	1.01	1.02	1.03	1.04	1.06	1.09	1.13	1.18	1.24	1.32	1.44	1.59
2.0	1.01	1.02	1.03	1.04	1.06	1.09	1.13	1.18	1.24	1.32	1.44	1.61

Table XXI

Coefficients K_b for the half-width correction, using formula (3.109)

$s/\Delta\nu_{1/2}$	0.10	0.15	0.20	0.25	0.30	0.35	0.40	0.45	0.50	0.55	0.60	0.65
a'_{max}						K_b						
0.2	0.99	0.98	0.96	0.93	0.91	0.88	0.84	0.80	0.75	0.69	0.63	0.56
0.3	0.99	0.98	0.96	0.94	0.91	0.88	0.85	0.80	0.75	0.69	0.63	0.56
0.4	0.99	0.98	0.96	0.94	0.92	0.88	0.85	0.81	0.76	0.70	0.64	0.57
0.5	0.99	0.98	0.96	0.94	0.92	0.88	0.85	0.81	0.76	0.70	0.64	0.58
0.6	0.99	0.98	0.96	0.94	0.92	0.89	0.85	0.81	0.76	0.71	0.65	0.59
0.7	0.99	0.98	0.96	0.94	0.92	0.89	0.86	0.82	0.77	0.71	0.66	0.59
0.8	0.99	0.98	0.97	0.94	0.92	0.89	0.86	0.82	0.77	0.72	0.66	0.60
0.9	0.99	0.98	0.97	0.95	0.92	0.89	0.86	0.82	0.78	0.72	0.67	0.61
1.0	0.99	0.98	0.97	0.95	0.93	0.89	0.86	0.83	0.78	0.73	0.67	0.61
1.2	0.99	0.98	0.97	0.95	0.93	0.90	0.87	0.83	0.79	0.74	0.69	0.62
1.4	0.99	0.98	0.97	0.95	0.93	0.90	0.87	0.84	0.80	0.75	0.70	0.63
1.6	0.99	0.98	0.97	0.95	0.93	0.91	0.88	0.84	0.81	0.76	0.71	0.64
1.8	0.99	0.98	0.97	0.95	0.93	0.91	0.88	0.85	0.81	0.77	0.73	0.65
2.0	0.99	0.98	0.97	0.96	0.93	0.92	0.88	0.86	0.82	0.78	0.74	0.66

Table XXII

Coefficients K_i for the integrated intensity correction, using formula (3. 110)

$s/\Delta\tilde{\nu}_{1/2}$	0.10	0.15	0.20	0.25	0.30	0.35	0.40	0.45	0.50	0.55	0.60	0.65
a'_{max}							K_i					
0.2	0.99	0.99	0.98	0.98	0.97	0.96	0.95	0.94	0.92	0.90	0.88	0.85
0.3	0.99	0.99	0.98	0.98	0.97	0.96	0.95	0.94	0.92	0.91	0.89	0.86
0.4	0.99	0.99	0.99	0.98	0.98	0.97	0.96	0.94	0.93	0.92	0.90	0.88
0.5	0.99	0.99	0.99	0.98	0.98	0.97	0.96	0.95	0.94	0.92	0.91	0.89
0.6	0.99	0.99	0.99	0.98	0.98	0.97	0.96	0.95	0.94	0.93	0.92	0.91
0.7	1.00	0.99	0.99	0.99	0.98	0.97	0.97	0.96	0.95	0.94	0.93	0.92
0.8	1.00	0.99	0.99	0.99	0.98	0.98	0.97	0.96	0.95	0.95	0.94	0.93
0.9	1.00	0.99	0.99	0.99	0.99	0.98	0.97	0.97	0.96	0.96	0.95	0.94
1.0	1.00	0.99	0.99	0.99	0.99	0.98	0.98	0.97	0.97	0.96	0.96	0.95
1.2	1.00	1.00	0.99	0.99	0.99	0.99	0.99	0.98	0.98	0.98	0.98	0.98
1.4	1.00	1.00	1.00	0.99	0.99	0.99	0.99	0.99	0.99	0.99	1.00	1.00
1.6	1.00	1.00	1.00	1.00	1.00	1.00	1.00	1.00	1.00	1.01	1.01	1.02
1.8	1.00	1.00	1.00	1.00	1.00	1.00	1.00	1.01	1.01	1.02	1.03	1.04
2.0	1.00	1.00	1.00	1.00	1.00	1.00	1.01	1.01	1.02	1.03	1.05	1.07

K_i for correction of the apparent integrated intensity, B, to give the true integrated intensity, A, was derived analogously

$$A = K_i \,.\, B. \tag{3.110}$$

The K_i values are given in Table XXII.

Correction formulae (3.108)–(3.110) can, however, only be used when the spectral slit width is substantially smaller than the band half-width. When the magnitudes of quantities s and $\Delta\tilde{\nu}_{1/2}$ are comparable, more complicated deconvolution methods, described in the next Section, must be used. When the spectral slit width is substantially larger than the true band half-width, the situation is hopeless, as the spectral curve shape is determined exclusively by the shape of the instrument function. It then follows that the measurement of the intensities and especially of the shapes of lines of rotational-vibrational transitions of gaseous compounds on commercial medium-dispersion instruments is virtually pointless.

3.3.3.3.4 Spectrum Deconvolution

Only exceptionally, with bands with extremely small half-widths (gases, crystals), is it necessary to reconstruct the dependence of the true transmittance on the wavenumber from the measured function $\tau'(\tilde{\nu})$, i.e. to carry out spectrum deconvolution. This practically means to solve integral equation (3.101), i.e. to determine function $\tau(\tilde{\nu})$ after the integral sign. It may seem that the solution follows from Eq. (3.103) and that it is sufficient to find inverse matrix $\mathbf{G}^{-1}$. However, this is not so, as matrix $\mathbf{G}$ is infinitely large and knowledge of a part of it is not sufficient for calculation of the corresponding elements of the inverse matrix.

It is apparently most suitable to combine deconvolution with band separation in the spectrum (see Section 3.3.4.3.3). During this procedure, the calculated spectrum is subjected to convolution according to Eq. (3.103) before comparing it with the experimental spectrum, and only this "perturbed" spectrum is employed for the calculation of the band parameter corrections.

If this procedure cannot be used for some reason, pseudodeconvolution, developed originally for x-ray spectroscopy but also suitable for infrared spectra, is applied. This is an iteration method, based on the assumption that the true spectrum, τ, is not very different from the spectrum perturbed by instrumental effects, τ'. Hence the measured spectrum, τ', is used as the initial estimate of the deconvoluted spectrum, τ_{est}. The calculation scheme is then as follows:

1. Using Eq. (3.103), convoluted spectrum τ_{conv} is calculated from estimate τ_{est}.

2. Differences $\tau_{conv} - \tau'$ are determined. If the criterion chosen*) for the differences is smaller than the present limiting value, the calculation is terminated and τ_{est} is considered as the best representation of τ.

3. A new corrected estimate, τ_{corr}, is calculated from the original estimate, τ_{est}, according to either the relationship

$$\tau_{corr} = \tau_{est.} + \tau' - \tau_{conv.} \tag{3.111}$$

or

$$\tau_{corr}(\nu) = \tau_{est.}(\nu) \cdot \tau'(\nu)/\tau_{conv.}(\nu). \tag{3.112}$$

4. The calculation is repeated starting from point 1 using the corrected estimate.

It should be pointed out that this algorithm is very sensitive to the input data quality. Even the lowest noise is amplified by this calculation scheme; therefore noise smoothing methods must also be used (see Section 3.2.2.3). As rounding off errors also represent a source of statistical noise during the calculation, it is recommended to repeat smoothing before each iteration (in step 1.). Even then it is very difficult to avoid unpleasant oscillations, especially close to the ends of the treated spectral region.**) A program for pseudodeconvolution is given in Appendix A.

3.3.3.3.5 Apparent Loss and Gain in Energy

In measurement of infrared spectra it is mostly assumed that the loss in the source radiation energy detected is caused by absorption of the infrared radiation in the sample. In this Section we should like to point out that this is practically never true; the measuring beam intensity can be attenuated not only by absorption, but also by other (chiefly optical) effects and the spectrometer records only the total loss. Information on relative losses in the intensity through absorption and through other mechanisms must be obtained by analysis of the conditions under which the measurement is performed. In some cases energy may apparently be gained rather than lost.

A certain part of energy loss is always caused by reflections, predominantly on the surfaces of the polished cell window materials. These

---

*) Either the sum of the squares of the deviations or the deviation with the largest absolute value.

**) For handling, the spectrum can be divided into smaller sections at points where there are no absorption bands.

losses can actually be considered as narrowing of the spectrometer photometric scale, as they are almost constant over a wide wavenumber range. They can be readily compensated by placing an optical material of the same kind in the reference beam. The decisive role here is played by the number of reflecting surfaces; for example, a cell with two NaCl windows 5 mm thick cannot be perfectly compensated for using a single window of double the thickness, but only by two windows (their thickness need not be identical with that of the cell windows, as the losses through absorption are minimal). However, interference occurs between windows of empty cells and consequently the compensation windows must be separated by a ring at least 1 cm thick instead of by a distance foil; the interference fringes are then too dense to be resolved by the instrument. If an empty cell of the same thickness as that of the measuring cell were used for the compensation, the recording would exhibit superposition of the spectrum (from the measuring cell) on interference bands (from the reference cell).

Loss in radiation through reflection also occurs with crystalline samples. If the conditions for reflections are satisfied (a large change in the refractive index at the reflecting interface, e.g. in a crystal-air system), the spectrum of the crystalline sample may be damaged by this mechanism. If e.g. the mean grain-size of the suspended material is comparable with the radiation wavelength, diffraction phenomena also occur and almost all the radiation energy may be lost. For example, in measurement on an "incorrectly" prepared suspension (the spectrum in Fig. 21), the spectrum background increases toward the short-wavelength region. This is a result of the effect of diffraction phenomena when radiation with a continuously decreasing wavelength passes through a sample with a constant particle distribution.

Further, sharp changes in the refractive index take place in the absorption band region; this effect also influences the sample reflectivity, as its refractive index substantially changes with respect to the constant refractive index of the medium. Superposition of the absorption and reflectance spectra causes apparent "distortion" of the absorption bands; bands suffering from this superposition (the Christiansen effect) cannot be handled unless the effect is eliminated. It seems that this effect can sometimes also occur in the spectra of liquids[1] and not only in those of crystals (cf. the reference on p. 123).

---

1 Horák M., Urban Š.: Difficulties in Interferometric Determination of the Thickness of Infrared Cells. *Chem. Listy 68*, 959 (1974).

Among instances of apparent gain in energy, two cases will be pointed out here. The first deals with scattered radiation. This is radiation incident on the detector simultaneously with the test signal, which, however, is not connected with the spectrometer optical scheme. The contribution from scattered radiation at a certain wavenumber is easy to determine, as it is recorded by the spectrometer on application of a totally absorbing filter for the given wavenumber. For more details on scattered radiation see Section 5.2.4.3.

The second effect is connected with the measurement of sample spectra at higher temperatures (above cca 200 °C). The heated sample, as well as its holder, a cell or a support, emit infrared radiation obeying similar laws to those governing a black body. The radiation may enter the spectrometer and complicate the radiating energy balance, especially if the heated source is placed only in the measuring beam. The higher the temperature, the more pronounced is the effect. The contribution from the radiation emitted by the apparatus must be determined by a test and, if necessary, compensated for (by introducing a reference device heated to the same temperature as the device placed in the measuring beam). Some modern spectrometers are, however, constructed so that the effect of emission from heated samples and cells is automatically eliminated during the measurement of spectra.

3.3.3.4 Molar Absorption Coefficient and Integrated Intensity

If monochromatic radiation with wavenumber $\tilde{\nu}$ passes through an absorbing medium, then the radiant power corresponding to a minute wavenumber interval, $\mathrm{d}\tilde{\nu}$, equals $\Phi_{\tilde{\nu}}(\tilde{\nu})\,\mathrm{d}\tilde{\nu}$. Let us assume that on the whole n oscillators capable of absorbing radiation with wavenumbers in the interval, $\langle\tilde{\nu}, \tilde{\nu} + \mathrm{d}\tilde{\nu}\rangle$, are present in a volume unit of the medium. The rate with which the radiant power will decrease along path l will evidently be proportional to the instantaneous radiant power and the frequency of occurence ("the concentration") of oscillators capable of interaction, i.e.,

$$\mathrm{d}[\Phi_{\tilde{\nu}}(\tilde{\nu}\,.\,l)\,\mathrm{d}\tilde{\nu}]/\mathrm{d}l = -\mathrm{const}\,.\,n\,.\,\Phi_{\tilde{\nu}}(\tilde{\nu}\,.\,l)\,\mathrm{d}\tilde{\nu}. \tag{3.113}$$

If this differential equation is integrated over the whole beam path length in the absorbing medium, from $l = 0$ to $l = d$, with the initial condition $\Phi_{\tilde{\nu}}(\tilde{\nu}, 0) = \Phi^0_{\tilde{\nu}}(\tilde{\nu})$ and simultaneously eliminating $\mathrm{d}\tilde{\nu}$, then

$$\Phi_{\tilde{\nu}}(\tilde{\nu}, d) = \Phi_{\tilde{\nu}}(\tilde{\nu})\,.\,\exp\left[-\mathrm{const}.\,n\,.\,d\right]. \tag{3.114}$$

The number of interacting oscillators, n, in the volume unit is proportional to the molar concentration,*) c_X, of compound X in the absorbing medium. If the corresponding substitution is carried out, the Lambert–Beer law is obtained in the form

$$\Phi_{\tilde{\nu}}(\tilde{\nu}, d) = \Phi_{\tilde{\nu}}^{0}(\tilde{\nu}) \exp\left[-(\ln 10) \,.\, \varepsilon(\tilde{\nu}) \,.\, c_X d\right] \tag{3.115}$$

or

$$\tau(\tilde{\nu}) = \Phi_{\tilde{\nu}}(\tilde{\nu}, d)/\Phi_{\tilde{\nu}}^{0}(\tilde{\nu}) = \exp\left[-(\ln 10) \,.\, \varepsilon(\tilde{\nu}) \,.\, c_X d\right] \tag{3.116}$$

or

$$D(\tilde{\nu}) = \log\left[1/\tau(\tilde{\nu})\right] = \varepsilon(\tilde{\nu}) \,.\, c_X d. \tag{3.117}$$

Quantity $\varepsilon(\tilde{\nu})$ is termed the molar linear absorption coefficient**) of the given compound at wavenumber $\tilde{\nu}$. If concentration c_X and cell thickness d are substituted into Eq. (3.117) in mol . 1^{-1} and cm, respectively, quantity $\varepsilon(\tilde{\nu})$ is then expressed in units with the dimension 1 . mol^{-1} . cm^{-1}.

It was assumed in the above derivation that the loss in the oscillators capable of interaction through transition from the ground state to an excited state is negligible; this condition is practically always satisfied with common infrared spectrometers, as the spectral radiant powers, $\Phi_{\tilde{\nu}}$, obtained from classical sources (globar, Nernst burner, etc.) are not large.

It simultaneously follows from the derivation procedure that the absorbances of various oscillators are additive, without respect to whether they belong to molecules of the same type or of various types (cf. Section 5.2.1).

As a consequence of the statistical distribution of various effects (see Section 2.2.2), the properties of a single type of oscillator change randomly; hence, not all of them absorb at the same wavenumber. If it is assumed that the probability of the absorption process is virtually independent of the magnitude of the absorbed energy quantum, $h\nu$, then statistically the most probable oscillator arrangements correspond to the band maximum wavenumber, $\tilde{\nu}_{max}$; the course of $\varepsilon(\tilde{\nu})$ for the given band is then analogous to the distribution function of the frequencies of these oscillators.

If function $\varepsilon(\tilde{\nu})$ is integrated over the whole spectral band, a value is obtained which is proportional to the mean probability of photon absorption by the given oscillator. This value is termed the integrated band intensity. It is evident that the relationship between the integrated intensity and the

---

*) More precisely, to the activity of a certain form of compound X, see Section 5.3.

**) This quantity was previously called the molar extinction coefficient or the molar absorptivity. The word "linear" is usually omitted.

oscillator inherent properties are much closer than with the maximum intensity.*)

A number of various mathematical definitions of the integrated intensity can be found in the literature, differing mostly only by a constant. The IUPAC Comission for Molecular Structure and Spectroscopy, in an effort to unify these definitions, has recommended the use to three of them.

The principal quantity is the absolute integrated intensity A_{abs}, defined by the relationship,

$$A_{abs} = \frac{2.3026}{n_X d} \int_{band} P(\nu)\, d\nu, \tag{3.118}$$

where d is the cell thickness, n_X is the number of molecules of X in a volume unit and ν is the frequency of the absorbed radiation. Quantity A_{abs} is thus related to a single molecule of the studied substance. If d is substituted in cm, n_x in molecules per cm^3 and ν in Hz, then the absolute integrated intensity unit has the dimension $cm^2 . s^{-1}$.

The secondary quantity, also related to a single molecule, is the integrated intensity A_{sec}, defined by

$$A_{sec} = \frac{2.3026}{n_X d} \int_{band} P(\nu)\, d(\ln \nu). \tag{3.119}$$

If n_X and d are substituted in the above units, then the dimension of the unit of A_{sec} is cm^2 and it has the significance of the molecule effective cross-section.

A practical quantity, used commonly especially in analytical practice, is integrated intensity A_{pract}, defined by the relationship,

$$A_{pract} = \frac{1}{f_X c_X d} \int_{band} P(\tilde{\nu})\, d\tilde{\nu}, \tag{3.120}$$

where c_X is the (analytical) concentration of substance X, f_X is the dimensionless activity coefficient and $\tilde{\nu}$ is the wavenumber of the absorbed radiation. If these quantities are substituted in common units $-c_X$ in mol . l^{-1} and $\tilde{\nu}$ in cm^{-1}, the dimension of the integrated intensity unit is l . mol^{-1} . cm^{-2}.

Two other definitions of integrated intensities can be encountered in the

---

*) Integrated intensity A_{abs}, corresponding to a vibrational transition, is proportional to $(\partial\mu/\partial q_i)^2$, where μ is the molecule dipole moment and q_i is the appropriate internal coordinate. Cf. Section 2.2 for the biatomic molecule approximation.

older literature. If the ln $(\Phi°/\Phi)$ scale is employed in place of the absorbance scale,*) the quantity

$$A_{RJ} = \frac{1}{c_X d} \int_{band} \ln(\Phi^0/\Phi)\, d\tilde{\nu} = \frac{2.3026}{c_X d} \int_{band} P(\tilde{\nu})\, d\tilde{\nu}, \qquad (3.121)$$

is obtained, the unit dimension being $1 \,.\, mol^{-1} \,.\, cm^{-2}$.

Further, if the integration is carried out on a wavelength scale λ, the relationship,

$$A_M = \frac{2.3026}{c_X d} \int_{band} P(\lambda)\, d\lambda. \qquad (3.122)$$

is obtained. If wavelength λ is expressed in cm, the unit of quantity A_M has the dimension $1 \,.\, mol^{-1}$.

The following relationships hold among the individual quantities given above**)

$$A_{abs} = [\tilde{\nu}_{max} c \,.\, A_{sec}] = 2.3026 \,.\, c \,.\, N_A^{-1} \,.\, A_{pract} = c \,.\, N_A^{-1} \,.\, A_{RJ} = = [\tilde{\nu}_{max}^2 \,.\, c \,.\, N_A^{-1} \,.\, A_M]; \qquad (3.123)$$

$$A_{see} = [\tilde{\nu}_{max}^{-1} \,.\, c^{-1} \,.\, A_{abs}] = [2.3026 \,.\, \tilde{\nu}_{max}^{-1} \,.\, N_A^{-1} \,.\, A_{pract}] = = [\tilde{\nu}_{max}^{-1} \,.\, N_A^{-1} \,.\, A_{RJ}] = [\tilde{\nu}_{max} \,.\, N_A^{-1} \,.\, A_M]; \qquad [3.124)$$

$$A_{pract} = 0.4343 \,.\, c^{-1} \,.\, N_A \,.\, A_{abs} = [0.4343 \,.\, \tilde{\nu}_{max} \,.\, N_A \,.\, A_{sec}] = = 0.4343 \,.\, A_{RJ} = [0.4343 \,.\, \tilde{\nu}_{max}^2 \,.\, A_M], \qquad (3.125)$$

where c is the velocity of light, N_A is Avogadro's constant and $\tilde{\nu}_{max}$ is the band maximum wavenumber. If the units specified with the definition formulae, (3.118)–(3.122), are used to express the individual integrated intensity types and $\tilde{\nu}_{max}$ is expressed in cm^{-1}, the numerical values of the conversion relationships will equal

$$A_{abs} = [2.998 \,.\, 10^{10} \,.\, \tilde{\nu}_{max} \,.\, A_{sec}] = 1.145 \,.\, 10^{-10} \,.\, A_{pract} = = 4.976 \,.\, 10^{-11} \,.\, A_{RJ} = [4.976 \,.\, 10^{-11} \tilde{\nu}_{max}^2 \,.\, A_M]; \qquad (3.126)$$

$$A_{sec} = [3.335 \,.\, 10^{-11} \,.\, \tilde{\nu}_{max}^{-1} \,.\, A_{abs}] = [3.820 \,.\, 10^{-21} \,.\, \tilde{\nu}_{max}^{-1} \,.\, A_{pract}] = = [1.660 \,.\, 10^{-21} \tilde{\nu}_{max}^{-1} \,.\, A_{RJ}] = [1.660 \,.\, 10^{-21} \,.\, \tilde{\nu}_{max} \,.\, A_M]; \qquad (3.127)$$

$$A_{pract} = 8.728 \,.\, 10^9 \,.\, A_{abs} = [2.617 \,.\, 10^{20} \tilde{\nu}_{max} \,.\, A_{sec}] = = 0.4343 \,.\, A_{RJ} = [0.4343 \,.\, \tilde{\nu}_{max}^2 \,.\, A_M]. \qquad [3.128)$$

---

*) In other words, if the decadic logarithm is replaced by the the natural logarithm in Eq. (2.14).

**) Formulae in square brackets are approximate.

3.3.3.4.1 Determination of the Band Maximum Absorbance and Molar Absorption Coefficient and of the Band Half-Width

While in the determination of the position of an isolated spectral band maximum the absolute value of the spectral background need not be known, this knowledge is necessary for determination of the band maximum absorbance. Disregarding the differential analysis at various concentrations or cell thicknesses, it is necessary to introduce an assumption concerning the profile function shape.

Simultaneous determination of all the parameters of the profile function will be dealt with in Section 3.3.4.3.3. Here only an approximate graphical method will be given (cf. Fig. 35).

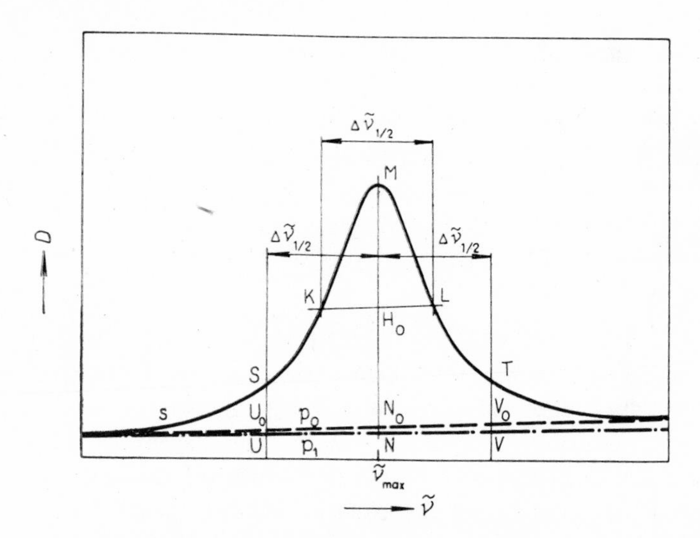

Fig. 35. Geometric construction of the band maximum absorbance. The initial background estimate is $p_0 = \overline{U_0N_0V_0}$, the corrected background is $p_1 = \overline{UNV}$; the intial absorbance estimate equals the length of $\overline{MN_0}$ and the corrected value is the length of $\overline{MN}$. For the explanation of the symbols see the text.

First, the background shape, p_0, is approximately assessed. A straight line perpendicular to the wavenumber axis is constructed at the band maximum (wavenumber $\tilde{\nu}_{max}$); this line intersects the spectral curve in point M and the estimated background curve in point N_0. Distance MN_0 is then the initial estimate of the maximum absorbance if the spectrum is recorded on the absorbance scale; Otherwise, the $D^{(0)}_{max}$ value is calculated from the appropriate transmittance value,

$$D^{(0)}_{max} = \log (\tau_{N_0}/\tau_M). \tag{3.129}$$

Point H_0 corresponding to absorbance $D_H = 1/2 \,.\, D^{(0)}_{max} + D_N$ is constructed and straight line p'_0 parallel with the background is drawn through it. The distance between points K and L is the initial estimate of the band half-width. Points S and T are then constructed at a distance of one half-width from the band maximum. Assuming the Cauchy profile, the absorbance at these points should be

$$D_S - D_U = D_T - D_V = 0.2 \,.\, D_{max}. \tag{3.130}$$

The background is thus corrected so that the lengths of $\overline{SU}$ and $\overline{TV}$ satisfy this requirement and the whole procedure is repeated with the corrected background until Eq. (3.130) is satisfied with sufficient accuracy.

The length of segment $\overline{PQ}$, constructed at three quarters of the band height, can be employed to check whether the band profile corresponds to the Cauchy curve; it should approximately equal $0.57 \,.\, \Delta\tilde{\nu}_{1/2}$. If a significant difference is found, some of the more complex functions given in Tables III and IV can be used for the approximation.

The band maximum molar absorption coefficient, ε_{max}, is obtained from the relationship,

$$\varepsilon_{max} = D_{max}/(c_X d). \tag{3.131}$$

If the quantities are read directly from an experimental spectrum, apparent values, distorted by the instrument function, are obtained. The corrections necessary to obtain the true values have been described in Section 3.3.3.3.3. The molar absorption coefficient dependences on the temperature and medium are discussed in Sections 3.4 and 5.2.2.

3.3.3.4.2 Determination of the Integrated Intensity

An ideal situation for the determination of the integrated intensity involves knowledge of the band profile function and the values of the function parameters. Analytical formulae can then be employed for calculation of the integral, or, in more complex cases, numerical integration can be carried out.

For bands with the Cauchy profile, (2.21), the integrated intensity equals

$$A_{pract} = \pi \,.\, \varepsilon_{max}/b_c. \tag{3.132}$$

For the Gaussian profile, (2.23),

$$A_{\text{pract}} = \sqrt{\pi} \, . \, \varepsilon_{\max}/b_g . \tag{3.133}$$

For a sum Cauchy–Gaussian curve, (2.28), the integrated intensity is obtained as the sum of the contributions from the appropriate curves. The integral of the product function, (2.29), is proportional to the real part of complex error function, erf.

Derivation of an analytical expression for the integrated intensity of a band described by a fractional rational function of type (2.30) is much more involved. Function (2.30) must first be converted into the form,

$$P(\tilde{\nu}) = \frac{\varepsilon_{\max} c_X d}{\{\sum_{j=0}^{n} \beta_{n-j}[i(\tilde{\nu} - \tilde{\nu}_{\max})]^j\}\{\sum_{j=0}^{n} \beta_{n-j}[i(\tilde{\nu}_{\max} - \tilde{\nu})]^j\}}, \tag{3.134}$$

where i is an imaginary unit. The expressions in the braces are the complex conjugate and hence the function is real and non-negative over the whole wavenumber range. Coefficients β_{n-j} are connected with width parameters b_i of function (2.30); the mutual relationship is found by carrying out the multiplication in the denominator in Eq. (3.134) and comparing with function (2.30). Coefficient β_n always equals unity ($\beta_n = 1$). Now determinants

$$H_n = \begin{vmatrix} \beta_1 i^{n-1} & \beta_3 i^{n-3} & \beta_5 i^{n-5} & \ldots & 0 \\ \beta_0 i^n & \beta_2 i^{n-2} & \beta_4 i^{n-4} & \ldots & 0 \\ 0 & \beta_1 i^{n-1} & \beta_3 i^{n-3} & \ldots & 0 \\ 0 & \beta_0 i^n & \beta_2 i^{n-2} & \ldots & 0 \\ \vdots & & & & \vdots \\ 0 & 0 & 0 & \ldots & \beta_n i^0 \end{vmatrix} \tag{3.135}$$

and

$$G_n = \begin{vmatrix} 0 & 0 & 0 & \ldots & \varepsilon_{\max} \\ \beta_0 i^n & \beta_2 i^{n-2} & \beta_4 i^{n-4} & \ldots & 0 \\ 0 & \beta_1 i^{n-1} & \beta_3 i^{n-3} & \ldots & 0 \\ 0 & \beta_0 i^n & \beta_2 i^{n-2} & \ldots & 0 \\ \vdots & & & & \vdots \\ 0 & 0 & 0 & \ldots & \beta_n i^0 \end{vmatrix}, \tag{3.136}$$

are defined. The integrated intensity value for polynomial function (3.134) then equals

$$A_{\text{pract}} = \frac{\pi i \, . \, G_n}{\beta_0 i^n \, . \, H_n} . \tag{3.137}$$

For polynomial function (2.30) or (3.134) of the second order ($n = 2$), substitution yields

$$A_{\text{pract}} = \frac{\pi \varepsilon_{\max}}{\beta_1} = \frac{\pi \varepsilon_{\max}}{[b_1^2 + b_2 \cdot 2]^{\frac{1}{2}}} \,. \tag{3.138}$$

For the third order ($n = 3$),

$$A_{\text{pract}} = \frac{\pi \beta_1 \varepsilon_{\max}}{\beta_1 \beta_2 - \beta_0} \,. \tag{3.139}$$

The following relationships then hold between coefficients b_i and β_j

$$\beta_0^2 = b_3^2/6, \tag{3.140}$$

$$\beta_1^2 - 2\beta_0\beta_2 = b_2^2/2 \tag{3.141}$$

and

$$\beta_2^2 - 2\beta_1 = b_1^2 \,. \tag{3.142}$$

If the profile function is not expressed analytically, common graphical or numerical methods can be employed for the integration. It is usually impossible to integrate over the whole wavenumber range; therefore the integration is stopped at a certain distance from the band maximum and a correction is made for the neglected contributions from the band wings. If the area under the integrated part of the band profile equals I, then the integrated intensity is obtained on multiplication by correction factor K_{wing},

$$A_{\text{pract}} = K_{\text{wing}} \cdot I. \tag{3.143}$$

The values of coefficient K_{wing} for bands with the Cauchy profile are given in Table XXIII.

If the basis for the determination of the integrated intensities is an experimental spectrum, apparent values, affected by the instrument function, are obtained. Assuming the Cauchy profile, the procedure given in Section 3.3.3.3.3 can be employed for corrections to obtain the true values.

If the band has not the Cauchy profile and if high precision of the determination of the integrated intensity is required, the Wilson – Wells method is employed, based on the assumption that apparent integrated intensity A'_{pract} approaches true intensity A_{pract} with decreasing band maximum absorbance. The relationship between the two quantities is given by

$$A_{\text{pract}} = A'_{\text{pract}}/(1 + \vartheta a'_{\max}). \tag{3.144}$$

For a detailed discussion of the derivation of this expression, see ref.[1]

1 Seshadri K. S., Jones R. N.: The Shapes and Intensities of Infrared Absorption Bands — A Review. *Spectrochim. Acta 19*, 1013 (1962).

Table XXIII

Coefficients K_{wing} for the integrated intensity correction, using formula (3.143)

$\frac{(\tilde{\nu}_2 - \tilde{\nu}_1)}{\Delta\tilde{\nu}_{1/2}}$	K_{wing}	$\frac{(\tilde{\nu}_2 - \tilde{\nu}_1)}{\Delta\tilde{\nu}_{1/2}}$	K_{wing}	$\frac{(\tilde{\nu}_2 - \tilde{\nu}_1)}{\Delta\tilde{\nu}_{1/2}}$	K_{wing}
4.0	1.185	7.0	1.099	10.0	1.068
4.2	1.175	7.2	1.096	10.5	1.064
4.4	1.166	7.4	1.094	11.0	1.061
4.6	1.158	7.6	1.091	11.5	1.058
4.8	1.151	7.8	1.088	12.0	1.056
5.0	1.144	8.0	1.086	12.5	1.053
5.2	1.138	8.2	1.084	13.0	1.051
5.4	1.132	8.4	1.081	13.5	1.049
5.6	1.127	8.6	1.079	14.0	1.047
5.8	1.122	8.8	1.078	14.5	1.046
6.0	1.118	9.0	1.076	15.0	1.044
6.2	1.113	9.2	1.074		
6.4	1.109	9.4	1.072		
6.6	1.106	9.6	1.070		
6.8	1.103	9.8	1.069		

$\tilde{\nu}_1, \tilde{\nu}_2$ — integration limits, placed symmetrically with respect to the band maximum.

Constant ϑ depends on the ratio of the instrument spectral slit width and the band half-width (see Table XXIV). Since complications may arise due to changes in the activity coefficient values of the studied compounds in dependence on changes in the concentration, it is more advantageous to

Table XXIV

Coefficient ϑ for the correction of the integrated intensity from the apparent to the real value, using Eq. (3.144)

$s/\Delta\nu_{1/2}$	ϑ	$s/\Delta\nu_{1/2}$	ϑ
0.1	—0.002	0.6	—0.028
0.2	—0.004	0.7	—0.036
0.3	—0.008	0.8	—0.046
0.4	—0.013	0.9	—0.057
0.5	—0.020	1.0	—0.070

extrapolate the integrated intensity values to zero absorbance by changing the cell thickness.

Generally, it can be observed that the integrated intensity is less sensitive to instrumental effects than the maximum absorbance and the band half-width.

3.3.4 Parameters of Overlapping Bands

The determination of the band maximum wavenumbers, band intensities and some other spectroscopic quantities such as the band width and symmetry is more complicated in real spectra than in idealized cases described in Sections 3.3.1 – 3.3.3. Complications stem from the fact that an isolated band is exceptional in real spectra and band separation must be undertaken before obtaining the band parameters. Only then can the methods and procedures described in the previous sections be applied.

Mutual overlapping of several bands is common in real spectra. Problems connected with the overlap can, however, be completely explained using a doublet; overlapping of a larger number of bands only makes the separation procedures more complex.

From a practical point of view, it is very important to have an indication, either experimental or theoretical, of band overlapping. With close band distances and low spectrometer resolution several bands may unresolvably merge into a single band. Difficulties then lie in the fact that any band can be described as a superposition of an arbitrary number of bands. An important role is played here by the spectrometer resolution, which will be discussed in the following section.

3.3.4.1 Spectrometer Resolution

Resolution R is one of the principal parameters determining the spectrometer quality. It is reported as distance $\Delta\tilde{\nu}$ between the maxima of two bands in the wavenumber scale, which the spectrometer can still differentiate at wavenumber $\tilde{\nu}$. Quantitatively it is given by

$$R = \tilde{\nu}/\Delta\tilde{\nu}. \tag{3.145}$$

The spectrometer resolution is predominantly determined by the quality of its monochromator. However, as a gradually narrower beam (in the energy wavenumber scale) is formed with increasing resolution, the radiant power incident on the detector decreases. Because of limitations due to noise (Section 3.3.3.1), the narrowing of the beam must be compensated for by increasing either the source power or the detector sensitivity, or both simultaneously. Hence the spectrometer resolution depends on the quality of all its component parts and also affects the instrument price. The maximum attainable resolution is actually a criterion of the spectrometer

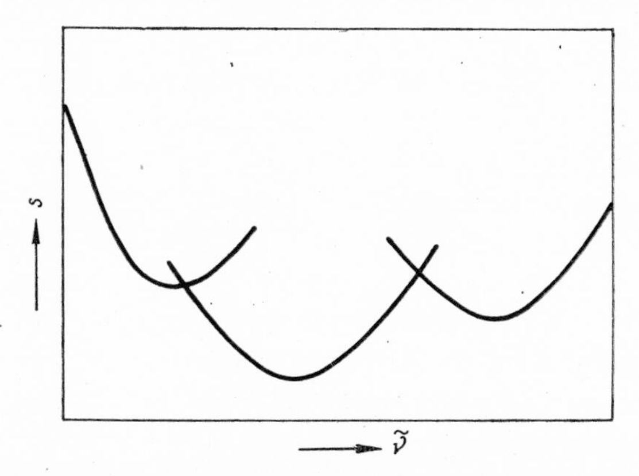

Fig. 36. Dependence of spectral slit width s on wavenumber $\tilde{\nu}$. A single grating and a single order cannot be used for the whole infrared spectrometer scale; this fact is expressed by the individual curves in the figure, as each is related to a certain order of a certain grating.

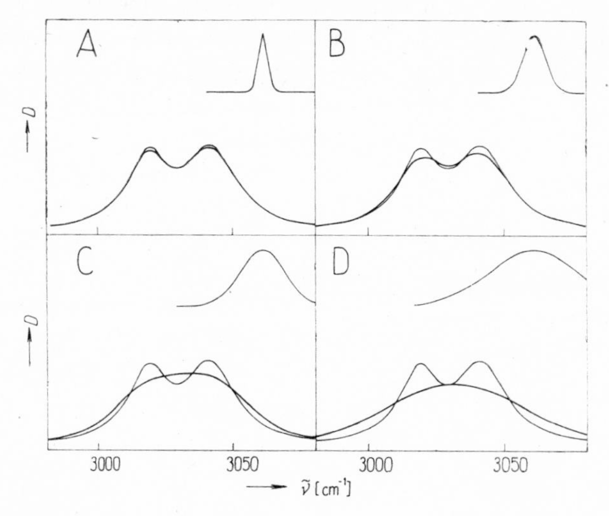

Fig. 37. Resolution of the doublets of the stretching vibrations of the C—H bonds for methanesulphonyl chloride (bands Nos. 1 and 2 in Table V) on a "spectrometer" with variable resolution; curves A—D were obtained by computer simulation using convolution with the instrument function depicted in the upper right corner of the figures.

quality, in the broadest sense; it can also serve as a criterion for classification of spectrometers into quality classes.

At present, diffraction gratings are almost exclusively employed for monochromatization in infrared and Raman spectrometers. Their resolution is given by the relationship,

$$R = Nn, \tag{3.146}$$

where N is the overall number of grooves on the grating and n is the interference order, by which the resolution is defined. It is evident that the resolution increases with increasing total size of the grating (as the number of grooves per unit length is dictated by the spectral region employed – here 90 – 30 grooves per millimetre) and with increasing interference order; it is independent of ambient conditions (e.g. the temperature), as no parameter in Eq. (3.146) is affected by them.

This spectrometer resolution is dependent not only on the quality of the grating but also on the quality of the whole photometric process. With the same quality grating, poorer results may be obtained when using a detector of low sensitivity, as it is necessary to compensate for the detector shortcomings by enlargening the geometric slit. The resolution also changes due to non-uniform contents of individual wavelengths in the source radiation, which must also be compensated for by changes in the geometric slit width; an example of a change in the spectrometer resolution stemming from this effect is given in Fig. 36 in dependence on the wavenumber.

The spectral slit width (Section 3.3.3.3.1) can be employed as an approximate measure of the spectrometer resolution. Figs. 37A – 37D depict how a change in the instrument resolution, expressed in terms of the values of the spectral slit width, can affect the appearance of an experimental spectrum. The figures were obtained by computer simulation from the parameters of bands 1 and 2 in the methanesulphonyl chloride spectrum (Table V) after subsequent convolution. The calculated spectrum (Fig. 37A), where the convolution was carried out for value $s = 4\ \text{cm}^{-1}$, is almost identical with the true spectrum (unaffected by the instrument parameters); an analogous situation is depicted in Fig.37 B, where $s = 10\ \text{cm}^{-1}$. In the third figure, where $s = 20\ \text{cm}^{-1}$, the doublet has already merged into an asymmetrical band and in the last figure ($s = 32\ \text{cm}^{-1}$) there is no trace of the existence of two bands.

Real spectrometer resolution must be determined experimentaly using components of rotational-vibrational spectra of gases, which can still be resolved with the given instrument adjustment. Here the resolution criteria should be mentioned.

The Rayleigh criterion, common in interference optics, according to which two interference patterns are considered resolved when the maximum of one falls in the first minimum of the other, has not given very good results in practical spectroscopy. The Sparrow criterion, according to which two interference patterns whose maxima are still separated by a perceptible minimum are considered resolved, is more frequently used in practice. However, all these definitions are problematic when the resolution of two bands of unequal intensities is to be evaluated.

In some cases, especially with high-resolution spectrometers, the instrument resolution can be estimated from band half-widths in rotational-vibrational transitions of gaseous samples. The method is based on the fact that the natural width of rotational-vibrational lines is very small, usually much smaller than the spectrometer spectral slit width (by one to three orders of magnitude) and thus the measured line width is chiefly proportional to the slit width. It is assumed that the resolution is cca 2/3 of the experimental half-width.

3.3.4.2 Simple Corrections of Parameters of Overlapping Bands

If a doublet with more or less resolved components is present in the studied spectrum, simple graphical-numerical methods can be employed for the determination of approximate values of the parameters of its components. The attainable precision is not particularly high, but suffices for many applications. An undisputable advantage of these methods is their rapidity and simplicity.

When determining the maxima positions for the components of a clearly resolved doublet (Fig. 38), it is assumed that the effect of one band on the farther wing of the other band is negligible and that this wing can consequently be considered as "background" superimposed on the first band. Let us assume that the spectrum is plotted on the absorbance scale. In the first step the maximum positions on the spectral curve are estimated roughly (points A and B) and a straight line perpendicular to the wavenumber axis is constructed at wavenumber $(\tilde{\nu}_A + \tilde{\nu}_B)/2 = \tilde{\nu}_s$. Point C is further found on the spectral curve so that $|\tilde{\nu}_C - \tilde{\nu}_A| = |\tilde{\nu}_A - \tilde{\nu}_B|$. Tangent DE to the spectral curve is constructed at point C. By transfering point D according to the symmetry axis, SE, to the other side, the estimated background direction, FE $= a_0$, is obtained. Lines $a_1 - a_5$ parallel to this straight line are then constructed, forming a number of chords of the bands, whose halving gives the band median, t_G. The latter intersects the spectral curve in point G. Wavenumber $\tilde{\nu}_G$ is considered as an estimate of the maximum position for one component of the doublet. The wavenumber for the maximum of the other component, $\tilde{\nu}_M$, is determined analogously.

Estimating of the maximum intensity and of the half-width is more complicated. As the first estimate of $D^{(0)}_{max,G}$, the difference*)

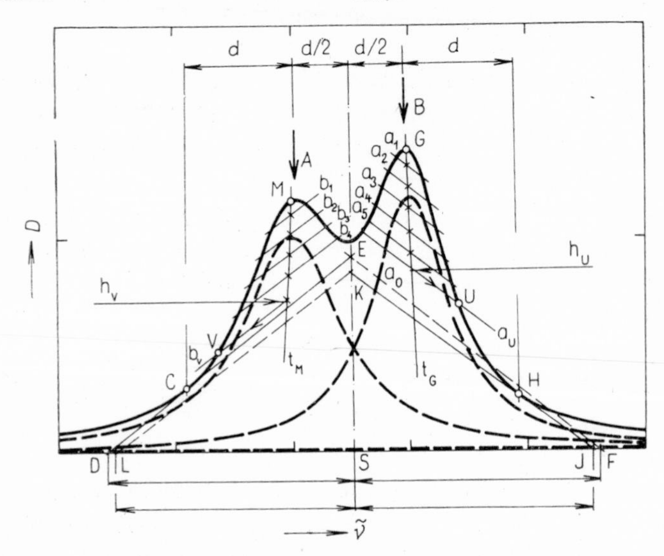

Fig. 38. Determination of the position and the intensity of the maxima of the components of a markedly resolved doublet. For the significance of the symbols, see the text.

$$D^{(0)}_{max,G} = D_G - D_C \tag{3.147}$$

is taken, and, analogously,

$$D^{(0)}_{max,M} = D_M - D_H . \tag{3.148}$$

If the transmittance scale is used directly, the previous formulae are converted into the form,

$$D^{(0)}_{max,G} = \log \tau_C/\tau_G \quad \text{or} \quad D^{(0)}_{max,M} = \log \tau_H/\tau_M . \tag{3.149}$$

These values are somewhat underestimated, as the effect of one band on the outer wing of the other band was neglected. Now lines parallel with the wavenumber axis are constructed at absorbance values $D_U = D_G - 0.5$.

*) If a maximum lies under the doublet and is substantially broadened with respect to the wavenumber axis, the values of D_G, D_C, D_M and D_H are first corrected for its contribution.

$. D^{(0)}_{max,G}$ or $D_V = D_M - 0.5 . D^{(0)}_{max,M}$; their intercepts with the spectral curve are denoted as U and V, respectively. The estimated band half-widths are then given by

$$\Delta\tilde{\nu}_{\frac{1}{2},G} = 2 . | \tilde{\nu}_G - \tilde{\nu}_U | \quad \text{or} \quad \Delta\tilde{\nu}_{\frac{1}{2},M} = 2 . | \tilde{\nu}_M - \tilde{\nu}_V |. \qquad (3.150)$$

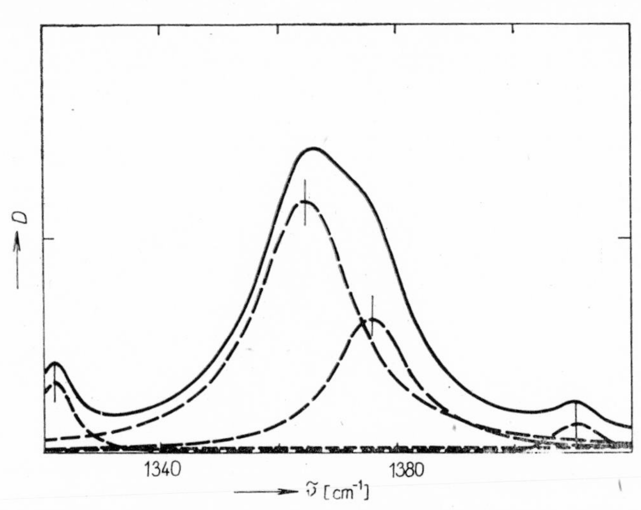

Fig. 39. The group of strongly overlapping bands in the region around 1400 cm^{-1} in the methanesulphonyl chloride infrared spectrum.

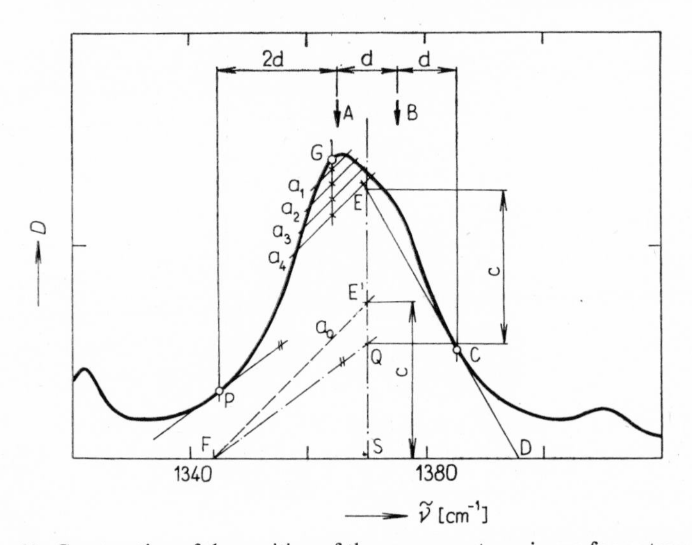

Fig. 40. Construction of the position of the component maximum for a strongly overlapping doublet (1st part). For the significance of the symbols, see the text.

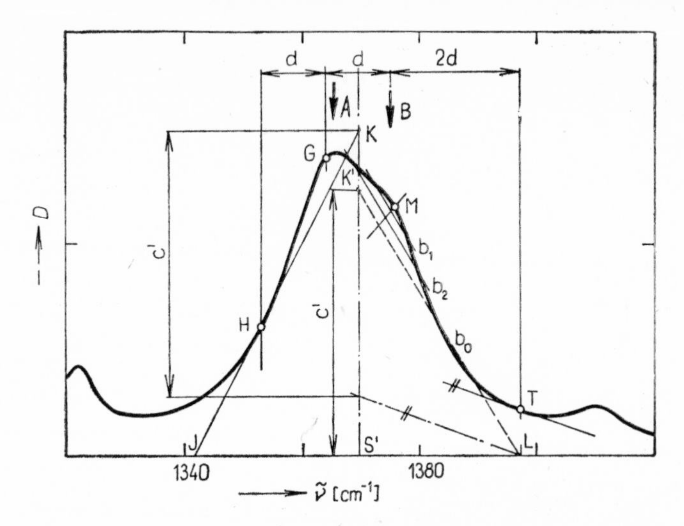

Fig. 41. Construction of the position of the component maxima for a strongly overlapping doublet (2nd part). For the significance of the symbols, see the text.

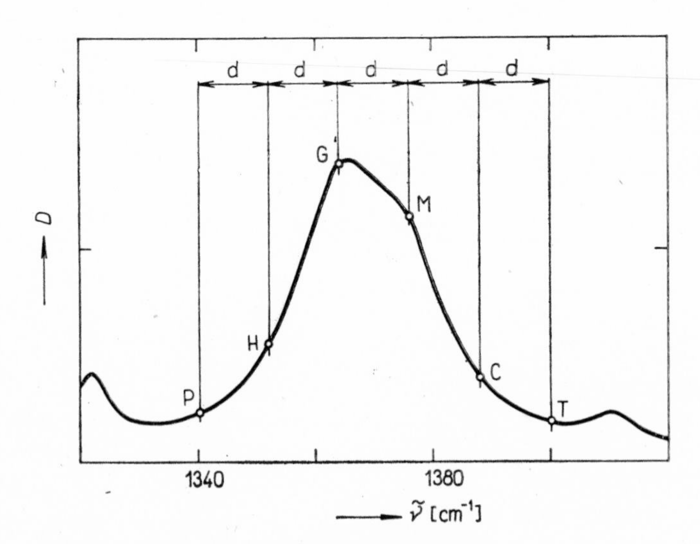

Fig. 42. Determination of the intensity values at the component maxima for a strongly overlapping doublet. For the significance of the symbols, see the text.

The procedure in the transmittance scale is analogous; the D_U and D_V values are converted into the corresponding transmittance values, $\tau_U = 10^{2-D_U}$, and lines parallel with the wavenumber axis are constructed at this value.

Table III (or IV) can now be used for refining the maximum intensity values. The correction for the effect of the wings is calculated for the selected profile; the corrected value of the maximum intensity then equals

$$D^{(1)}_{\max,G} = D^{(0)}_{\max,G} + P(2 \,.\, |\,\tilde{\nu}_G - \tilde{\nu}_C\,|)/\Delta\tilde{\nu}_{\frac{1}{2},G}). \qquad (3.151)$$

The correction for the other doublet component is analogous. The refined values of $D^{(1)}_{\max,G}$ and $D^{(1)}_{\max,M}$ obtained can then be employed for simultaneous refinement of the band half-widths.

If the doublet components strongly overlap (Fig. 39), the situation is more complex. Here distortion of band outer wings in the doublet must be considered even in the first determination. The geometric construction for maximum location is evident from Figs. 40 and 41. Points A and B are again obtained by initial estimation and points G and M by refined estimation of the maximum position.

In order to assess the maximum intensities, more complex formulae must be employed (see Fig. 42),

$$D^{(0)}_{\max,G} = D_G - D_{G,p} - (D_C - D_{C,p}) + D_P - D_{P,p}, \qquad (3.152)$$

where subscript $_p$ denotes the corresponding background.

The construction of points for location of the half-width values is quite analogous to the previous case. The maximum intensities and half-widths obtained can again be further refined.

If, however, the parameters obtained are used as initial estimates for numerical separations (Section 3.3.4.3.3), there is no need for further refining; experience has shown that even very rough estimates are sufficient.

3.3.4.3 Spectral Band Separation

It was assumed in Section 2.3.1 that a spectrum expressed as the absorbance dependence on the wavenumber, $D(\tilde{\nu})$, can be written as the sum of the background function, $P_0(\tilde{\nu})$, and a finite number of profile functions $P_i(\tilde{\nu})$. This transcription is suitable in spectroscopic practice for separation of spectra into individual bands and for the determination of the parameters of the corresponding profile functions.

As intensity additivity is valid only in the absorbance scale, it is necessary to convert the spectrum from the more common transmittance scale into the absorbance scale before the treatment. If band separation is expected before measuring the spectrum, it is advantageous to record it directly on the absorbance scale; this will be assumed in the ensuing discussion.

3.3.4.3.1 Graphical Methods

Graphical separation methods, the oldest and simplest, have certain advantages as they require no assumption concerning the analytical shape of the profile functions and no complex technical equipment. However,

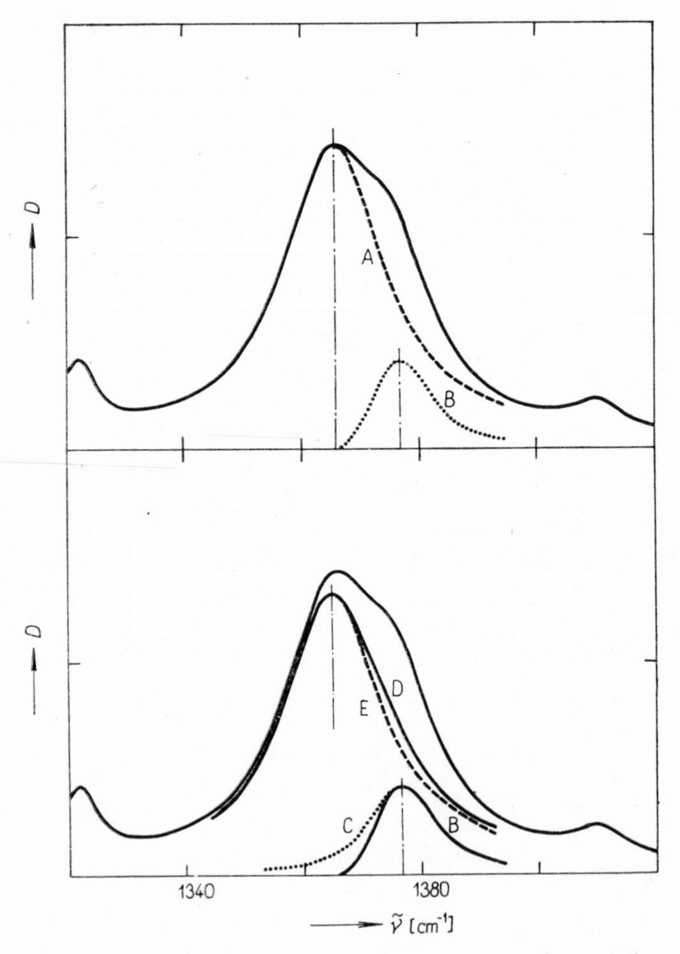

Fig. 43. Procedure for graphical separation of the components of a strongly overlapping doublet: A—E, curves constructed gradually for the separated components. For the significance of the symbols, see the text.

graphical methods are unsuitable for separations of larger numbers of bands and hence their application is mostly limited to the determination of the parameters of band pairs.

Graphical methods are based on gradual symmetrization of all multiplet components. They assume that band profiles are symmetrical in the wavenumber scale, which is actually sufficiently satisfied with most infrared bands. A further assumption requires a linear background, parallel with the wavenumber axis; if this requirement is not met in the experimental spectrum, the contribution from the background must be subtracted prior to the separation itself.

The following scheme is observed in graphical separation of doublet bands (Fig. 43):

1. The band with the greater integrated intensity is selected;
2. the position of the maximum of this band is assessed and a line perpendicular to the wavenumber axis is constructed at this point, forming the symmetry axis of the band;
3. the half of the selected band located on the opposite side to the overlap is rotated along the symmetry axis "in" between the two bands;
4. the band thus symmetrized is graphically subtracted from the experimental spectrum; the result is called the residual spectrum;
5. if the residual spectrum represents a sufficiently symmetrical band, the graphical separation is finished;
6. otherwise, the maximum is located on the residual spectrum curve and a perpendicular line is constructed though this point as the symmetry axis of the other doublet component;
7. operations sub 3–5 are carried out with the other doublet component. In this way the first band is corrected and the cyclic corrections are continued if necessary. Three or four iterations usually suffice for satisfactory solution.

An analogous procedure is adopted for triplets. It is advantageous when the most intense band is not located in the middle. However, if one of the bands is located further from the other two, the symmetrization process is started with it without respect to its intensity.

3.3.4.3.2 Separation on Analog Computers

Prior to mass application of digital computers, analog computers, either universal or single-purpose (analyzers from the Du Pont Company), were employed for separations of overlapping bands. These instruments are rather costly, but the quality of the results obtained does not correspond to the expense and effort. Four to five bands can be reliably resolved, but the parameters obtained must often be refined numerically.

3.3.4.3.3 Numerical Separation

If the profile functions can be expressed analytically, digital computers can be used for band separation in the spectra. The question of profile functions has been discussed (Section 2.4.1); it follows from practice that for most common applications with spectra measured in the liquid state, the Cauchy (Lorentz) function [Eq. (2.21)] suffices for band description. Therefore, in this section attention will be centred principally upon this function.

The simplest case, but rarely occuring in practice, is that when the positions and half-widths of all bands in the studied spectral region are known with sufficient precision and only the individual band maximum intensities ($D_{max,i}$) are to be determined. The spectrum is a linear function of these parameters; the process of finding the best fitting values is mathematically multidimensional linear regression (Section 3.2.2.2.2).

If the spectrum is specified in terms of the experimental absorbance values in N points and of an m-th order polynomial ($m = 0$ corresponds to a constant background), then general functions $a_i(x)$ from relationship (3.51) assume the forms,

$$\left.\begin{aligned} a_i(\tilde{\nu})_j &= 1/[1 - b_{c,i}^2(\tilde{\nu}_j - \tilde{\nu}_{max,i})^2], & i \in \langle 1, n \rangle \\ a_{n+k}(\tilde{\nu}_j) &= (\tilde{\nu}_j - \tilde{\nu}_{centre})^{k-1}, & k \in \langle 1, m+1 \rangle \end{aligned}\right\} j \in \langle 1, N \rangle, \quad (3.153)$$

where $\tilde{\nu}_{centre}$ is an arbitrarily selected wavenumber value, which is set as the "beginning" of the polynomial function. For numerical reasons it is suitable to choose $\tilde{\nu}_{centre}$ in the centre of the studied spectrum interval. The values of the right-hand sides of Eqs. (3.51) are experimentally determined absorbances at wavenumbers $\tilde{\nu}_j$, $j \in \langle 1, N \rangle$

$$y_j = D(\tilde{\nu}_j). \quad (3.154)$$

The obtained (overdetermined) system of linear equations is solved using the procedure described in Section 3.2.2.2.2.

However, all parameters of overlapping bands must usually be determined simultaneously, i.e. the maximum positions and band half-widths must be found in addition to the maximum intensities. The problem is nonlinear with respect to these parameters and the iteration methods described in Section 3.2.2.2.3 must be applied in order to solve it.

A spectrum with n bands and with a background described by an m-th order polynomial is then represented by a $(3n + m + 1)$-dimensional parameter vector, $\boldsymbol{p}$. If the parameters of the individual bands are grouped by three in the order, the maximum position, $\tilde{\nu}_{max,i} = p_{3i-2}$, the width parameter, $b_{c,i} = p_{3i-1}$, and the maximum intensity, $a_{max,i} = p_{3i}$, the corresponding Jacobian elements are given by the relationship

$$\left.\begin{aligned} J_{j,3i-2} &= \frac{2D_{max,i}\, b_{c,i}^2(\tilde{\nu} - \tilde{\nu}_{max,i})}{[1 + b_{c,i}^2(\tilde{\nu} - \tilde{\nu}_{max,i})^2]^2} \\ J_{j,3i-1} &= \frac{2D_{max,i} b_{c,i}(\tilde{\nu} - \tilde{\nu}_{max})^2}{[1 + b_{c,i}^2(\tilde{\nu} - \tilde{\nu}_{max,i})^2]^2} \\ J_{j,3i} &= 1/[1 + b_{c,i}^2(\tilde{\nu} - \tilde{\nu}_{max,i})^2] \end{aligned}\right\} \begin{aligned} & j \in \langle 1, N \rangle, \\ & i \in \langle 1, n \rangle, \end{aligned} \qquad (3.155)$$

while the Jacobian elements corresponding to the background are

$$J_{j,3n+k} = (\tilde{\nu}_j - \tilde{\nu}_{centre})^{k-1}, \qquad k \in \langle 1, m + 1 \rangle, \qquad j \in \langle 1, N \rangle. \quad (3.156)$$

On the right-hand side of the equations of the linearized system appears the difference between the experimental absorbances and those calculated from the best estimate of the spectral parameters:

$$y_j = D(\tilde{\nu}_j) - \Sigma P_i(\tilde{\nu}), \qquad j \in \langle 1, N \rangle. \qquad (3.157)$$

The initial estimates of the parameter vector $\boldsymbol{p}$ elements are obtained using the approximate methods described in the previous sections.

It was observed elsewhere (Section 3.3.3) that the error in the measurement of transmittance τ is virtually constant within a relatively wide transmittance range. It would thus seem advantagous to carry out the calculation in the transmittance scale, not in absorbances. It must, however, be kept in mind that recalculation of absorbances to transmittances requires raising of base 10 to a rational exponent; this operation, which is involved in the calculation of each Jacobian element, places great demands on the computing time. It is therefore better to circumvent this obstacle by employing statistical weights, which reflect the non-uniformity in the error distribution in the absorbance scale and, moreover, involve acknowledgement of the fact that the individual points in the spectrum are *a priori* subject to various errors. As has been pointed out in Section 3.2.2.6, the weight values, w_j, should be inversely proportional to the overall variance (the square of the standard deviation) for the measurement of the individual points. If the standard deviation for stationary transmittance measurement (i.e. at a constant wavenumber) is σ_τ and the error of the wavenumber determination*) is $\sigma_{\tilde{\nu}}$, then, according to the law of error propagation, the overall transmittance standard deviation is given by

$$\sigma_{\tau,\,\mathrm{tot}} = \left[\sigma_\tau^2 + (\mathrm{d}\tau/\mathrm{d}\tilde{\nu})^2\,\sigma_{\tilde{\nu}}^2\right]^{1/2}. \tag{3.158}$$

Applying the same law, the total absorbance value is

$$\sigma_{D,\,\mathrm{tot}} = \mathrm{d}D/\mathrm{d}\tau\,.\,\sigma_{\tau,\mathrm{tot}} = 0.4343\,.\,\sigma_{\tau,\mathrm{tot}}/\tau, \tag{3.159}$$

i.e. it holds for the weight of the j-th point that

$$w_j = \tau^2/\{0.1886\,.\,[\sigma_\tau^2 + (\mathrm{d}\tau/\mathrm{d}\tilde{\nu})^2\,\sigma_{\tilde{\nu}}^2]\}. \tag{3.160}$$

During the calculation, derivative $\mathrm{d}\tau/\mathrm{d}\tilde{\nu}$ can be replaced with sufficient precision by the fraction of the increments of τ and $\tilde{\nu}$ between neighbouring points in the spectrum.

The iteration procedure, described in Section 3.2.2.2.3 is terminated once the sum of the squares of the deviations between the calculated and experimental spectrum is less than a pre-set limit. If statistical weighing is not employed, the resultant sum of the squares of the deviations can be replaced by the estimate of the square of the mean standard deviation for the determination of the absorbance (between 10^{-4} to 10^{-6} for common spectra), multiplied by the number of experimental points, N; with weighing, the expected value of the sum of the squares of the deviations equals $N-1$.

A program for separation of bands expressed by Cauchy (Lorentz)–type profile functions and by fractional rational function (2.30) is given in the Appendix.

If band separation is to be combined with study of the effect of the finite spectral slit width, the transmittance scale must be used. Then the Jacobian elements will be given by (the order of the parameters is identical with that in Eqs. (3.155)–(3.157))

$$\left.\begin{aligned}
J_{j,3i-2} &= -\ln 10\,\frac{2D_{\mathrm{max},i}b_{c,i}^2(\tilde{\nu} - \tilde{\nu}_{\mathrm{max},i})}{[1 + b_{c,i}^2(\tilde{\nu} - \tilde{\nu}_{\mathrm{max},i})^2]^2}\,\tau_{\mathrm{calc}}(\tilde{\nu}_j)\\
J_{j,3i-1} &= \ln 10\,\frac{2D_{\mathrm{max},i}b_{c,i}(\tilde{\nu} - \tilde{\nu}_{\mathrm{max},i})^2}{[1 + b_{c,i}^2(\tilde{\nu} - \tilde{\nu}_{\mathrm{max},i})^2]^2}\,\tau_{\mathrm{calc}}(\tilde{\nu}_j)\\
J_{j,3i} &= \ln 10\,.\,\tau_{\mathrm{calc}}(\tilde{\nu}_j)/[1 + b_{c,i}^2(\tilde{\nu} - \tilde{\nu}_{\mathrm{max},i})^2]
\end{aligned}\right\}\quad \begin{aligned} j &\in \langle 1, N\rangle\\ i &\in \langle 1, n\rangle\end{aligned} \tag{3.161}$$

and for the background,

$$J_{j,3n+k} = \ln 10\,.\,(\tilde{\nu}_j - \tilde{\nu}_{\mathrm{centre}})^{k-1}\tau_{\mathrm{calc}}(\tilde{\nu}_j),\qquad \begin{aligned} j &\in \langle 1, N\rangle\\ k &\in \langle 1, m+1\rangle,\end{aligned} \tag{3.162}$$

---

*) Both values can be considered constant in wide ranges of τ and $\tilde{\nu}$.

where $\tau_{calc}(\tilde{\nu}_j)$ is the calculated transmittance value,

$$\tau_{calc}(\tilde{\nu}_j) = 10^{-D_{calc}(\tilde{\nu}_j)}, \qquad j \in \langle 1, N \rangle, \tag{3.163}$$

without the effect of the instrument function. However, the right-hand side expresses the difference between the experimental and the calculated spectrum, after carrying out the convolution (Section 3.3.3.3), e.g. using Eq. (3.102).

3.3.5 Parameters of Raman Lines

With certain exceptions, Raman spectra are handled similarly to infrared spectra. This fact follows from the formal similarity of the description of infrared bands and Raman lines by profile functions. The profile of Raman lines satisfies the Gauss or Cauchy profile function or some combination thereof; however, the parameters of these functions have different significance for infrared bands and for Raman lines.

The principal parameter of the profile function, the band maximum wavenumber, has the same definition in both cases. Its determination is also identical for infrared and Raman spectra. The band half-width is similar in this respect, as far as its definition and the method of determination are considered. However, for many reasons there is no close relationship between the true width of the infrared band and the width of the Raman line for vibrations which are active in both kinds of vibration spectrum on the basis of the selection rules. Many bands which are very broad in infrared spectra (e.g. those of vibrations of hydrogen-bonded groups and bonds) may have narrow counterparts in the corresponding Raman lines and, on the contrary, narrow infrared bands may be replaced by broader Raman lines. While the infrared maximum wavenumber and the Raman line wavenumber can generally be used alternatively for the description of the transition magnitude between two vibrational levels, data concerning the band (line) width cannot be used in this way.

3.3.5.1 Raman Line Intensities

The problem of the Raman line intensity is quite specific. The Raman spectrum is a spectrum of scattered radiation and hence the basis of the Raman line intensity is quite different in character from that for absorption infrared bands. Radiation is scattered by a molecule in all directions (generally non-uniformly), so that the intensity would have to be monitored in the whole space in order to determine the scattered Raman radiation intensity. This is, of course, not carried out for practical reasons and hence the Raman line intensity value is a partial quantity. It is thus clear that the Raman line intensity can be given only as a relative quantity, related e.g. to a suitably selected standard under certain experimental conditions.

The quantity recorded directly during measurement of Raman spectra is the radiant power

$$\Phi_r(\tilde{\nu}) \doteq \Phi_{\tilde{\nu}}(\tilde{\nu})\,\Delta\tilde{\nu}, \tag{3.164}$$

scattered by the sample in a direction forming angle ϑ with the incident beam of the excitation radiation, within a small wavenumber range, $\Delta\tilde{\nu}$, separated by the monochromator from the overall radiant power Φ_r. The measured scattered beam exhibits a certain finite divergence (determined basically by the inlet slit of the monochromator unit and the overall instrument geometry), which can be described by the magnitude of the appropriate solid angle, $\Delta\Omega$. If the magnitudes of the spectral radiant power of the Raman radiation, $\Phi_{\tilde{\nu}}(\tilde{\nu})$, are related to a unit solid angle (steradian), the quantity,

$$\Phi_{\tilde{\nu}}(\tilde{\nu}, \vartheta) = \mathrm{d}\Phi_{\tilde{\nu}}(\tilde{\nu})/\mathrm{d}\Omega \doteq \Phi_r(\tilde{\nu})/(\Delta\tilde{\nu}\,\Delta\Omega), \tag{3.165}$$

is obtained, termed the spectral radiant power of the sample at the given wavenumber.

The absolute magnitude of the Raman scattering intensity, I_{abs}, for a given spectral transition could be obtained by double integration of the spectral radiant power, $\Phi_{\tilde{\nu}}(\tilde{\nu}, \vartheta)$, over the whole space around the sample and over the entire band of the studied transition and by relating it to the overall number of scattering molecules, n_x,

$$I_{abs} = (1/n_X) \int_{band} \int_{\Omega} \Phi_{\tilde{\nu}}(\tilde{\nu}, \vartheta)\,\mathrm{d}\tilde{\nu}\,\mathrm{d}\Omega. \tag{3.166}$$

The quantity thus defined has the significance of the effective molecular cross-section for Raman scattering and has the dimension of area.

Experiments in which the absolute magnitude of the Raman scattering intensity was determined have shown that, similar to classical scattering, it is inversely proportional to the fourth power of the wavelength of the excitation radiation. It then follows that the Raman line intensity depends on an adjustable parameter; although it is not always advantageous for other reasons, Raman effect excitation sources with the shortest possible wavelength should be employed to attain the maximum intensity of Raman lines. The use of the red excitation line of laser sources is not in accordance with this requirement and only the very high intensity of these sources helps to overcome this handicap.

It can be shown in a quasi-classical approximation that the magnitude of Raman scattering in the Stokes spectral region (the molecule effective cross-section) equals

$$I_{\mathrm{St,i}} = \mathrm{const} \,.\, (\tilde{\nu}_0 - \tilde{\nu}_i)^4 / \{\mu_i \tilde{\nu}_i [1 - \exp(-hc\tilde{\nu}_i/kT)]\}, \qquad (3.167)$$

where $\tilde{\nu}_i$ and $\tilde{\nu}_0$ are the wavenumbers of the vibrational transition and of the excitation radiation, respectively, and μ_i is the reduced mass of the (biatomic) harmonic oscillator. The proportionality constant depends chiefly on the partial derivatives of the elements of the molecule polarizability tensor with respect to the appropriate normal coordinate; this term is proportional to $(\partial\alpha/\partial q_i)^2$ for isotropic molecules.

The ratio of the intensities of anti-Stokes and Stokes lines can be expressed as

$$I_{\mathrm{aSt,i}}/I_{\mathrm{St,i}} \doteq [(\tilde{\nu}_0 + \tilde{\nu}_i)/(\tilde{\nu}_0 - \tilde{\nu}_i)]^4 \,.\, \exp(-hc\tilde{\nu}_i/kT). \qquad (3.168)$$

However, in practice, Raman line intensities are always given for a certain angle ϑ between the scattered and the incident radiation. Basically, three procedures are employed. The first is based on selecting the most intense Raman line in the spectrum (on the $\Phi_{\tilde{\nu}}(\tilde{\nu}, \vartheta)$ scale) and attributing unit intensity to it. The intensity of the other lines is then expressed relatively with respect to this standard intensity. The intensity value is specified in brackets after the line wavenumber, e.g. 1705 cm^{-1} (0.3) means that a Raman line with wavenumber 1705 cm^{-1} attains 30 % of the intensity of the strongest line in the spectrum, which must be given explicitly; the selection of the strongest line need not always be identical when spectra are recorded on various spectrometers.

Another procedure involves the determination of the "standard intensity", $I^{\mathrm{i}}_{\mathrm{std}}$, with respect to an external standard, which is usually, tetrachloromethane. The standard intensity is introduced by the relationship

$$\begin{aligned} I^{\mathrm{i}}_{\mathrm{std}} = {} & (I^{\mathrm{i}}/I^{\mathrm{std}})\,(1 + \varrho^{\mathrm{std}}/1 + \varrho^{\mathrm{i}})\,(n/n^{\mathrm{std}})^2\,(\sigma^{\mathrm{std}}/\sigma^{\mathrm{i}})\,(M/d)\,. \\ & .\,(d^{\mathrm{std}}/M^{\mathrm{std}})\,(\Delta\tilde{\nu}^{\mathrm{i}}/\Delta\tilde{\nu}^{\mathrm{std}})\,[(\tilde{\nu}^{\mathrm{i}} - \tilde{\nu}^{\mathrm{std}})/(\tilde{\nu}^{\mathrm{i}} - \Delta\tilde{\nu}^{\mathrm{i}})]^2\,. \\ & .\,[1 - \exp.\,(1.44\,\Delta\tilde{\nu}^{\mathrm{i}}/T)]\,[1 - \exp.\,(1.44\,\Delta\tilde{\nu}^{\mathrm{std}}/T], \end{aligned} \qquad (3.169)$$

where

$$I_{\mathrm{i}} = \int_{\mathrm{band}} \Phi_{\tilde{\nu}}(\tilde{\nu}, \vartheta)\,\mathrm{d}\tilde{\nu} \qquad (3.170)$$

is the integrated intensity of i-th Raman line and I^{std} is the integrated intensity of a standard Raman line (e.g. the line of the totally symmetrical stretching vibration, ν_1, of tetrachlormethane at 459 cm^{-1}, which is most frequently used as the standard). With all quantities, the superscript ^{std} is related to the standard line, ϱ denotes the depolarization factors, n the refractive indices, σ the spectral sensitivity (a parameter whose magnitude depends on the experimental technique of the spectrum measurement), M the molecular mass, d the density, $\Delta\tilde{\nu}$ the Raman shift in wavenumbers and T the absolute temperature. The complexity of the formula indicates the problems connected with obtaining the standard magnitude of Raman line intensities and with their accuracy and precision.*)

Complications connected with the use of an external standard have led to the introduction of an internal intensity standard. The intensity is determined in very much the same way as that described on the previous page, but the intensity of the lines of admixtures or solvents added directly to the test compound sample is employed as the standard intensity. In addition to tetrachloromethane, chloroform, dichloromethane, tetrachloroethylene and other compounds have been used for this purpose.

3.3.5.2 The Depolarization Factor

Raman spectra can also be monitored with respect to polarization of the excitation or scattered radiation, often yielding very valuable information about the symmetry of molecules and of normal vibrations. The source of information proper is the information about the distribution of the scattered radiation polarization within the space around the sample.

The possibility of using information on the polarization of scattered radiation follows from the fact that, on elastic interaction of molecules with radiation, an electric dipole moment is induced in the molecules, whose direction is related to the direction of the electric vector of the excitation radiation. An important role is played by the symmetric properties of polarizability, described by the polarizability tensor. The polarizability has a spherical part α,

$$\alpha = 1/3(\alpha_{xx} + \alpha_{yy} + \alpha_{zz}) \quad (3.171)$$

---

*) In determining Raman line intensities limitations stemming from the experimental technique must also be borne in mind. While infrared spectrometers of various makes permit work in roughly the same set of conditions and yield data of approximately the same quality, Raman spectrometers from various manufacturers are so different that comparison of measured values is impossible. As the scattered radiation intensities are further exceptionally sensitive to errors in sample preparation (e.g. to the presence of impurities, fluorescence, etc.), experimentally determined Raman line intensities have a rather problematic quality.

and an anisotropic part β^2, whose magnitude follows from the relationship,

$$\beta^2 = 1/2[(\alpha_{xx} - \alpha_{yy})^2 + (\alpha_{yy} - \alpha_{zz})^2 + (\alpha_{zz} - \alpha_{xx})^2 + \\ + 6(\alpha_{xy}^2 + \alpha_{yz}^2 + \alpha_{zx}^2)]. \quad (3.172)$$

For molecules with isotropic polarizability distribution, e.g. CH_4, CCl_4, etc., the direction of the induced electric dipole moment coincides with the direction of the electric vector of the excitation radiation at any orientation of the molecule. On the other hand, with anisotropic molecules the direction of the induced dipole moment coincides with the direction of the electric vector only when the direction of this vector coincides with one of the axes of the polarizability ellipsoid describing the polarizability distribution in space. If the scattered radiation is then observed in a plane perpendicular to the excitation beam, it is found that, with isotropic molecules, the scattered radiation is completely polarized, even without respect to whether the excitation radiation was polarized or not. With anisotropic molecules the scattered radiation observed in this plane is not completely polarized and the degree of its depolarization also depends on whether the excitation radiation was polarized or not.

For description of the polarizability of the scattered radiation, the depolarization factor is used, determined for natural (non-polarized) radiation ϱ_n or for linearly polarized radiation ϱ_1. The value of these factors is determined as the ratio of the intensities of the scattered radiation observed perpendicularly, $I_\perp$, and parallel, $I_\parallel$, to plane XY (the axes are denoted in the same way as in the description of the tensor of molecule polarizability); in the measurement the relationship (perpendicularity, parallel orientation) with respect to the spectrometer slit is followed. It holds for the Rayleigh (frequency non-modulated) scattering of natural radiation that

$$\varrho_n = I_\perp/I_\parallel = 6\beta^2/(45\alpha^2 + 7\beta^2), \quad (3.173)$$

where α and β are the quantities from Eqs. (3.171) and (3.172). However, in the Raman (frequency modulated) scattering, not the polarizability alone but changes during molecular vibration play a role. If thus the elements of polarizability tensor α_{xx} and the others in Eqs. (3.171) and (3.172) are replaced by their partial derivatives with respect to the appropriate normal coordinate q_i, i.e. $\partial\alpha_{xx}/\partial q_i$, Eq. (3.173) can be employed for calculation of the depolarization factor for the i-th Raman line. For non-symmetrical vibrations quantity α equals zero*) and the ϱ_n value attains a maximum value of 6/7 (0.857). For symmetrical vibrations α is non-zero and ϱ_n may

*) With Rayleigh scattering, it always holds that $\alpha > 0$.

attain any value between 6/7 and zero. Raman lines with $\varrho_n = 0$ are called fully polarized,*) with $\varrho_n \in (0, 6/7)$ partially polarized and with $\varrho_n = 6/7$ depolarized.

In application of linearly polarized excitation radiation, practically identical conclusions can be drawn as those given as an explanation of Eq. (3.173); however, for ϱ_1 it holds that

$$\varrho_1 = I_{\perp}/I_{\parallel} = 3\beta^2/(45\alpha^2 + 4\beta^2), \tag{3.174}$$

so that $\varrho_1 \in \langle 0, 3/4 \rangle$.

Readers interested in the problems of polarization of scattered radiation are referred to the appropriate chapters of monographs dealing with the Raman effect (for references see Chapter 2).

The method of determining the depolarization factor with individual types of Raman spectrometers is described in detail in manuals for operation

Table XXV

The values of depolarization factors ϱ for Raman lines of standard liquids

Compound	Wavenumber[a] $\tilde{\nu}$ (cm^{-1})	ϱ
Tetrachloromethane	459 317	0.013 0.857
Tetrachlorosilane	424 221	0.013 0.857
Tetramethylsilane	2962 598	0.857 0.000

[a] Raman line maximum wavenumber, for which the depolarization factor value ϱ is given

of the instruments. Several standard values are given in Table XXV for control of the accuracy of the experimental depolarization factors.

In practice, depolarization factors are determined from the ratio of the Raman line maximum intensities under conditions (specified for each instrument) where $I_{\perp}$ and $I_{\parallel}$ are measured. Some spectrometers from the JEOL Company permit recording of Raman spectra as differences of two

*) The experimentally determined depolarization factor is minutely larger than zero even with fully polarized lines, due to perturbations in the symmetry of molecules owing to their mutual interactions.

spectra monitored perpendicularly and in parallel to the slit. The depolarization factor at any wavenumber can then be determined from these spectra.

3.4 Dependence of Band Parameters on External Conditions

The parameters of infrared bands and Raman lines are determined predominantly by the inherent properties of vibrating molecules, such as bond strengths, electric dipole moment contributions or bond polarizability, etc. However, these parameters are also affected to a certain extent by forces involved in mutual molecular interactions. Both effects are manifest simultaneously in experimental spectra of macroscopic samples of test compounds. Section 3.4 should point out the direction and magnitude of the effect of intermolecular interactions on vibrational spectra, so that these effects could be corrected for or at least considered, e.g. in attempts to spectroscopically evaluate the bond properties.

The intensity of the effect of intermolecular interactions on the band parameters varies greatly. Non-polar molecules affect one another very little and the nature of the interaction is physical. On the other hand, molecules of polar compounds or molecules containing an acidic hydrogen often interact strongly, sometimes with perturbations of a chemical nature – complex formation. The state of aggregation (whose change is also manifested in changes in the band parameters) and the effect of the pressure and temperature, i.e. quantities that determine or change intermolecular interactions, also markedly affect the spectra.

Most applications of vibrational spectroscopy methods do not consider the direction and magnitude of the intermolecular interaction effects and these effects actually interfere in the spectrum analysis. Therefore, an experimental arrangement, where the effects of intermolecular interactions are eliminated, is preferred. This situation can be attained in practice only in dilute gases, where, however, molecular rotation can cause unpleasant complications. Then it is necessary to measure the spectra of compounds in dilute solutions in a cooled inert gas matrix; under these experimental conditions the effect of intermolecular forces is, however, not eliminated, but only "standardized".

3.4.1 Effect of the State of Aggregation

The state of aggregation of the test compound samples affects the appearance of infrared and Raman spectra and the band parameters very characteristically. In each state of aggregation, the molecules are exposed to a somewhat different form of mutual interaction, which is then indirectly reflected in the vibrational spectra.

Our considerations should start from the dilute gas state. In this state the molecules affect one another very little. Especially the molecules of non-polar compounds spend most of the time travelling uniformly in a straight line through space; the time during which they are engaged in collision states (in collisions with other molecules or with the cell walls) is incomparably shorter. Hence, in dilute gases the spectra of virtually "free" molecules are measured, as the assumption that these molecules are not exposed to the effect of any external force is satisfied relatively well. However, an increase in the gas pressure and in the molecule polarity impose a limit on the independence of the molecules. In compressed gases the collision probability increases and the time the molecule spends in collision states increases. Very dense gases resemble liquids in this respect and the spectra obtained for them (i.e. at pressures approaching the critical pressure) strongly resemble the spectra of liquids.

Strong interactions or even molecular association may occur even in dilute gases, when compounds with permanent electric dipole moments and especially those with acid hydrogens are involved. A typical example is carboxylic acids, which form equilibrium mixtures of the monomeric form with the cyclic dimer even in a dilute gas. Gaseous hydrogen fluoride also forms equilibrium mixtures containing oligomers up to the octamer, in addition to the monomer. Molecules of different compounds may also associate; e.g. hydrogen halides, nitric acid, etc. form complexes with ethers and other organic bases in the gaseous state.

In condensed phases the molecules are close to one another and their mutual interaction is thus stronger. Interactions in a non-polar liquid can be described in a simplified form considering a pair of immediately neighbouring molecules. Both molecules spend a certain time vibrating around the equilibrium position (intermolecular vibration in which the distance between the centres of gravity of the two molecules varies periodically); after the end of this partnership both molecules immediately contact other molecules (in a "dense" medium) and the process is repeated with new partners. It seems that molecules in the liquid state are never free, as the time of movement between two collisions is extremely short.

In non-polar liquids, density fluctuations reflecting the movement and mutual interactions of molecules can be observed. The aggregates formed are distributed according to size. Orientation effects are not very marked with non-polar liquids. Often, especially with compounds with molecules having low moments of inertia, molecular rotation typical for the gaseous state survives in the liquid state. Molecular rotation slowly ceases with cooling of the samples and occurs only exceptionally in the crystalline state*) (e.g. with crystalline hydrogen). Examples of manifestations of residual rotation of CO and NO molecules, followed in solutions for the sake of lucidity, are given in Fig. 44. Residual rotation in liquids is manifested in

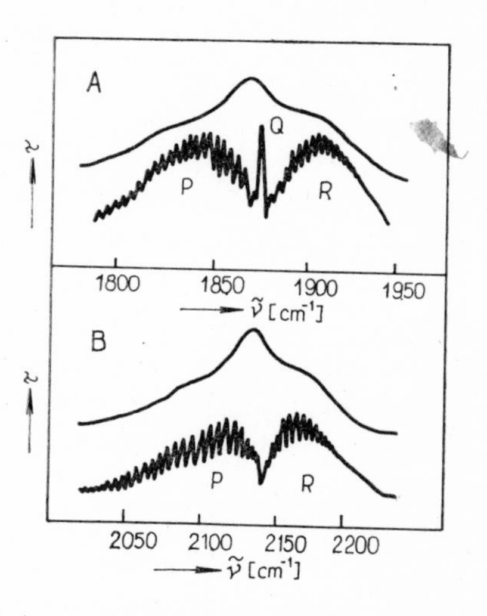

Fig. 44. Infrared absorption spectra of tetrachloromethane solutions of: A — nitrogen oxide NO, and B — carbon monoxide CO. The spectra of the two oxides in the gaseous state (the lower curves with resolved rotational structure) are given for comparison.

broad unresolved bands (transitions from P- and R-branches) on both sides of the main band (transitions from the Q-branch). If the spectra of liquids do not exhibit these satellite bands, they can be reliably described as purely vibrational spectra.

Compounds with a permanent dipole moment or with an acid hydrogen form aggregate in the liquid state, in which the molecules are kept together by oriented effects (e.g. dipole-dipole interactions, hydrogen bonds, charge-transfer complexes, coordination effects). The life-time of these aggregates is quite long (compared with the life-times of "coupled" molecules in non-polar liquids), and thus complexes with a greater number of particles, linearly or cyclically interconnected, are formed. The internal structure of

*) Rotation of the ions in ionic salt crystals is encountered much more frequently, especially if the ions have a high symmetry and are located in a crystal field of high symmetry. Rotation of the ions in a crystal lattice ceases at a certain temperature, typical for each crystal; at this temperature the ions only slightly turn along the rotation axes, which is called libration.

liquids of this type is very complicated; various types of complexes are in equilibrium and the dynamics of the processes taking place depend on the temperature and the pressure. Hence a polar liquid can be defined as an equilibrium system of strongly associated aggregates. The only exception are liquids whose molecules cannot mechanically form strongly bonded aggregates because of steric hindrance to association (e.g. 2,6-ditert. butyl-phenol). Very broad bands appear in the spectra of strongly associated polar liquids; the composition of the equilibrium mixtures (e.g. the size and structure of aggregates) cannot be reliably determined from the spectra.

Even more complex systems are formed when a single molecule contains two or more polar groups or groups with acid hydrogens. Then aggregates with inter- and intramolecular bonds among molecules can be formed in the liquid state (e.g. ethylene glycol). The number of possible forms in the liquid state is extremely large and attempts to analyze the spectra are pointless. The spectra of compounds of this type must necessarily be analyzed in dilute solutions or under similar conditions, where the interactions of polar molecules are limited as much as possible.

Mutual interactions of particles in crystals are of a different quality. In the ordered crystal structure there are basic building units, crystal cells,

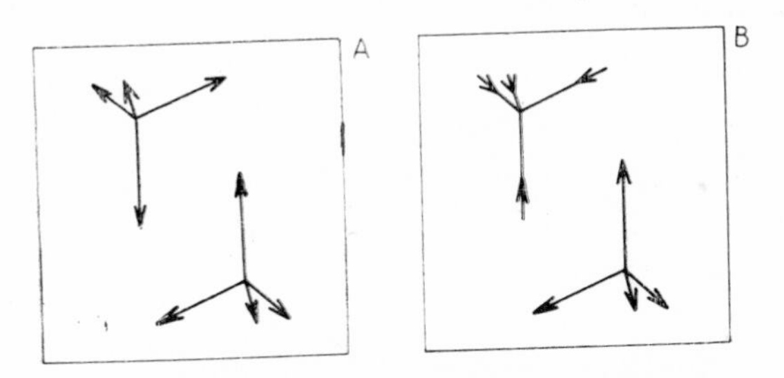

Fig. 45. Schematic representation of coupled vibrations in a crystal unit cell: A — vibrations in phase, B — vibrations with opposite phases. Both vibrations originate from totally symmetrical vibrations of tetrahedral molecules forming the crystal unit cell.

which determine not only the external geometric parameters of crystals but also their vibrational spectra. In order to interpret spectra of crystals correctly*) it is necessary to know not only the chemical composition of the

*) In this Section only polycrystalline samples are discussed.

molecules and ions forming the crystal, but also the unit cell structure and symmetry.

Forces operative in the crystal lattice among molecules and especially among ions can be very large in some instances, so that they may even cause coupling of vibrations of the particles forming the unit cell (as happens with bonding of atoms in molecules). Manifestations of this coupling make the spectra of crystals more complicated than those of liquid samples of the same compounds. This coupling is depicted symbolically in Fig. 45. Identical vibrations of two ions of the same type present in a unit cell may proceed in phase or with opposite phases and both vibrations may appear in the vibrational spectra, provided that the selection rules for the given unit cell symmetry permit it. Splitting of this kind is termed correlation field splitting. The coupling of vibrational modes and its spectral manifestations are further complicated with an increasing number of unit cell building particles, according to its symmetry.

Modes non-existent in the gaseous and liquid states are encountered in the spectra of crystals, especially in the low wavenumber region. These are translation modes, where the unit cell building blocks (even monoatomic ions) vibrate periodically around equilibrium positions and libration modes, which can be described as the turning of polyatomic ions or molecules around rotation axes, in equilibrium positions in the unit cell. Further there are acoustic modes (phonons), representing propagation of certain deformations through the crystal, etc. The background for the analysis of crystal spectra is given in the special literature cited at the end of Chapter 5.

The determination of the spectra of crystals by the unit cell structure is most pronounced with polymorphous compounds. Various crystalline forms of the same compound can yield very different spectra (Fig. 46A, B); in other cases, the spectra are rather similar (Fig. 46C, D). The extent of the differences among the spectra of various crystalline forms can be assessed only if the crystallographic description of the forms compared and the unit cell composition are known. Differences in the spectra appear only when the measurement is performed on crystals; of course, the spectra of solutions of polymorphous forms are identical. The experimental technique employed must also be evaluated carefully when measuring the spectra of polymorphous forms. For example, the use of high energies in milling and pressing in the KBr technique may cause conversion of one crystalline form into another (which, moreover, need not be quantitative).

In the measurement of crystal spectra, the temperature dependence of processes in the crystal lattices must also be considered. For example, the

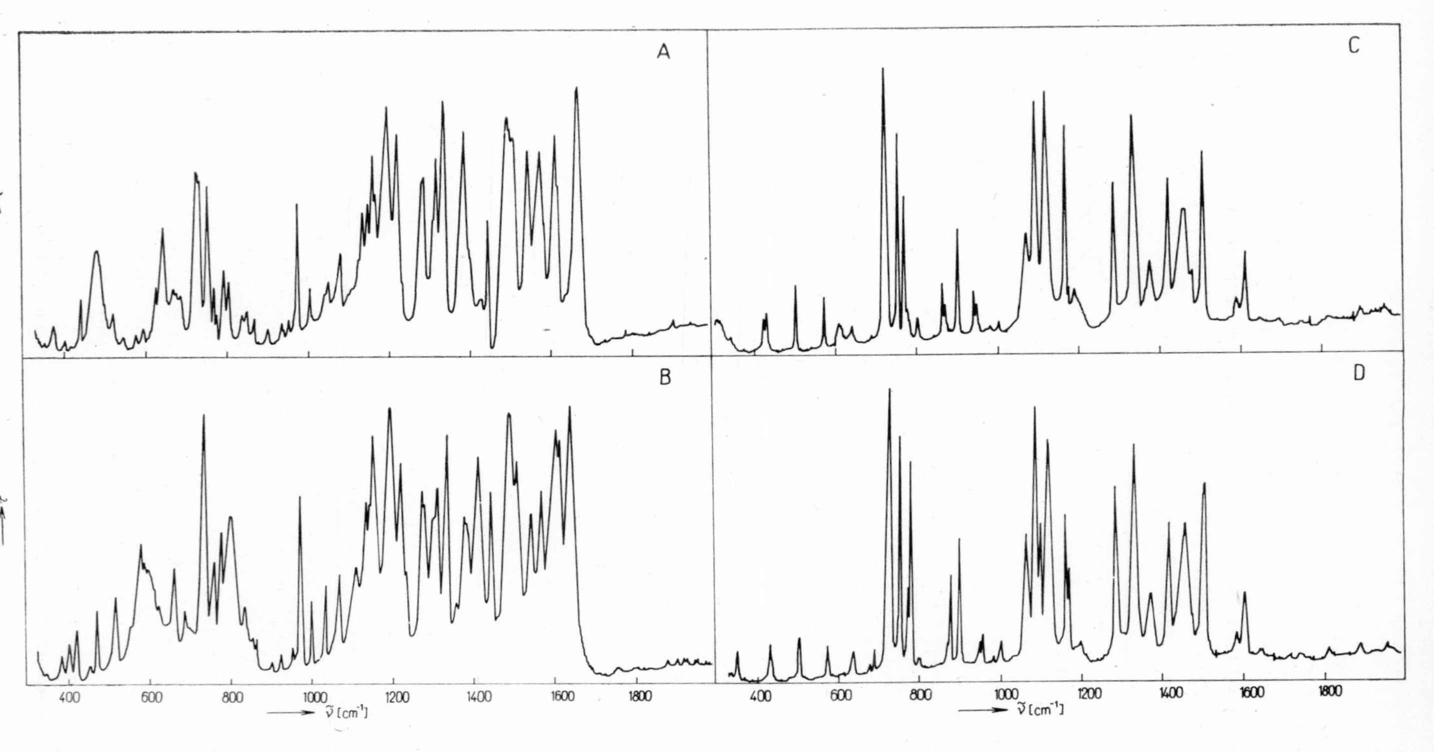

Fig. 46. Infrared absorption spectra of nujol mulls of the following crystal modifications: A — α-form of ostacene yellow, B — β-form of ostacene yellow (spectra A and B differ significantly), C — α-form of Cu-phthalocyanine, D — β-form of Cu-phthalocyanine (spectra C and D are very similar).

existence of individual polymorphous forms is connected with certain temperature intervals. Thus, if the temperature of a phase transition is exceeded, another modification and a different spectrum are obtained. Further, with compounds forming mixtures of rotational isomers in the liquid state, certain forms may be "frozen-out" at low temperatures. A super-cooled crystal is then formed by the molecules of the thermodynamically most stable rotational isomer alone and hence its spectrum is simpler than the spectrum of a liquid sample. However, on rapid supercooling crystallization frequently does not occur at all, the supercooled sample

Table XXVI

Vibrational band wavenumbers for some simple molecules from infrared or Raman spectra in the gaseous, liquid and solid state (cm^{-1})

Compound	Vibration	Gas	Liquid	Crystal
CH_4	ν_1	2916	2906	2905
	ν_2	1534	1535	1538
	ν_3	3019	3020	3010
	ν_4	1306	1300	1303
CO_2	ν_1	1388	1388	1388
	ν_2	667	—	656
	ν_3	2349	—	2288
HCl		2866	2785	2768
H_2O	ν_1	3657	3428[a]	3256[a]
	ν_2	1595	1637	1644
	ν_3	3756	3428[a]	3256[a]

[a] The bands of stretching vibrations ν_1 and ν_3 merge in the condensed phases.

remains amorphous (the so-called "glassy" modification) and its spectrum does not differ from that of the liquid sample, as the rotational isomers are not completely frozen-out under these conditions.

For various reasons, it is often necessary to compare the band parameters obtained from the spectra of gaseous, liquid and crystalline samples. An example of changes in the band parameters for some typical vibrations, obtained for various states of aggregation, is given in Table XXVI.

The comparison of the band wavenumbers can be direct, as the concept of the band "centre" has the same significance for rotational-vibrational spectra of gases, liquids and solids. The wavenumbers of stretching vibration bands usually decrease when going from a gas to a liquid and then to a solid; with non-polar molecules these are small changes of the order of units or at the most tens of cm^{-1}, but with polar molecules and those containing acid hydrogens the interactions are much stronger and the bands may shift by hundreds or even thousands of cm^{-1}. Deformation vibration bands, on the other hand, move to higher wavenumbers with the same change in the state of aggregation; no simple description is possible concerning changes of degenerate vibrations. Usually each band in the spectrum changes its position differently (both in direction and in the magnitude of the shift); the greatest effects are mostly found with vibration bands for bonds which participate directly in the intermolecular interactions, especially those of the X–H type (where X=O, N), hydrogen-bonded in the condensed phases.

In transfer from the gaseous state to the liquid and to the crystal, the integrated band intensity usually increases; changes in it may be as large as orders of magnitude with strongly associating molecules. The band maximum intensity need not always grow; if very strong interactions take place among the molecules in the liquid state and in the crystal, the growth of the integrated intensity may be predominantly given by band broadening and the band maximum intensity may even decrease. Bands with half-widths of the order of hundreds of cm^{-1} have been found for some strong hydrogen bonds; a band is sometimes broadened so much that it merges with the spectral background.

Most difficulties arise when band widths (half-widths) are to be compared in spectra measured in various states of aggregation. In no case may the width of even an unresolved rotational-vibrational band be compared for a gaseous and a liquid sample, as the bands in the gaseous and liquid states manifest physical processes incomparable in this respect. The band width represents very sensitively the strength and mechanism of molecular interactions, rotational movements of molecules and their parts and other effects.

3.4.2 Solvent Effects

The infrared and Raman spectra of compounds are often measured in solution. In dilute solutions in non-polar liquids, the association of polar compound molecules is suppressed, work with solutions is easy and in many cases the measurement cannot be carried out otherwise (e.g. in a number of quantitative analyses).

In any case, interactions between the solute and solvent molecules occur in solutions. In Section 3.4.2 some aspects of solution interactions will be summarized, especially those which must be considered in the evaluation of experimental spectra.

Solutions of non-polar solutes in non-polar solvents need not be discussed; the molecular interaction is weak in these systems and, moreover, this type of system is rarely studied.

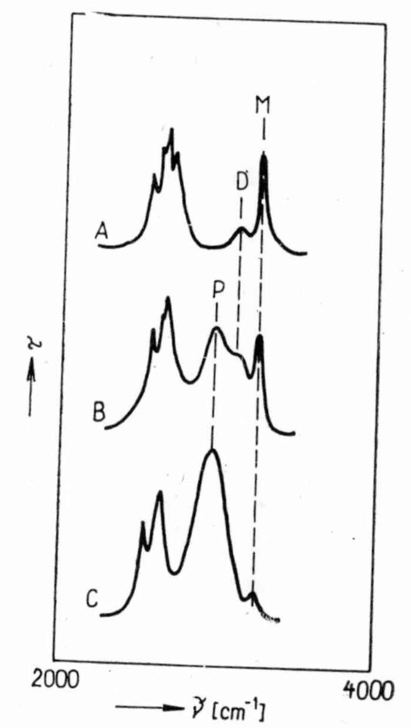

Fig. 47. Infrared spectra of tetrachloromethane solutions of methanol: A — 0.01 % vol. methanol (0.1 cm cell), B — 0.1 % vol. (0.01 cm), C — 1 % vol. (0.002 cm). M — monomer band, D — dimer band, P — polymer band.

The most important combination is a polar solute in a non-polar solvent. Polar solutes dissolved in non-polar liquids associate; e.g. nitromethane dissolved in tetrachloromethane can form higher concentrations of the dimer in addition to the monomer, which appear as typical bands in the infrared spectrum. Solutions of nitromethane in tetrachloromethane (or other non-polar liquids) are thus equilibrium mixtures of the monomer and the dimer, the contents of the two forms depending on the total nitromethane concentration and the temperature; the equilibrium shifts in favour of the monomer with increasing temperature and decreasing concentration.

Compounds with an acid hydrogen also associate when dissolved in non-polar liquids. The monomers are present only at high dilutions (e.g. alcohols and phenols in CCl_4 at concentrations lower than 1 mmole . l^{-1}), and even then are not always found; e.g. carboxylic acids associate so strongly that they are not completely converted into the monomeric forms by any possible dilution. With increasing concentration the molecules of alcohols and phenols first dimerize (the dimer yields a typical $\nu(O-H)$ band) and at higher concentrations trimers, oligomers and finally polymers are formed. An extremely high concentration is actually the liquid state, in which polymeric forms predominate. Fig. 47 shows the changes in the methanol spectrum with increasing methanol concentration; the bands of the monomer, dimer and a polymer are perceptible. Carboxylic acids exhibit the dimer (in addition to the monomer) both in the gaseous state and in dilute solutions in non-polar liquids; temperatures around 200 °C are required for the dissociation of the dimeric form in the gaseous state. The same holds for solutions. Carboxylic acid monomers and dimers have different spectra; hence all changes in the monomer-dimer equilibrium appear as changes in the spectra. Amines also associate in solution (the association of primary amines is a very specific problem); amides and lactams associate very strongly and lactams also yield cyclic dimers in addition to monomers in dilute solutions.

The degree of association of polar solutes dissolved in non-polar liquids is also strongly affected by the solvent used. For example, the degree of association is always higher in liquids of low dielectric permittivity than in those with a higher permittivity.

Molecular interaction is very complicated when strong solute-solvent interactions can occur (e.g. hydrogen bonding). This happens e.g. with solutions of compounds containing an acid hydrogen (alcohols, carboxylic acids) in organic bases, e.g. ethers or amines, where specific solute-solvent complexes are formed.

In qualitative description of the effects operative in solutions, the effect on the equilibrium content of rotational isomers and tautomeric forms cannot be omitted. With compounds whose molecules are capable of forming various rotational isomers owing to hindered rotation, the equilibria are shifted in solutions of variously polar liquids. Generally, the content of the more polar rotational isomer increases with increasing solvent polarity; the bands in the spectra measured under these conditions increase if they represent vibrations of the more polar form or decrease in the opposite case. The polarity also has an effect on keto-enol tautomerism of compounds in solution. The enol-form usually predominates in less polar solvents and the keto-form in more polar ones.

Further, the mechanism of molecular interactions in solutions will be discussed, especially from the point of view of spectral manifestations of these interactions. The explanation of the basic laws of intermolecular effects in solutions is rather difficult since no satisfactory theory of liquids is forthcoming at present. Therefore we will primarily deal with experimental facts which should help us to understand certain rules of this part of physics, at least in idealized cases.

A great contribution to the understanding of molecular interactions in solutions was the study of ternary mixtures; one component of the mixture was a polar solute and the other two were miscible solvents.

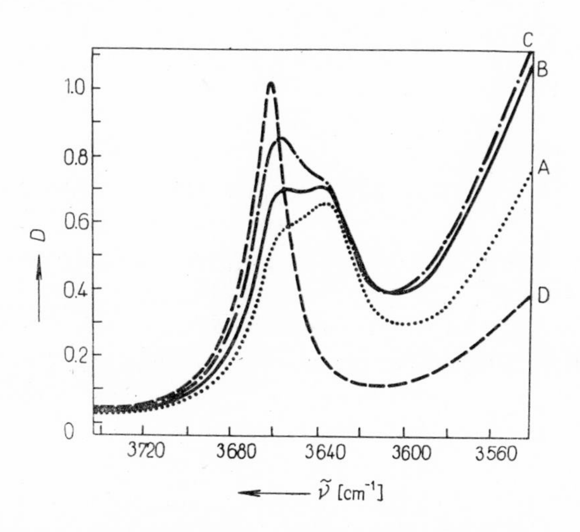

Fig. 48. The bands of collision complexes methanol — hexafluorobenzene and methanol — carbon disulphide in the infrared absorption spectra of ternary mixtures methanol — hexafluorobenzene — carbon disulphide. The amounts of hexafluorobenzene in the mixtures (% vol.): A — 20, B — 50, C — 70, D — methanol solution in pure hexafluorobenzene.

Surprisingly enough, doublets of the "free" hydroxyl band were found in the spectra of polar solutes (alcohols, phenols, carboxylic acid monomers) in pairs of non-polar solvents (Fig. 48); the intensity of the doublet branches depends on the ratio of the two solvents in the mixtures. This experiment indicates that specific interactions, i.e. complex formation, occur in polar solute solutions in non-polar solvents among the molecules. The complexes formed have short life-times as the interaction energy is relatively low; they are formed when the solute molecules are surrounded by a large number of solvent molecules, so that immediately after breakage of the complex molecule its active component, the polar solute molecule, can form a new complex with a solvent molecule of the same kind (or with a molecule of the other solvent in a ternary mixture). Effects of this type cannot be distinguished in binary solute-solvent mixtures, although these weak complexes, called collision complexes, also exist there.

It is also possible to form ternary mixtures containing a polar and a non-polar solvent. The results obtained with a ternary mixture of phenol-tetrachloromethane-acetonitrile are given in Fig. 49. Phenol in pure tetrachloro-

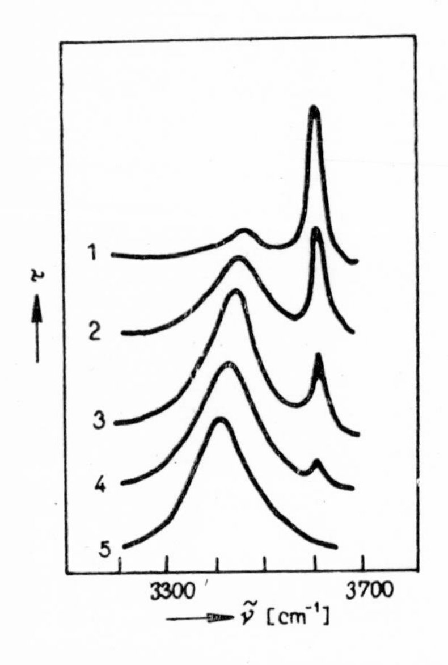

Fig. 49. The bands of the phenol — tetrachloromethane collision complex and the phenol — acetonitrile donor-acceptor complex in the infrared absorption spectra of phenol-acetonitrile-tetrachloromethane ternary mixtures. The amounts of acetonitrile in the mixtures (% vol.): 1 — 0.2, 2 — 1, 3 — 3, 4 — 6 and 5 — phenol solution in pure acetonitrile.

methane yields a band corresponding to the collision complex (phenol-tetrachloromethane) at 3611 cm^{-1}. On addition of acetonitrile, another band appears in the spectrum at lower wavenumbers, which belongs to the donor-acceptor phenol-acetonitrile complex (hydrogen-bonded). On increasing the acetonitrile concentration, the intensity of the donor-acceptor complex band increases, while that of the collision complex band decreases. The donor-acceptor complex is much more stable than the collision complex; on addition of 6 % vol. acetonitrile virtually all the phenol is bound in the former complex.

The donor-acceptor complex exhibits a 1 : 1 stoichiometric ratio of phenol and acetonitrile; the excess solvent molecules participate in secondary solvation of the complex molecules. In agreement, the donor-acceptor complex band shifts from 3463 cm^{-1} (mixtures with small amounts of acetonitrile) to 3409 cm^{-1} (phenol solution in pure acetonitrile). The frequency shifts of the complex band can be described in terms of the Onsager reaction field theory. A molecule of the phenol-acetonitrile complex represents a point dipole placed in a spherical cavity of the continuum

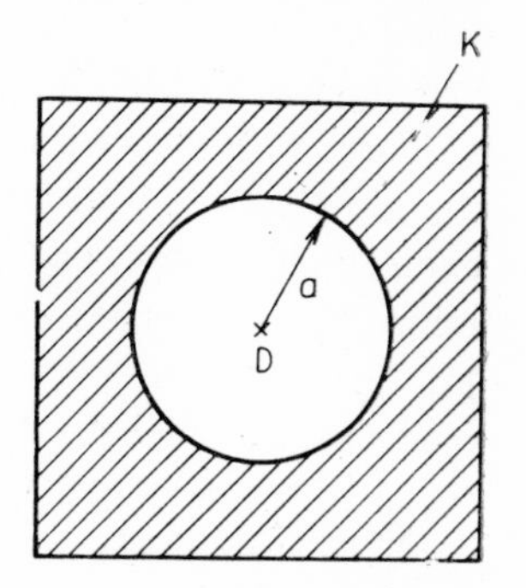

Fig. 50. Schematic representation of solute-solvent interactions according to the reaction field theory. D — a point dipole (approximating the solute molecule) in a spherical cavity of radius *a* and in continuum K formed by the solvent molecules.

formed by solvent molecules (Fig. 50). Assuming that the changes in the solution composition do not affect the complex stoichiometry but only the continuum properties, it holds that reaction field F,

$$F = g \cdot \mu, \tag{3.175}$$

acts upon the complex molecule with point electric dipole moment μ in a medium of relative permittivity*) ε_r, where g is defined by the relationship,

$$g = (2/a^3) \cdot f(\varepsilon_r); \tag{3.176}$$

and a is the radius of the spherical cavity (in the continuum) and function f is generally defined as

$$f(\xi) = (\xi - 1)/(2\xi + 1). \tag{3.177}$$

Here, quantity ξ is equal to permittivity ε_r and it will follow from the text that it may also equal the square of the refractive index, n^2.

In the described case the force acting upon the dipole (the phenol-acetonitrile complex) is proportional to the medium polarizibility, which can be described by its macroscopic permittivity ε_r. It has been derived classically that the relative change in the vibrating dipole wavenumber, $(\tilde{\nu}_0 - \tilde{\nu}_s)/\tilde{\nu}_0$, caused by a change in the continuum properties is given by

$$(\tilde{\nu}_0 - \tilde{\nu}_s)/\tilde{\nu}_0 = C \cdot f(\varepsilon_r); \tag{3.178}$$

*) The concept of permittivity is used in place of the earlier term dielectric constant.

where $\tilde{\nu}_0$ and $\tilde{\nu}_s$ are the wavenumbers for the vibrating dipole when the dipole is subject to a reaction field continuum of relative permittivity 1 (i.e. a vacuum) and in the given medium, respectively, and C is a proportionality constant given by the properties of the vibrating dipole. This relationship, named after its discoverers the Kirkwood–Bauer–Magat relationship (the KBM relationship), was incorrectly tested on frequency shifts of solute vibration bands in various solvents, in binary solute-solvent mixtures; its validity was then doubted because of disagreement of experimental wavenumbers with theoretical. However, the dipole cannot be considered to be the same entity in all binary systems (virtually in each system a different complex is present) and thus only experiments with ternary mixtures permitted the correct testing of the relationship. It has then been found that Eq. (3.178) can be employed, within the limits given by the validity of the model used, for the description of changes in the wavenumber of the phenol-acetonitrile complex band in the ternary mixture of phenol-tetrachloromethane-acetonitrile with varying contents*) of the two solvents.

New relationships have been derived for relative bond shifts on the basis of quantum-mechanical considerations, assuming that factor g consists of the contributions from two terms for polar solvents,

$$g_\varepsilon = (2/a^3) \,.\, f(\varepsilon_r) \tag{3.179}$$

and

$$g_n = (2/a^3) \,.\, f(n^2), \tag{3.180}$$

where n is the medium refractive index. For frequency shifts in non-polar solvents, the relationship

$$(\tilde{\nu}_0 - \tilde{\nu}_s)/\tilde{\nu}_0 = \sum_{i=0}^{m} C_i \,.\, [f(\varepsilon_r)^i] \tag{3.181}$$

was proposed; for those in polar solvents,

$$(\tilde{\nu}_0 - \tilde{\nu}_s)/\tilde{\nu}_0 = \sum_{i=0}^{m} \sum_{j=0}^{k} C_{i,j}[f(\varepsilon_r)]^i \,.\, [f(n^2)]^j. \tag{3.182}$$

These relationships, tested under the same conditions as the KBM relation, were also doubted.

However, in the Onsager reaction field theory, the effect of complex formation is not included and hence it cannot be verified in cases when

*) The phenol concentration was maintained constant throughout.

complexes are evidently formed during solvation*). If the same complex is, however, followed during the whole experiment, Eqs. (3.181) and (3.182) are suitable for the description of induced wavenumber changes. For the system described here, quantity $\tilde{\nu}_0$ is then the wavenumber of band $\nu(O-H)$ for the phenol-acetonotrile complex in a medium with relative permittivity equal to unity, i.e. in the gaseous state. As this quantity cannot be obtained experimentally (the phenol-acetonitrile complex cannot exist in the gaseous state), a value of $\tilde{\nu}_0 = 3530\ cm^{-1}$ was calculated from Eqs. (3.178) or (3.182). If the wavenumber of band $\nu(O-H)$ of phenol in the gaseous state, $3654\ cm^{-1}$, that of band $\nu(O-H)$ for the phenol-acetonitrile complex in the gaseous state, $3530\ cm^{-1}$, and that of band $\nu(O-H)$ for the phenol-acetonitrile complex in pure acetonitrile, $3409\ cm^{-1}$, are considered, it can be seen that the decrease in the wavenumber on complex formation is roughly the same as that caused by secondary solvation of the complex by acetonitrile. It is thus evident that both specific (complex formation) and non-specific (secondary solvation) effects play a role in determining the solvent effect on polar solutes. It should be pointed out that the two effects need not always be comparable; e.g. some basic solvents can cause large specific effects (e.g. tertiary amines), but their non-specific effect is weak because of their low polarity. Sometimes complex formation may increase rather than decrease the band wavenumber; this occurs e.g. with the nitrile group band, whose wavenumber increases with increasing complex formation interaction, or with the NO group in X–NO compounds, etc. This type of shift is called "blue" in contrast to "red" shifts with which the wavenumber decreases on complex formation. A more detailed description of these problems can be found in a series of papers on the solvent effect, especially in refs.[1–4].

*) It must be borne in mind that polar solutes form collision complexes even in non-polar liquid solutions. It has been shown experimentally that shifts of the bands of these weak complexes caused by changes in the continuum (experiments with mixtures) can be explained analogously as shifts of bands of strong complexes.

1 Horák M., Plíva J.: Studies of Solute-Solvent Interactions. I. General Considerations. *Spectrochim. Acta 21*, 911 (1965).

2 Horák M., Moravec J., Plíva J.: Studies of Solute-Solvent Interactions. II. Formation of Collision Complexes in Solutions of Some Phenols. *Spectrochim. Acta 21*, 919 (1965).

3 Horák M., Poláková J., Jakoubková M., Moravec J., Plíva J.: Studies of Solute-Solvent Interactions. III. Solvation of Donor-Acceptor Complexes of Phenols with Basic Solvents. *Collect. Czechoslov. Chem. Commun. 31*, 622 (1966).

4 Horák M., Moravec J.: Studies of Solute-Solvent Interactions. V. The Solvent Induced Frequency Shifts of the $\nu(C=O)$ and $\nu(C\equiv N)$ Infrared Bands. *Collect. Czechoslov. Chem. Commun. 36*, 2757 (1971).

Many authors attempted to order solvents in a series according to the decrease in the band wavenumbers for certain vibrations in their solutions. It has been found that it is impossible to make a single series of solvents; this is understandable in view of the independent mechanisms of solvent-solute interactions, i.e. specific and non-specific. There is a certain sense in classifying solvents in two series, the first suitable for the description of the ν(X–H) wavenumbers and the other for ν(X–O) wavenumbers. The

Table XXVII

The wavenumbers of the stretching vibration bands for X—H and C=O bonds in selected models in solutions of various solvents (cm^{-1})

Solvent	Pyrrole ν(N—H)	Phenol ν(O—H)
—[a]	3530	3654
n-Hexane	3506	3623
Tetrachloromethane	3500	3611
Chloroform	3485	3594
Carbon sulphide	3481	3593
Benzene	3458	3553
Acetonitrile	3422	3409
Diethyl ether	3352	3344
Pyridine	3219	3330

Solvent	Acetone ν(C=O)	Acetophenone ν(C=O)
—[a]	1737	1709
n-Hexane	1722	1697
Diethyl ether	1721	1694
Tetrachloromethane	1719	1692
Carbon sulphide	1717	1690
Benzene	1717	1690
Acetonitrile	1713	1687
Pyridine	1713	1687
Chloroform	1713	1683

[a] The wavenumbers of the corresponding bands from the spectra of gaseous samples are given for the sake of comparison

solvent series and examples of band shifts and wavenumber changes are given in Table XXVII.

For very similar oscillator types such as e.g. ν(N–H) in pyrrole and ν(O–H) in phenol, very good agreement in the solvent-induced relative

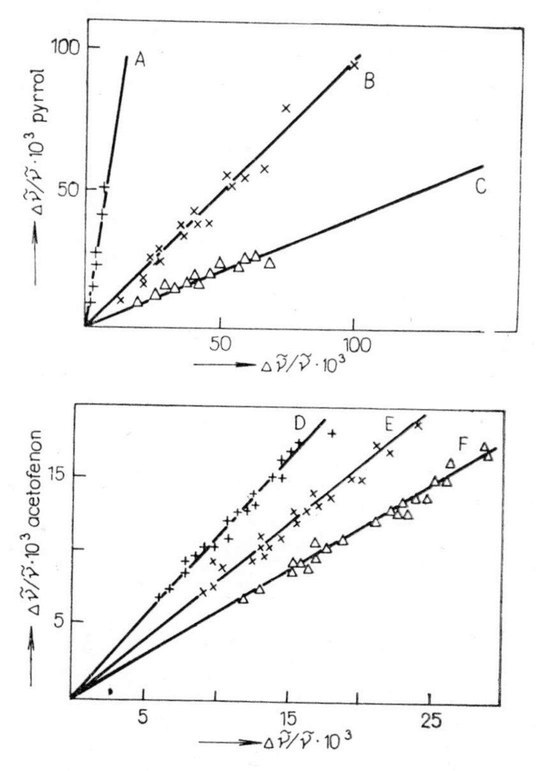

Fig. 51. Correlations of the relative solvent shifts, $(\tilde{\nu}_0 - \tilde{\nu}_s)\,\tilde{\nu}_0$, of the stretching vibration bands for some bonds with the relative shifts of the stretching vibration band of the N—H bond in pyrrole (A — C) and the stretching vibration band of the C=O bond in acetophenone (D — F):
A — stretching vibration of the B—H bond in decaborane, B — stretching vibration of the O—H bond in methanol, C — fundamental frequency of the HCl molecule, D — stretching vibration of the C=O bond in benzophenone, E — stretching vibration of the C=O bond in methyl acetate, F — stretching vibration of the C=O bond in dimethylformamide.

wavenumber shifts, $(\tilde{\nu}_0 - \tilde{\nu}_s/\tilde{\nu}_0)$, has been found. The dependences for some oscillators, referred to pyrrole as a standard, are depicted in Fig. 51. Compounds containing relatively more acid hydrogens give steeper correlations; no good agreement was observed for weakly acid compounds.

Table XXVIII

Liquids used as solvents and their properties

Solvent	Relative molecular mass M_r	Melting point t_t (°C)	Boiling point			Density		Relative permit-tivity ε_r	Refractive index		Values of function [a]	
			t_v (°C)	$dt_v/dp \cdot 10^5$ (K/Pa)	$dt_v/dp \cdot 10^5$ (K/Torr)	ϱ (g/cm^3)	$d\varrho/dt \cdot 10^3$ (g/cm^3.K)		n	$dn_D/dt \cdot 10^3$ (K^{-1})	$f(\varepsilon_r)$	$f(n_D^2)$
n-Pentane	72.15	—130	36	0.2892	38.56	0.6214	—0.975	1.836	1.3547	—0.530	0.1789	0.1788
n-Hexane	86.17	—95	69	0.3144	41.91	0.6548	—0.891	1.882	1.3723	—0.542	0.1851	0.1853
n-Heptane	100.20	—91	98	0.3361	44.81	0.6795	—0.840	1.917	1.3851	—0.508	0.1897	0.1898
Cyclohexane	84.16	7	81	0.3285	43.80	0.7739	—0.921	2.015	1.4325	—0.540	0.2018	0.2030
Benzene	78.11	6	80	0.3203	42.70	0.8737	—1.051	2.274	1.4979	—0.640	0.2296	0.2267
Diethyl ether	74.12	—116	34	0.279	37.2	0.7078	—1.147	4.235	1.3499	—0.560	0.3416	0.1770
Tetrahydrofuran	72.10	—108	67			0.8892		7.291	1.4057	—0.14	0.4037	0.1971
1,4-Dioxan	88.10	12	101	0.324	43.2	1.0269	—1.376	2.209	1.4202	—0.411	0.2231	0.2020
Dichloromethane	84.94	—97	40	0.323	43.0	1.3168	—1.801	8.893	1.4214	—0.632	0.4202	0.2024
Chloroform	119.39	—64	61	0.302	40.3	1.4799	—1.857	4.718	1.4426	—0.598	0.3563	0.2094
Tetrachloro-methane	153.84	—23	77	0.324	43.2	1.5842	—1.927	2.228	1.4576	—0.271	0.2251	0.2142
Dibromomethane	173.86	10	97			2.4842	—2.56	7.22	1.5419	—0.05	0.4028	0.2393
Bromoform	252.77	8	150	0.368	49.0	2.8775	—2.593	4.337	1.5948	—0.570	0.3449	0.2536

Table XXVIII - continued

Solvent	Relative molecular mass M_r	Melting point t_t (°C)	Boiling point t_v (°C)	Boiling point $dt_v/dp \cdot 10^5$ (K/Pa)	Boiling point $dt_v/dp \cdot 10^5$ (K/Torr)	Density ϱ (g/cm^3)	Density $d\varrho/dt \cdot 10^3$ (g/cm^3.K)	Relative permittivity ε_r	Refractive index n	Refractive index $dn_D/dt \cdot 10^3$ (K^{-1})	Values of function [a] $f(\varepsilon_r)$	Values of function [a] $f(n_D^2)$
Diiodomethane	257.87	6	180			3.3105	—2.98	5.316	1.7378	—0.64	0.3710	0.2869
Tetrachloro-ethylene	165.85	—22	121	0.375	50.0	1.6146	—1.714	2.30	1.5030	—0.530	0.2321	0.2282
Carbon Disulphide	76.14	—112	46	0.315	42.0	1.2556	—1.484	2.629	1.6240	—0.780	0.2603	0.2610
Acetonitrile	41.05	—46	82	0.2932	39.09	0.7768	—1.079	36.7	1.3416	—0.450	0.4798	0.1739
Nitromethane	61.04	—28	101	0.330	44.0	1.1312	—1.349	36.7	1.3795	—0.480	0.4798	0.1879
Acetone	58.09	—95	56	0.2846	37.95	0.7851	—1.142	20.70	1.3561	—0.500	0.4646	0.1794
Nitrobenzene	123.11	6	211	0.360	48.0	1.1984	—0.988	34.82	1.5506	—0.400	0.4788	0.2418
Ethyl acetate	88.10	—84	77	0.301	40.1	0.8946	—0.120	6.02	1.3698	—0.490	0.3850	0.1844
Triethylamine	101.19	—115	89	0.30	40	0.7235	—0.904	2.42	1.4010	—0.002	0.2432	0.1955
Pyridine	78.10	—42	116	0.315	42.0	0.9778		12.3	1.5067	—0.500	0.4414	0.2292

Melting points are given for normal pressure, i.e. 1.01325×10^5 Pa (760 Torr); density, relative permittivity, refractive index and the values of functions $f(\varepsilon_r)$ and $f(n_D^2)$ for normal pressure and temperature 25 °C (298 K)

[a] The function is defined by Eq. (3.177).

Table XXVIII contains some spectroscopically important data for most common solvents, including the relative permittivities and refractive indices together with the values of function (3.177) for the permittivity and the square of refractive index (for Eqs. (3.179) or (3.180)).

Finally it should be pointed out that the effect of solvents on polar molecules whose active groups are strongly sterically protected, e.g. by bulky alkyl-groups, may be quite atypical.

Solvent-induced changes in the band intensities and half-widths are much more complex than wavenumber changes. Very approximately it can be said that the integrated band intensity and mostly also the band half-width increase with increasing solvent polarity. However, each individual case requires more detailed study.

3.4.3 Effect of Temperature and Pressure

Variations in the temperature and pressure in samples of studied compounds are reflected in vibrational spectra to a greater or lesser degree. The two effects are also different in different states of aggregation.

With molecules in the gaseous state an increase in the temperature leads to an increase in the number of molecules in higher rotational quantum states (see Section 2.2.2). The spectra of a gaseous sample measured at two different temperatures may therefore differ substantially. In the spectrum obtained at a higher temperature, bands for transitions from higher quantum states which were not populated at the lower temperature are visible provided that the resolution is sufficient and the rotational-vibrational band is actually "broadened" (it contains more transitions) and also lowered (the band integrated intensity is practically independent on a change in the temperature). A change in the temperature also does not affect the position of the centre of a rotational-vibrational band, which is thus temperature-independent.

When referring to gaseous samples, dilute gases are most often considered. On increasing the gas pressure, the required degrees of freedom are lost, molecules spend more and more time in collision states; in "dense" gases (approaching the critical pressure above the critical temperature) the gases do not differ very much from liquids (also from the point of view of their spectra).

Changes in the temperature of liquid samples cause changes in intermolecular effects. An increase in the temperature is usually accompanied by a decrease in the band molar absorption coefficient (due to decreasing intermolecular forces) and by band broadening (due to increasingly more energetic collisions). With compounds whose molecules occur in the form of various rotational isomers, thermodynamically less stable forms are more populated with increasing temperature. The mean aggregate size decreases with increasing temperature with associating molecules; however, the free form never appears even at very high temperatures (e.g. with alcohols; strongly sterically hindered molecules are exceptional).

Liquids are quite incompressible and thus increasing the pressure has practically no effect and the vibrational spectra remain unchanged. However, if liquids are exposed to very high pressures of the order of 10^6 Pa, deep changes in their structure and consequently also in the spectra occur. Liquids, the molecules of which are brought closer by a high pressure, are often converted into crystalline forms and sometimes they react chemically (e.g. polymerize); this gives rise to considerable changes in the vibrational spectra. It has been shown recently that, under the effect of high pressures on solutions of compounds forming various rotational isomers, the content of these isomers can be varied; thermodynamically less stable forms are more populated under higher pressures.

Crystalline forms of polymorphous compounds change with a change in the temperature, provided that the phase-transition temperature is passed. Other transitions have also been found with certain compounds in the crystalline state. For example, the spectra of cycloalkanes measured at temperatures slightly above and below the melting point are virtually identical. Only on cooling to a certain temperature, different for each cycloalkane, do the spectra change considerably. The cycloalkane molecules apparently change their conformation at these temperatures and this change is reflected in the spectra.

On cooling samples of compounds, whose molecules form rotational isomers, to very low temperatures, thermodynamically less stable forms are often frozen-out during crystallization. The process is usually complete. From the point of view of crystal structure it is more advantageous if they consist of a single type of molecule, which are then packed better in the crystal lattice.

Cooling liquids or mixtures must be used for cooling samples for the measurement of infrared or Raman spectra at temperatures below 0 °C. Liquid nitrogen is most frequently employed for cooling in common spectroscopic practice; its boiling point, −196 °C (77 K), is the lowest temperature which can then be attained in the cell. However, such a low temperature is rarely attained because of temperature losses in the cryostat;

if the exact temperature of the cooled sample must be known, it must be determined experimentally (e.g. by a calibrated thermocouple placed directly on the cell transparent window). Liquid methane is suitable for attaining a temperature of −162 °C (111 K) and a liquid acetone-solid CO_2 bath for −78 °C (195 K). Cryogenic experiments with cooled matrices require much lower temperatures, obtained by using hydrogen with the boiling point, −253 °C (20 K) and especially liquid helium boiling at −269 °C (4 K).

Work with cooling liquids is rather dangerous; liquids such as methane, hydrogen and mixtures of organic liquids with CO_2 yield flammable and some-times explosive vapour mixtures and work with cryostats cooled by them requires special safety precautions. Work with liquid helium is also dangerous; because of its low heat capacity, even small overheating of a liquid sample leads to rapid evaporation and explosion.

Polymorphous compounds can also be converted from one form into another by pressure; the modification obtained is often incapable of existence at normal pressures. This effect also sometimes appears unexpectedly, e.g. during pressing of KBr pellets with samples of polymorphous compounds.

3.4.4 Isolated Matrices

At the end, the possibility of the study of the spectra of compounds transferred into matrices of another materials should be mentioned. Experiments of this kind are used for various purposes, chiefly according to the matrix type and the measuring conditions.

Inorganic polyatomic ions can be studied in crystal lattices of matrices of e.g. the alkali metal halides. For example, from fused potassium chloride, contaminated by traces of potassium sulphate, a KCl single-crystal doped by sulphate ions can be grown. In the absorption spectrum of a plate cut from the single crystal, the bands of vibrations of the sulphate anion placed in the potassium chloride crystal field appear. Sulphate anions can analogously be introduced into other potassium halides (KF, KBr, KI) and into halides of other alkali metals (sodium, lithium, etc.); the spectra then yield information about the forces to which the sulphate anion is subject in matrices of various materials. According to experiments carried out with some anions, the effect of both cations and anions is accompanied by considerable changes in the vibration band wavenumbers and the integrated intensities.

However, the spectra of compounds in so-called cooled isolated matrices of inert gases are studied much more frequently. The experimental technique of this method is very demanding, as work at very low temperatures (in liquid hydrogen or helium) is usually required.

Principally, a gaseous mixture of the studied compound and an inert gas (chiefly argon, but nitrogen, carbon monoxide, sulphur hexafluoride, etc. can also be used) is cooled rapidly, so that it deposits as a solid film on the transparent cell window. The molecules of the test compound, if present at a low concentration, are dispersed in the inert gas crystal lattice and thus exposed to only very weak intermolecular forces. Thus the technique approaches the measurement of spectra in the gaseous state, but with the advantage that, in contrast to the spectra of gases, spectra obtained from matrices are "purely" vibrational.

The advantages of this experimental technique are undisputable. The spectra of strongly polar compounds can be measured in the almost inert medium of cooled matrices under conditions where their molecules do not associate. By an increase in the test compound concentration, imperfect distribution within the matrix can be attained and the formation of dimers, trimers, etc. can be followed under simple and unambiguous conditions. The study of a spectrum can sometimes distinguish the crystal field effect; for example, the glycine molecule, which exists in the crystal and in solutions only as the zwitterion, $H_3N^+CH_2COO^-$, is present in a matrix in the neutral form, H_2NCH_2COOH. Matrix measurements also reflect unusual forms of intermolecular interactions; for example, a mixture of hydrogen fluoride with ammonia in a matrix forms a hydrogen-bonded complex FH ... NH_3 and not ammonium fluoride NH_4F as under other conditions.

Details on this complicated experimental technique, the analysis of the spectra obtained and examples of molecules studied can be found in many works and especially in the monographs[1,2].

---

1 Meyer B.: Low Temperature Spectroscopy. Optical Properties of Molecules in Matrices, Mixed Crystals and Frozen Solutions. Elsevier, Amsterdam 1971.

2 Hallam H. E.: Vibrational Spectroscopy of Trapped Species. Wiley, Chichester 1973.

Chapter 3 — References

Experimental Technique

1. Brügel W.: Einführung in die Ultrarotspektroskopie. Steinkopf, Darmstadt 1957.
2. Kössler I.: Methoden der Infrarot-Spektroskopie in der chemischen Analyse. Quantitative Analyse. SNTL, Prague 1960 (translation — Akad. Verlag, Leipzig 1966).
3. Conn G. K. T., Avery D. G.: Infrared Methods. Academic Press, New York 1960.
4. Potts W. J., Jr.: Chemical Infrared Spectroscopy. Vol. I, Techniques. Wiley, New York 1963.
5. Geppert G.: Experimentelle Methoden der Molekülspektroskopie. Akademie Verlag, Berlin 1964.
6. Martin A. E.: Infrared Instrumentation and Techniques. Elsevier, Amsterdam 1966.
7. Stine K. E.: Modern Practices in Infrared Spectroscopy. Technical Information Sect., Sci. Instrum. Div., Beckman Instruments, Fullerton (California) 1970.
8. Stewart J. E.: Infrared Spectroscopy. Dekker, New York 1970.
9. May L. (ed.): Spectroscopic Tricks. Vol. I., II. Plenum Press, New York 1971.
10. Miller R. G. J., Stace B. C. (eds.): Laboratory Methods in Infrared Spectroscopy. Heyden, London 1972.
11. Zýka J. (ed.): Analytical Handbook, SNTL, Prague 1972.

Statistics, Experiment Evaluation

1. Rényi A.: Wahrscheinlichkeitsrechnung mit einem Anhang über Informationstheorie. Deutscher Verlag der Wissenschaften, Berlin 1962.
2. Korn G. A., Korn T. M.: Mathematical Handbook for Scientists and Engineers. McGraw—Hill, New York 1961.
3. Eadie W. T., Dryard D., James F. E., Roos M., Sadoulet B.: Statistical Methods in Experimental Physics. North-Holland, Amsterdam 1971.
4. Linnik J. V.: Method Naimenshikh Kvadratov i Osnovy Teorii Obrabotki Nabliudenii. Gos. Izdat. Fiziko-Matematicheskoi Literatury, Moscow 1962.
5. Sachs L.: Statistische Methoden. Springer, Berlin 1970.
6. Janko J.: Statistical Tables. Publ. House of Czechoslovak Academy of Sciences, Prague 1958.
7. Ahlberg J. H., Nilson E. N., Walsh J. L.: The Theory of Splines and Their Applications. Academic Press, New York 1967.

Intermolecular Effects

1. Pimentel G. C., McClellan A. L.: The Hydrogen Bond. Freeman, San Francisco 1960.
2. Bakhshiev N. G.: Spektroskopiya Mezhdumolekularnykh Vzaimodeistvii. Nauka, Leningrad 1972.
3. Bulanin M. O. (ed.): Spektroskopiya Vzaimodeistvuyushchikh Molekul. Leningrad University, Leningrad 1970).
4. Yarwood J.: Spectroscopy and Structure of Molecular Complexes. Plenum Press, New York 1973.

CHAPTER 4

Analysis of Vibrational Spectra and Band Assignment

In Chapter 3, procedures were discussed for obtaining accurate and sufficiently precise values of the band maximum wavenumbers and intensities and the values of half-widths and integrated intensities from infrared bands and Raman lines. Sources of error arising in the measurement and processing of spectra and the forms of their elimination (description) were also described. The formal procedure for spectrum processing is quite general, i.e. it can be used for any band in infrared spectra or for any Raman line.

Chapters 4 and 5 will deal with vibrational spectra from another point of view, namely, utilization of information stored in them for solution of chemical problems. Information contained in the spectrum is quite specifically distributed in the recording and the data concerning the "location" of its storage must be well known in spectrum handling. Chapter 4 discusses the procedures employed in the analysis of spectra (i.e. in the extraction of useful information from vibrational spectra) and Chapter 5 describes the type of task which is solved using vibrational spectra.

Vibrational spectra can be successfully utilized in chemistry only because certain molecular vibrations have group (local) character. Generally the nuclei of all atoms in a molecule take part in any normal vibration, but their contributions are not equal. In any normal vibration some nuclei contribute more, others less and the contributions from still others can be neglected. Thus any normal vibration is connected with vibrations of the nuclei in certain bonds or groups, i.e. certain localities in molecules; structural elements, in which the specificity of individual molecules is manifested, appear in vibrational spectra as certain characteristic features.

It was shown in Chapter 2 that each normal vibration is connected with the existence of certain vibrational quantum levels. Transitions among these levels for all normal vibrations – provided that they are permitted by the selection rules are reflected in the vibrational spectra. Hence any normal molecular vibration is connected with absorption (scattering) in a certain region of the infrared (Raman) spectrum, with exceptions determined by the selection rules. If this statement is combined with the information in the previous paragraph, it can be seen that the bands in spectra can be assigned to vibrations of atomic nuclei in certain parts (bonds, groups) of molecules. The band parameters are then the source of information on which correlations of this type are based. In assigning bands in spectra, not only the atomic nuclei contributing to the given normal vibration, but also the way in which these nuclei vibrate, are determined. A single group of atomic nuclei can participate in several normal vibrations and hence can be connected with the occurrence of a greater number of bands in various regions of vibrational spectra.

Spectra can, of course, also be analyzed without considering the group character of localized vibrations. This approach is taken e.g. in the analyses of spectra before calculating force constants. The purpose of these analyses is to determine the fundamental frequency bands (the complete set of $1 \leftarrow 0$ transitions), whose wavenumbers serve as a basis for calculation of force constants. In calculating force conditions in a molecule, information on the group character of the vibrations has no importance; it can at most facilitate the spectrum analysis. This also indicates that data on group localized vibrations have less importance in physical applications than in chemical ones.

The content of Chapter 4 is closely related to the theory of vibrational spectra; as this theory is not discussed in the present book, the interested reader is referred to the more general monographs cited at the end of Chapter 2.

Finally it should be pointed out that errors in the assignment of bands in infrared spectra and lines in Raman spectra are gross errors and, like all mistakes, lead to failure of all constructions built on this basis. This warning is important, because not all bands in the spectra of polyatomic molecules can be reliably and unambiguously assigned and not all molecular vibrations are discernible in the spectra.

4.1 Classification of Molecular Vibrations According to Symmetry

It follows from the theory of molecular vibrations that individual molecular vibrations can be classified on the basis of their behaviour under the effect of symmetry operations. Table XXIX gives an example of a classification of molecular vibrations, carried out for the methanesulphonyl chloride molecule which we studied. The band maximum wavenumbers are given, together with the description of the corresponding normal vibrations; this procedure is usually called band assignment. The table helps to classify

Table XXIX

Wavenumbers of the fundamental frequency bands for the CH_3SO_2Cl molecule (from the infrared spectra of a liquid sample) and their assignment

Symmetry	Vibration	Wavenumber (cm^{-1})	Assignment
	ν_1	3018	stretching C—H
	ν_2	2932	stretching C—H
	ν_3	1410	deformation CH_3
	ν_4	1321	deformation CH_3
	ν_5	1173	stretching SO_2
A'	ν_6	967	rocking CH_3
	ν_7	749	stretching C—S
	ν_8	540	scissoring SO_2
	ν_9	494	wagging SO_2
	ν_{10}	377	stretching S—Cl
	ν_{11}	258	deformation C—S—Cl
	ν_{12}	3040	stretching C—H
	ν_{13}	1376	deformation CH_3
	ν_{14}	1364	stretching SO_2
A''	ν_{15}	967	rocking CH_3
	ν_{16}	520	rocking SO_2
	ν_{17}	235	twisting SO_2
	ν_{18}	?	torsion H—C—S—Cl

the manner in which molecular vibrations are classified; this procedure is based on the application of group theory and hence the principles of the latter must be explained.

4.1.1 Symmetry Elements and Operations

If a polyatomic molecule contains a larger number of atoms of the same kind, it can sometimes be converted, by rotating it through a certain angle in space, by reflecting it in a certain plane or through some other geometric operation, into another configuration, which is indistinguishable from the original configuration, provided, of course, that atoms of the same kind are not distinguished. If plane σ is constructed through the atoms of chlorine, sulphur, carbon and one hydrogen atom from the methyl group in a methanesulphonyl chloride molecule in the equilibrium configuration, then reflection in this plane results in the remaining two hydrogen atoms and the two oxygen atoms exchanging respective places with formation of a configuration indistinguishable from the original (see Fig. 52). Analogously, when

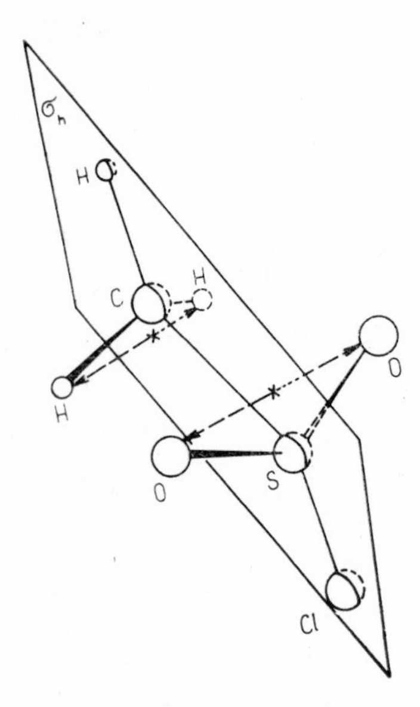

Fig. 52. Reflection operation $\boldsymbol{\sigma}_{\mathbf{h}}$ through reflection plane σ_h, demonstrated on the methanesulphonyl chloride molecule.

a phosgene molecule is rotated through 180° around the C_2 axis (see Fig. 53), the chlorine atoms exchange their places and the configuration obtained is indistinguishable from the original one.

Both these geometric operations are examples of symmetry operations and the plane σ and axis C_2 are examples of symmetry elements. The individual symmetry elements and the corresponding symmetry operations will now be discussed systematically.

The first large group of symmetry elements consists of n-fold axes of rotational symmetry, denoted by symbol C_n, $n \in \langle 1; \infty \rangle$. The molecule can be rotated n-times by angle $2\pi/n$ (i.e. $360°/n$) with respect to such an axis. On each rotation one indistinguishable configuration is converted into another. If these elementary rotations are mutually combined, symmetry operations $\mathbf{C}_n^1, \mathbf{C}_n^2, \ldots, \mathbf{C}_n^n$, generally $\mathbf{C}_n^k$, $k \in \langle 1; n \rangle$, corresponding to rotation through angle $2\pi k/n$, are obtained. It is evident that operation $\mathbf{C}_n^n$ converts the configuration back to the original position; this operation is trivial, is termed the identity operation and is denoted **E** or ***I***. This operation can be performed on any molecule, even if it has no symmetry element. It should further be pointed out that no symmetry operation other than the identity operation is connected with one-fold rotational axis C_1; for this reason the alternative symbol $\mathbf{C}_1^1$ can sometimes be encountered.

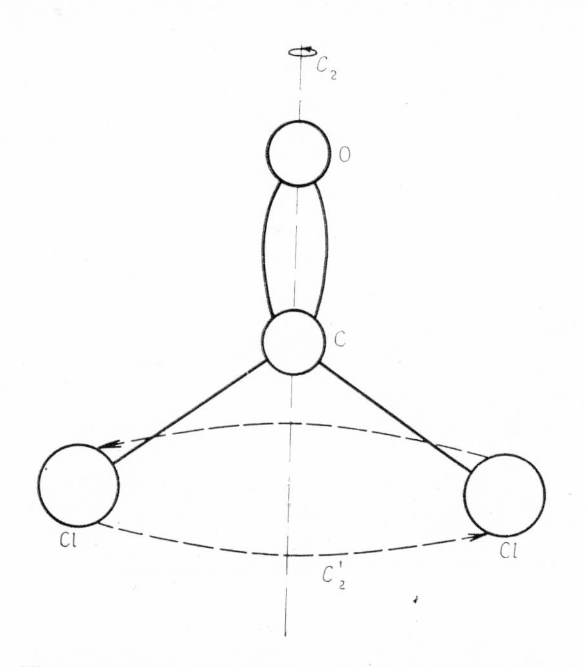

Fig. 53. Rotation operation $\mathbf{C}_2$ around two-fold axis C_2, demonstrated on the phosgene molecule.

The second group of symmetry elements contains reflection planes σ; the corresponding operation $\boldsymbol{\sigma}$ is called reflection and is denoted by the same symbol. It is evident that an even number of reflections in the same plane is again an identity operation. Individual reflection planes (and the operations of reflection connected with them) are distinguished by various subscripts, such as $_h$ (horizontal) for the plane perpendicular to the rotational axis with the highest order, $_v$ (vertical) for the plane in which the axis with the highest order lies and $_d$ (diagonal) for the plane in which the rotational axis with the highest order lies and which simultaneously halves the angle between the other two rotational axes. Individual planes are further specified according to the axes of the selected coordination system in the molecule to which they are related, e.g., σ_{xy}, etc.

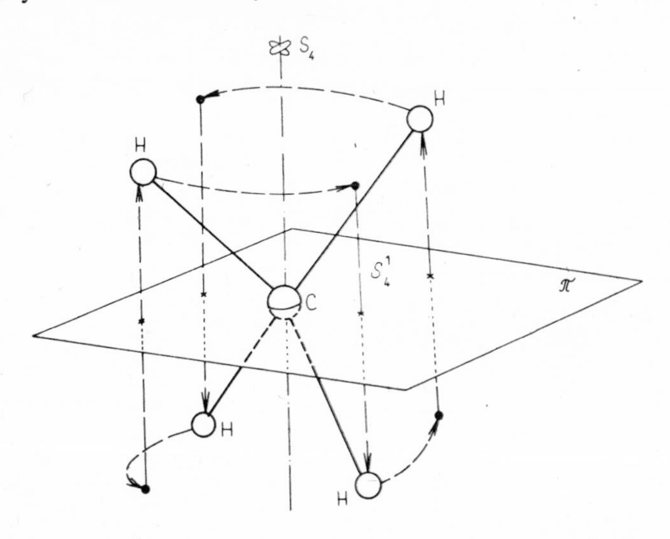

Fig. 54. Symmetry operation $\mathbf{S_4}$ with four-fold rotation-reflection axis S_4, demonstrated on the methane molecule. π is the auxiliary reflection plane.

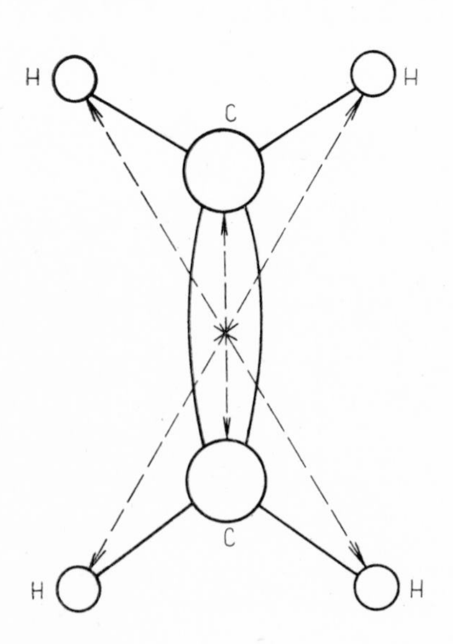

Fig. 55. Inversion operation **i** around symmetry centre *i*, demonstrated on the ethylene molecule.

Another group of elements is formed by n-fold rotation-reflection axes S_n. The symmetry operations corresponding to them can be described as follows: First, the system is rotated through angle $2\pi/n$ around the appropriate axes S_n and then, on each rotation, is reflected in the plane π perpendicular to axis S_n (Fig. 54). It can be seen that, for odd values of n, all operations $\mathbf{S}_n^k$, $k \in \langle 1, 2n - 1 \rangle$, can be replaced either by simple rotations $\mathbf{C}_n^k$, $k \in \langle 1, n - 1 \rangle$, or by rotations $\mathbf{C}_n^k$, $k \in \langle 1, n \rangle$, followed by reflection $\boldsymbol{\sigma}_h$.

The last type of symmetry elements is the centre of symmetry, i, connected with the inversion operation, **i**. Here the equivalent pairs of atoms are connected by straight lines, whose centres lie in the centre of symmetry (Fig. 55).

Table XXX

Survey of equivalences of elementary symmetry operations

Symmetry operation **R**	Equivalent operation[a] **R′**										
$\mathbf{C}_n^k$	n \ k	1	2	3	4	5	6	7	8	9	10
	1	E	E	E	E	E	E	E	E	E	E
	2	*	E	C_2^1	E	C_2^1	E	C_2^1	E	C_2^1	E
	3	*	*	E	C_3^1	C_3^2	E	C_3^1	C_3^2	E	C_3^1
	4	*	C_2^1	*	E	C_4^1	C_2^1	C_4^3	E	C_4^1	C_2^1
	5	*	*	*	*	E	C_5^1	C_5^2	C_5^3	C_5^4	C
	6	*	C_3^1	C_2^1	C_3^2	*	E	C_6^1	C_3^1	C_2^1	C_3^2

Symmetry operation **R**	Equivalent operation[a] **R′**										
$\mathbf{S}_n^k$	n \ k	1	2	3	4	5	6	7	8	9	10
	1	σ	E	σ	E	σ	E	σ	E	σ	E
	2	i	E	i	E	i	E	i	E	i	E
	3	*	C_3^2	σ	C_3^1	*	E	S_3^1	C_3^2	σ	C_3^1
	4	*	C_2^1	*	E	S_4^1	C_2^1	S_4^3	E	S_4^1	C_2^1
	5	*	C_5^2	*	C_5^4	σ	C_5^1	*	C_5^3	*	E
	6	*	C_3^1	i	C_3^2	*	E	S_6^1	C_3^1	i	C_3^2
	8	*	C_4^1	*	C_2^1	*	C_4^3	*	E	S_8^1	C_4^1

[a] The asterisk * denotes operations which cannot be written in a simpler manner.

In the notation of symmetry operations $\mathbf{C}_n^1$ and $\mathbf{S}_n^1$, the superscript, ^{1}, is often omitted. A number of operations $\mathbf{C}_n^k$ and $\mathbf{S}_n^k$ are identical with other operations. Table XXX summarizes the data on these equivalencies and simultaneously enables selection of the simplest notation.

4.1.2 Symmetry Point Groups

The individual elementary symmetry operations, described in the previous section, can be successively applied to a given molecule. A sequence of two or more successive operations is termed the product of symmetry operations. In contrast to the common arithmetical product, the commutative law does not hold here; if the order of the operations is changed, the result need not be identical. In any case, the product of symmetry operations always yields an operation of symmetry applicable to the given molecule. If a set of all operations of symmetry corresponding to a certain molecule is formed (this set must always contain the identity operation, **E**), then the product of an arbitrary series of operations from this set again yields an operation from this set and, vice versa, each operation from the set can be expressed as the product of other operations. From this point of view, the set is mathematically a group of operations. The number of elementary symmetry operations forming a group is termed the order of the group, g.

Symmetry point groups are employed for the description of isolated rigid molecules (those which are not exposed to the effect of external forces). These are called point groups because there is always at least one point in the molecule whose position does not change on carrying out any symmetry operation. Symmetry point groups are, of course, unsuitable for molecules with almost free internal rotation (e.g. nitromethane, ferrocene and others); permutation-inversion symmetry groups have been derived[1] for this purpose on different principles. For the description of crystal structures, symmetry space groups are used, which are closely related to symmetry point groups, but contain, in addition, e.g. translation operations describing the periodical occurrence of particles (molecules, ions) in the crystal lattice (e.g. a shift in the direction of some crystallographic axis by an integral multiple of the appropriate dimension of the crystal unit cell).

In this book, only symmetry point groups will be discussed. Two systems, Schoenflies' and Hermann–Mauguin's (also called the international

---

1 Hougen J. T.: The Vibrational Problem in Molecules with Nearly Free Internal Rotation. *Pure Appl. Chem. 11*, 481 (1965).

system), have basically become established in the literature for their notation. We shall adhere to the former, which is common in spectroscopy, while the latter has found use chiefly in crystallography for the description of space groups.

A system of point groups can be built gradually, by adding further and further operations to the groups. This is illustrated well by Table XXXI, which shows the similarity among point groups proper (i.e. groups consisting exclusively of $\mathbf{C}_n^k$-type operations, i.e. mere rotations).

Table XXXI

Subgroups of proper point groups

Point groups	Subgroups										
$\mathscr{O}$	$\mathscr{C}_1$	$\mathscr{C}_2$	$\mathscr{C}_3$	$\mathscr{C}_4$		$\mathscr{D}_2$	$\mathscr{D}_3$	$\mathscr{D}_4$		$\mathscr{T}$	$\mathscr{O}$
$\mathscr{T}$	$\mathscr{C}_1$	$\mathscr{C}_2$	$\mathscr{C}_3$			$\mathscr{D}_2$				$\mathscr{T}$	
$\mathscr{D}_6$	$\mathscr{C}_1$	$\mathscr{C}_2$	$\mathscr{C}_3$		$\mathscr{C}_6$	$\mathscr{D}_2$	$\mathscr{D}_3$		$\mathscr{D}_6$		
$\mathscr{D}_4$	$\mathscr{C}_1$	$\mathscr{C}_2$		$\mathscr{C}_4$		$\mathscr{D}_2$		$\mathscr{D}_4$			
$\mathscr{D}_3$	$\mathscr{C}_1$	$\mathscr{C}_2$	$\mathscr{C}_3$				$\mathscr{D}_3$				
$\mathscr{D}_2$	$\mathscr{C}_1$	$\mathscr{C}_2$				$\mathscr{D}_2$					
$\mathscr{C}_6$	$\mathscr{C}_1$	$\mathscr{C}_2$	$\mathscr{C}_3$		$\mathscr{C}_6$						
$\mathscr{C}_4$	$\mathscr{C}_1$	$\mathscr{C}_2$		$\mathscr{C}_4$							
$\mathscr{C}_3$	$\mathscr{C}_1$		$\mathscr{C}_3$								
$\mathscr{C}_2$	$\mathscr{C}_1$	$\mathscr{C}_2$									
$\mathscr{C}_1$	$\mathscr{C}_1$										

The simplest and most important are cyclic groups $\mathscr{C}_n$, consisting of $g = n$ operations, namely, $\{\mathbf{E}, \mathbf{C}_n^1, \mathbf{C}_n^2, \ldots, \mathbf{C}_n^{n-1}\}$, i.e. groups describing molecules with a single symmetry element, n-fold rotational axis C_n. The simplest point group of all, i.e. the $\mathscr{C}_1$ group, which contains a single symmetry operation, identity operation $\mathbf{E}$, also belongs here. Although the frequency, n, of a rotational axis can be arbitrary in point groups,*) there are only a few kinds of molecules with five-, seven- or more-fold axes of symmetry.

If there are n two-fold axes C_2 perpendicular to C_n in a molecule, in addition to n-fold rotational axis C_n, groups of order $g = 2n$ are formed,

*) On the other hand, with space groups only rotational axes with orders 1, 2, 3, 4 and 6 are permissible for purely geometric reasons.

termed dihedral and denoted as $\mathscr{D}_n$. If $n = 2$, a special case is obtained, as it is impossible to decide which of the three mutually perpendicular two-fold axes is primary; then symbol $\mathscr{V}$ (from German Vierergruppe) is also used, besides $\mathscr{D}_2$. It can be shown that group $\mathscr{D}_1$ is identical with group $\mathscr{C}_2$.

If, on the other hand, a horizontal reflection plane σ_h, perpendicular to C_n, is combined with the n-fold axis, $\mathscr{C}_n$, groups of order $g = 2n$ are again obtained. The addition of the reflection operation necessitates introduction of n-fold rotation-reflection axis S_n, lying on axis C_n. For even values of n, another element appears, namely, the centre of symmetry, i. Group $\mathscr{C}_{1h}$, consisting only of operations $\{\mathbf{E}, \boldsymbol{\sigma}_h\}$, is usually denoted as $\mathscr{C}_s$.

Analogously, groups $\mathscr{D}_n$ can be extended to groups $\mathscr{D}_{nh}$ by adding horizontal planes σ_h. As in the previous case, here the product of operations $\mathbf{C}_n^k$ and $\boldsymbol{\sigma}_h$ also yields operations of the $\mathbf{S}_n^k$-type (for odd k's). Furthermore, the combination of reflection in plane σ_h and rotation around secondary two-fold axes C_2 is equivalent to reflection in vertical planes σ_v perpendicular to σ_h and containing axes C_2. Centre of symmetry i appears for even n. The overall order of these groups is $g = 4n$. Group $\mathscr{D}_{2h}$ is denoted as $\mathscr{V}_h$ and group $\mathscr{D}_{1h}$ is identical with group $\mathscr{C}_{2v}$.

Groups of the $\mathscr{C}_{vn}$-type are obtained from groups $\mathscr{C}_n$ by adding vertical reflection planes σ_v. The order of these groups is $g = 2n$. Group $\mathscr{C}_{1v}$ is identical with group $\mathscr{C}_s$.

However, a similar extension of groups of the $\mathscr{D}_n$-type does not lead to a new group type, as groups $\mathscr{D}_{nh}$ contain both horizontal and vertical reflection planes, σ_h and σ_v, respectively. A new type of group can, however, be formed by adding vertical diagonal reflection planes σ_d, passing through the n-fold rotational axis and halving the angles between secondary two-fold rotational axes C_2. Another symmetry element simultaneously appears, namely, rotation-reflection axis S_{2n} and, with odd n, also a centre of symmetry, i. The order of these groups, denoted by symbols $\mathscr{D}_{nd}$, is $g = 4n$. Group $\mathscr{D}_{2d}$ is also denoted as $\mathscr{V}_d$ and group $\mathscr{D}_{1d}$ is identical with group $\mathscr{C}_{2h}$.

The last systematically ordered class of groups are groups $\mathscr{S}_n$ with order $g = n$ for even n; if n is odd, these groups are identical with groups $\mathscr{C}_{nh}$ and their order is $g = 2n$. Groups with even n which is not a multiple of four contain a centre of symmetry, i (it holds that $\mathbf{S}_2^1 \equiv \mathbf{S}_6^3 \equiv \mathbf{S}_{10}^5 \equiv \mathbf{S}_{14}^7 \equiv \ldots \equiv \mathbf{i}$). As group $\mathscr{S}_2$ consists of only two operations, $\{\mathbf{E}, \mathbf{S}_2^1\} \equiv \{\mathbf{E}, \mathbf{i}\}$, i.e. identity and inversion, it is often denoted as group $\mathscr{C}_i$.

Groups denoted as $\mathscr{T}$ are connected with the symmetry of a regular tetrahedron. Group $\mathscr{T}$ (the group order is $g = 12$) consists of all rotational symmetry operations for the tetrahedron (four three-fold and three two-fold axes), group $\mathcal{T}_d$ (group order, $g = 24$) also has all the reflection planes and group $\mathcal{T}_h$ (group order, $g = 24$), a centre of symmetry. Group $\mathcal{O}$ (group order, $g = 24$) is analogously formed by all rotational symmetry operations for an octahedron or a cube; group $\mathcal{O}_h$ (group order, $g = 48$) further contains a centre of symmetry and all other symmetry operations for an octahedron. The highest point groups, $\mathcal{I}$ (group order, $g = 60$) and $\mathcal{I}_h$ (group order, $g = 120$), correspond to the symmetry of a regular icosahedron or dodecahedron.

Any non-linear molecule can be assigned to some of the above symmetry groups. Linear molecules without a centre of symmetry belong to group $\mathcal{C}_{\infty v}$ and those with a centre of symmetry to group $\mathcal{D}_{\infty h}$. Examples of assignment of molecules to important symmetry point groups are given in Table XXXII.

Table XXXII

The most important point groups of symmetry and examples of molecules belonging to them

Point group	Molecule
$\mathcal{C}_1$	HCFClBr, camphor quinone, cholesterol
$\mathcal{C}_2$	HOOH, HSSH
$\mathcal{C}_s$	CH_3SO_2Cl, benzaldehyde, chlorocyclohexane
$\mathcal{C}_{2v}$	H_2O, $Cl_2C{=}O$, C_6H_5Cl, 2,5-dichlorothiophene
$\mathcal{C}_{3v}$	CH_3Cl and $CHCl_3$, NH_3, $CH_3C{\equiv}CCl$, 1-chloroadamantane
$\mathcal{C}_{\infty v}$	HF, HCN, OCS, $HC{\equiv}CCl$
$\mathcal{D}_{2h} = \mathcal{V}_h$	$H_2C{=}CH_2$, diborane
$\mathcal{D}_{3h}$	BF_3, anion NO_3^-, 1,3,5-trichlorobenzene
$\mathcal{D}_{6h}$	benzene
$\mathcal{D}_{\infty h}$	H_2, CO_2, $HC{\equiv}CH$
$\mathcal{D}_{2d} = \mathcal{V}_d$	$H_2C{=}C{=}CH_2$
$\mathcal{D}_{3d}$	cyclohexane, regularly distorted hexabromobenzene
$\mathcal{T}_d$	methane, anion SO_4^{2-}, hexamethylenetetramine, cation NH_4^+
$\mathcal{O}_h$	hexafluorides SF_6, UF_6 etc.
$\mathcal{I}_h$	anion $[Mo(CN)_8]^{2-}$, anion $[B_{12}H_{12}]^{2-}$

4.1.3 Irreducible Representations

If a coordinate system describing the positions of the atoms in a molecule is selected, each symmetry operation can be described by a linear transformation of the coordinates, transcribed e.g. in the form of a transformation matrix. All linear transformations corresponding to the symmetry operations of the corresponding point group also form a group, termed the representation of the given symmetry point group. The concrete form of the matrices naturally depends on the selected coordinate system, but the properties of the groups of representations are the same as those of the symmetry point groups*) whatever the method used in selecting the coordinate system.

In the study of molecular vibrations, the description of changes in the positions of the atomic nuclei in which the individual vibrations are manifested is of prime importance. These are most frequently described as changes in the internal coordinates, i.e. internuclear distances and valence angles. In methanesulphonyl chloride, 7 bonds and 13 angles are present; as two of the angles are redundant,**) the number of deformation coordinates is eleven.

The methanesulphonyl chloride molecule belongs to point group $\mathscr{C}_s$, formed by operations $\{\mathbf{E}, \boldsymbol{\sigma}_h\}$. The identity operation, **E**, converts each coordinate into itself; hence the matrix transcription of this operation is always a diagonal unit matrix (the matrix whose diagonal terms equal unity and non-diagonal terms are zero). A more complex situation occurs with the reflection operation, $\boldsymbol{\sigma}_h$. A change in the length of one C−H bond which does not lie in the σ_h plane is reflected in the other symmetrically located C−H bond and the internuclear distances in the S=O bonds and the deformation of some angles change similarly. Therefore, there will be non-zero non-diagonal elements in the matrix notation. If, however, each normal vibration of the molecule is described by a suitable linear combination of changes in the internal coordinates (generally all the atoms

*) The groups are said to be isomorphous.

**) The coordinates employed for the description of normal vibrations of molecules must, of course, be independent; the planar and space angles in molecules, however, do not satisfy this condition. The triangle can serve as an example of the redundance of angular coordinates; only two angles in a planar triangle are independent, the third being the complement to 180°. Molecules are similar; for example, in the planar phosgene molecule only two angles of the three are independent (ClCCl and ClCO), of the six angles in the methane molecule only five are independent, etc.

in the molecule contribute to normal vibrations), which is termed the normal coordinate, q_i, of the i-th normal vibration, the transformation matrix for all symmetry operations can normally be simplified, i.e. "reduced", to the diagonal form. If no further simplification of the matrix representation of a point group is possible, the representation is termed fully reduced.

However, it is often impossible to reduce a matrix fully to the diagonal form. It may happen with groups of higher symmetry that certain pairs or groups of three normal coordinates cannot be separated and these normal coordinates are mixed in certain symmetry operations. This appears as the presence of non-diagonal elements in the matrix representation. The corresponding normal vibrations are also indistinguishable and vibration degeneration occurs. If two coordinates are mixed, the vibrations are doubly degenerate; with three mixed coordinates they are triply degenerate. Degenerate vibrations are physically characterized by the same frequency.

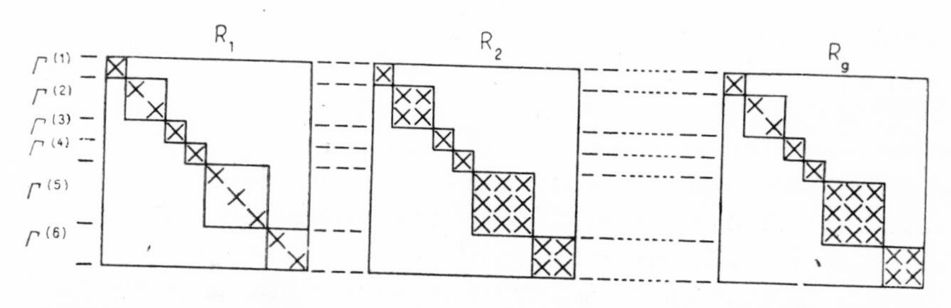

Fig. 56. Scheme of the decomposition of the transformation matrices $\mathbf{R}_i$ of a fully reduced representation to irreducible representations $\Gamma^{(\gamma)}$. The crosses denote the non-zero elements of the transformation matrices.

Even this imperfect simplification enables separation of transformation matrices into blocks (submatrices), among which there are no bonds for all g symmetry operations $\mathbf{R}_i$, $i \in \langle 1, g \rangle$, with the given symmetry point group. The mutually corresponding submatrices for the same vibration but for various symmetry operations form the irreducible representations $\Gamma^{(\gamma)}$. This is schematically depicted in Fig. 56; here irreducible representations $\Gamma^{(1)}$, $\Gamma^{(3)}$ and $\Gamma^{(4)}$ correspond to non-degenerate vibrations, $\Gamma^{(2)}$ and $\Gamma^{(6)}$ to doubly degenerate vibrations and $\Gamma^{(5)}$ to the triply degenerate vibration. If the irreducible representations of a concrete molecule are compared, it can be seen that they can be repeated. Generally, the fully reduced representation of a symmetry point group for a given molecule consists of a certain number of irreducible representations $\Gamma^{(\gamma)}$. The number of occurrences of a given irreducible representation depends on the molecular structure (see Section 4.1.4).

Introduction of the irreducible representation simplifies the complicated mathematical notation of each symmetry operation for non-degenerate normal vibrations to a single number, which is the corresponding diagonal element of the appropriate transformation matrix (actually transformation submatrix with dimension 1×1). This number can attain only two values, $+1$ and -1. In the former case the normal coordinate does not change its sign during the given symmetry operation (it is said to be symmetrical with respect to the given symmetry operation); in the latter its sign changes (the vibration is antisymmetrical). Two normal vibrations of the methane-sulphonyl chloride molecule can serve as an example: here the nuclei of the O atoms and of the SO_2 group vibrate in phase (I) and out of phase (II):

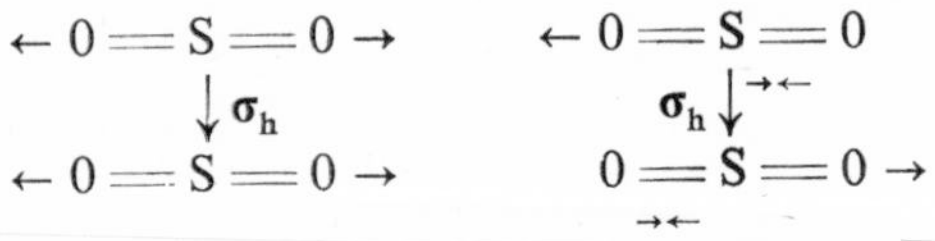

$$\begin{array}{ccc} \leftarrow 0=S=0 \rightarrow & \qquad & \leftarrow 0=S=0 \\ \downarrow \sigma_h & & \sigma_h \downarrow \;{\rightarrow\leftarrow} \\ \leftarrow 0=S=0 \rightarrow & & 0=S=0 \rightarrow \\ & & {\rightarrow\leftarrow} \\ \text{I} & & \text{II} \end{array}$$

During vibration in phase (I), the direction of the vibrational movement of the nuclei does not change after performing symmetry operation $\boldsymbol{\sigma}_h$; hence the vibration is symmetrical. In out-of-phase vibration (II), a change occurs and consequently the vibration is antisymmetrical.

With degenerate vibrations, complete information about a symmetry operation for a irreducible representation cannot be conveyed by a single number, but submatrices with the size 2×2 and 3×3 for doubly and triply degenerate vibrations, respectively, are required. For evaluation of the character of a given irreducible representation, however, the sum of the diagonal elements of the transformation submatrix is decisive. This number, denoted as $\chi_R^{(\gamma)}$, is termed the character of irreducible representation $\Gamma^{(\gamma)}$ for symmetry operation $\mathbf{R}$.

Tables of the character of irreducible representations have been drawn up for all symmetry points groups using the methods of group theory and are given, together with other data on groups, in Table XXXIII. If the tables of characters were actually written down for all symmetry operations in each group, it would be found that some columns in the tables occur repeatedly. This reflects a property of groups, namely, that their elements form certain classes. All symmetry operations belonging to the same class have, in addition to other common properties, the same characters of

Table XXXIII

Characters of irreducible representations of the most important point groups of symmetry

Point groups $\mathscr{G}$ (group order g)			Classes of symmetry operations[a]			
Irreducible representations[b] $\Gamma^{(\gamma)}$			Character[c] $\chi_j^{(\gamma)}$			
$\mathscr{C}_1$ $(g = 1)$			E			
	$\mathscr{C}_s \equiv \mathscr{C}_{1h}$ $(g = 2)$		E	σ_h		
		$\mathscr{C}_i$ $(g = 2)$	E	i		
A 3 3 IR Ra	A' 2 1 IR Ra	A_g 0 3 Ra	1	1		
...	A'' 1 2 IR Ra	A_u 3 0 IR	1	—1		
$\mathscr{C}_2$ $(g = 2)$			E	C_2^1		
	$\mathscr{C}_{2h}$ $(g = 4)$		E	C_2^1	i	σ_h
A 1 1 IR Ra	A_g 0 1 Ra	...	1	1	1	1
...	A_u 1 0 IR	...	1	1	—1	—1
B 2 2 IR Ra	B_g 0 2 Ra	...	1	—1	1	—1
...	B_u 2 0 IR	...	1	—1	—1	1
$\mathscr{C}_3$ $(g = 3)$			E	C_3^1/C_3^2		
	$\mathscr{C}_{3h}$ $(g = 6)$		E	C_3^1/C_3^2	σ_h	S_3^1/S_3^5
		S_6 $(g = 6)$	E	C_3^1/C_3^2	i	S_6^5/S_6^1
A 1 1 IR Ra	A' 0 1 Ra	A_g 0 1 Ra	1	1	1	1
...	A'' 1 0 IR	A_u 1 0 IR	1	1	—1	—1
E 2 2 IR Ra	E' 2 0 IR Ra	E_g 0 2 Ra	2	—1	2	—1
...	E'' 0 2 Ra	E_u 2 0 IR	2	—1	—2	1

Table XXXIII — continued

Point groups $\mathscr{G}$ (group order g)			Classes of symmetry operations[a]					
Irreducible representations[b] $\Gamma^{(\gamma)}$			Character[c] $\chi_j^{(\gamma)}$					
$\mathscr{C}_4$ ($g = 4$)			**E**	$\mathbf{C_4^1/C_4^3}$	$\mathbf{C_2^1}$			
	$\mathscr{C}_{4h}$ ($g = 8$)		**E**	$\mathbf{C_4^1/C_4^3}$	$\mathbf{C_2^1}$	**i**	$\mathbf{S_4^1/S_4^3}$	σ_h
		S_4 ($g = 4$)	**E**	$\mathbf{S_4^1/S_4^3}$	$\mathbf{C_2^1}$			
A 1 1 IR Ra	A_g 0 1 Ra	A 0 1 Ra	1	1	1	1	1	1
...	A_u 1 0 IR	...	1	1	1	—1	—1	—1
B 0 0 Ra	B_g 0 0 Ra	B 1 0 IR Ra	1	—1	1	1	—1	1
...	B_u 0 0 —	...	1	—1	1	—1	1	—1
E 2 2 IR Ra	E_g 0 2 Ra	E 2 2 IR Ra	2	0	—2	2	0	—2
...	E_u 2 0 IR	...	2	0	—2	—2	0	2
$\mathscr{C}_5$ ($g = 5$)			**E**	$\mathbf{C_5^1/C_5^4}$	$\mathbf{C_5^2/C_5^3}$			
	$\mathscr{C}_{5h}$ ($g = 10$)		**E**	$\mathbf{C_5^1/C_5^4}$	$\mathbf{C_5^2/C_5^3}$	σ_h	$\mathbf{S_5^1/S_5^9}$	$\mathbf{S_5^3/S_5^7}$
A 1 1 IR Ra	A' 0 1 Ra	...	1	1	1	1	1	1
...	A'' 1 0 IR	...	1	1	1	—1	—1	—1
E_1 2 2 IR Ra	E_1' 2 0 IR	...	2	m	—p	1	m	—p
...	E_1'' 0 2 Ra	...	2	m	—p	—1	—m	p
E_2 0 0 Ra	E_2' 0 0 Ra	...	2	—p	m	1	—p	m
...	E_2'' 0 0 —	...	2	—p	m	—1	p	—m

Table XXXIII — continued

Point groups $\mathscr{G}$ (group order g)			Classes of symmetry operations[a]							
Irreducible representations[b] $\Gamma^{(\gamma)}$			Character[c] $\chi_j^{(\gamma)}$							
$\mathscr{C}_6$ ($g = 6$)			E	C_6^1/C_6^5	C_3^1/C_3^1	C_2^1				
	$\mathscr{C}_{6h}$ ($g = 12$)		E	C_6^1/C_6^5	C_3^1/C_3^2	C_2^1	i	S_3^1/S_3^5	S_6^1/S_6^5	σ_h
A 1 1 IR Ra	A_g 0 1 Ra	...	1	1	1	1	1	1	1	1
...	A_u 1 0 IR	...	1	1	1	1	—1	—1	—1	—1
B 0 0 —	B_g 0 0 —	...	1	—1	1	—1	1	—1	1	—1
...	B_u 0 0 —	...	1	—1	1	—1	—1	1	—1	1
E_1 2 2 IR Ra	E_{1g} 0 2 Ra	...	2	1	—1	—2	2	1	—1	—2
...	E_{1u} 2 0 IR	...	2	1	—1	—2	—2	—1	1	2
E_2 0 0 Ra	E_{2g} 0 0 Ra	...	2	—1	—1	2	2	—1	—1	2
...	E_{2u} 0 0 —	...	2	—1	—1	2	—2	1	1	—2
$\mathscr{D}_2 \equiv \mathscr{V}$ ($g = 4$)			E	C_{2z}	C_{2y}	C_{2x}				
	($\mathscr{D}_{2v} \equiv \mathscr{V}_h$ $g = 8$)		E	C_{2z}	C_{2y}	C_{2x}	i	σ_{xy}	σ_{zx}	σ_{yz}
		$\mathscr{C}_{2v}$ ($g = 4$)	E	C_2	σ_{zx}	σ_{yz}				
A 0 0 Ra	A_g 0 0 Ra	A_1 1 0 IR Ra	1	1	1	1	1	1	1	1
...	A_u 0 0 —	...	1	1	1	1	—1	—1	—1	—1
B_1 1 1 IR Ra	B_{1g} 0 1 Ra	A_2 0 1 Ra	1	1	—1	—1	1	1	—1	—1
...	B_{1u} 1 0 IR	...	1	1	—1	—1	—1	—1	1	1
B_2 1 1 IR Ra	B_{2g} 0 1 Ra	B_1 1 1 IR Ra	1	—1	1	—1	1	—1	1	—1

Table XXXIII — continued

Point groups $\mathscr{G}$ (group order g)			Classes of symmetry operations[a]
Irreducible representations[b] $\Gamma^{(\gamma)}$			Character[c] $\chi_j^{(\gamma)}$
… B_3 1 1 IR Ra …	B_{2u} 1 0 IR B_{3g} 0 1 Ra B_{3u} 1 0 IR	B_2 1 1 IR Ra … …	1 —1 1 —1 —1 1 —1 1 1 —1 —1 1 1 —1 —1 1 1 —1 —1 1 —1 1 1 —1
$\mathscr{D}_3$ ($g = 6$)			E $2C_3$ $3C_2$
	$\mathscr{D}_{3d}$ ($g = 12$)		E $2C_3$ $3C_2$ i $2S_6$ $3\sigma_d$
		$\mathscr{C}_{3v}$ ($g = 6$)	E $2C_3$ $3\sigma_v$
A_1 0 0 Ra … A_2 1 1 IR … E 2 2 IR Ra …	A_1' 0 0 Ra A_1'' 0 0 — A_2' 0 1 — A_2'' 1 0 IR E' 2 0 IR Ra E'' 0 2 Ra	A_1 1 0 IR Ra … A_2 0 1 — … E 2 2 IR Ra …	1 1 1 1 1 1 1 1 1 —1 —1 —1 1 1 —1 1 1 —1 1 1 —1 —1 —1 1 2 —1 0 2 —1 0 2 —1 0 —2 1 0
$\mathscr{D}_4$ ($g = 8$)			E $2C_4$ C_2^1 $2C_2'$ $2C_2''$
	$\mathscr{D}_{4h}$ ($g = 16$)		E $2C_4$ C_2^1 $2C_2'$ $2C_2''$ i $2S_4$ σ_h $2\sigma_v$ $2\sigma_d$
		$\mathscr{C}_{4v}$ ($g = 8$)	E $2C_4$ C_2^1 $2\sigma_v$ $2\sigma_d$
A_1 0 0 Ra	A_{1g} 0 0 Ra	A_1 1 0 IR Ra	1 1 1 1 1 1 1 1 1 1

Table XXXIII — continued

Point groups $\mathscr{G}$ (group order g)			Classes of symmetry operations[a]									
Irreducible representations[b] $\Gamma^{(\gamma)}$			Character[c] $\chi_j^{(\gamma)}$									
...	A_{1u} 0 0 —	...	1	1	1	1	1	−1	−1	−1	−1	−1
A_2 1 1 IR	A_{2g} 0 1 —	A_2 0 1 —	1	1	1	−1	−1	1	1	1	−1	−1
...	A_{2u} 1 0 IR	...	1	1	1	−1	−1	−1	−1	−1	1	1
B_1 0 0 Ra	B_{1g} 0 0 Ra	B_1 0 0 Ra	1	−1	1	1	−1	1	−1	1	1	−1
...	B_{1u} 0 0 —	...	1	−1	1	1	−1	−1	1	−1	−1	1
B_2 0 0 Ra	B_{2g} 0 0 Ra	B_2 0 0 Ra	1	−1	1	−1	1	1	−1	1	−1	1
...	B_{2u} 0 0 —	...	1	−1	1	−1	1	−1	1	−1	1	−1
E 2 2 IR Ra	E_g 0 2 Ra	E 2 2 IR Ra	2	0	−2	0	0	2	0	−2	0	0
...	E_u 2 0 IR	...	2	0	−2	0	0	−2	0	2	0	0
$\mathscr{D}_4$ ($g = 8$)			E	$2C_4$	C_2^1	$2C_2'$	$2C_2''$					
	$\mathscr{D}_{2d} \equiv \mathscr{V}_d$ ($g = 8$)		E	$2S_4$	C_2^1	$2C_2'$	$2C_2''$					
		C_{4v} ($g = 8$)	E	$2C_4$	C_2^1	$2\sigma_v$	$2\sigma_d$					
A_1 0 0 Ra	A_1 0 0 Ra	A_1 1 0 IR Ra	1	1	1	1	1					
A_2 1 1 IR	A_2 0 1 —	A_2 0 1 —	1	1	1	−1	−1					
B_1 0 0 Ra	B_1 0 0 Ra	B_1 0 0 Ra	1	−1	1	1	−1					
B_2 0 0 Ra	B_2 1 0 IR Ra	B_2 0 0 Ra	1	−1	1	−1	1					
E 2 2 IR Ra	E 2 2 IR Ra	E 0 0 Ra	2	0	−2	0	0					

Table XXXIII — continued

Point groups $\mathscr{G}$ (group order g)			Classes of symmetry operations[a]							
Irreducible representations[b] $\Gamma^{(\gamma)}$			Character[c] $\chi_j^{(\gamma)}$							
	$\mathscr{D}_{4d}$ ($g = 16$)		E	$2S_8$	$2C_4$	$2S_8^3$	C_2^1	$4C_2'$	$4\sigma_d$	
		S_8 ($g = 8$)	E	S_8^1/S_8^7	C_4^1/C_4^3	S_8^3/S_8^5	C_2^1			
...	A_1 0 0 Ra	A 0 1 Ra	1	1	1	1	1	1	1	
...	A_2 0 1 —	...	1	1	1	1	1	—1	—1	
...	B_1 0 0 —	B 1 0 IR	1	—1	1	—1	1	1	—1	
...	B_2 1 0 IR	...	1	—1	1	—1	1	—1	1	
...	E_1 2 0 IR	E_1 2 2 IR	2	$\sqrt{2}$	0	$-\sqrt{2}$	—2	0	0	
...	E_2 0 0 —	E_2 0 0 Ra	2	0	—2	0	0	0	0	
...	E_3 0 2 IR	E_3 0 0 Ra	2	$-\sqrt{2}$	0	$\sqrt{2}$	—2	0	0	
$\mathscr{D}_5$ ($g = 10$)			E	$2C_5$	$2C_5^2$	$5C_2$				
	$\mathscr{D}_{5d}$ ($g = 20$)		E	$2C_5$	$2C_5^2$	$5C_2$	i	$2S_{10}^3$	$2S_{10}$	$5\sigma_d$
		$\mathscr{D}_{5h}$ ($g = 20$)	E	$2C_5$	$2C_5^2$	$5C_2$	σ_h	$2S_5$	$2C_5^3$	$5\sigma_v$
A_1 0 0 Ra	A_{1g} 0 0 Ra	A_1' 0 0 Ra	1	1	1	1	1	1	1	1
...	A_{1u} 0 0 —	A_1'' 0 0 —	1	1	1	1	—1	—1	—1	—1
A_2 1 1 IR	A_{2g} 0 1 —	A_2' 0 1 —	1	1	1	—1	1	1	1	—1
...	A_{2u} 1 0 IR	A_2'' 1 0 IR	1	1	1	—1	—1	—1	—1	1
E_1 2 2 IR Ra	E_{1g} 0 2 Ra	E_1' 2 0 IR	2	m	$-p$	0	2	m	$-p$	0
...	E_{1u} 2 0 IR	E_1'' 0 2 Ra	2	m	$-p$	0	—2	$-m$	p	0

Table XXXIII — continued

Point groups $\mathscr{G}$ (group order g) / Irreducible representations[b] $\Gamma^{(\gamma)}$			Classes of symmetry operations[a] / Character[c] $\chi_j^{(\gamma)}$											
E_2 0 0 Ra	E_{2g} 0 0 Ra	E_2' 0 0 Ra	2	—p	m	0	2	—p	m	0				
...	E_{2u} 0 0 —	E_2'' 0 0 —	2	—p	m	0	—2	p	—m	0				
$\mathscr{D}_5$ ($g = 10$)			**E**	**2C$_5$**	**2C$_5^2$**	**5C$_2$**								
	$\mathscr{C}_{5v}$ ($g = 10$)		**E**	**2C$_5$**	**2C$_5^2$**	**5σ$_v$**								
A_1 0 0 Ra	A_1 1 0 IR Ra	...	1	1	1	1								
A_2 1 1 IR	A_2 0 1 —	...	1	1	1	—1								
E_1 2 2 IR Ra	E_1 2 2 IR Ra	...	2	m	—p	0								
E_2 0 0 Ra	E_2 0 0 Ra	...	2	—p	m	0								
$\mathscr{D}_6$ ($g = 12$)			**E**	**2C$_6$**	**2C$_3$**	**C$_2^1$**	**3C$_2'$**	**3C$_2''$**						
	$\mathscr{D}_{6h}$ ($g = 24$)		**E**	**2C$_6$**	**2C$_3$**	**C$_2^1$**	**3C$_2'$**	**3C$_2''$**		**2S$_3$**	**2S$_6$**	**σ$_h$**	**3σ$_d$**	**3σ$_v$**
		$\mathscr{D}_{3h}$ ($g = 12$)	**E**		**2C$_3$**		**3C$_2$**		**σ$_h$**		**2S$_3$**		**3σ$_v$**	
A_1 0 0 Ra	A_{1g} 0 0 Ra	A_1' 0 0 Ra	1	1	1	1	1	1	1	1	1	1	1	1
...	A_{1u} 0 0 —	A_1'' 0 0 —	1	1	1	1	1	1	—1	—1	—1	—1	—1	—1
A_2 1 1 IR	A_{2g} 0 1 —	A_2' 0 1 —	1	1	1	1	—1	—1	1	1	1	1	—1	—1
...	A_{2u} 1 0 IR	A_2'' 1 0 IR	1	1	1	1	—1	—1	—1	—1	—1	—1	—1	1
B_1 0 0 —	B_{1g} 0 0 —	...	1	—1	1	—1	1	—1	1	—1	1	—1	1	—1
...	B_{1u} 0 0 —	...	1	—1	1	—1	1	—1	—1	1	—1	1	—1	1
B_2 0 0 —	B_{2g} 0 0 —	...	1	—1	1	—1	—1	1	1	—1	1	—1	—1	1

Table XXXIII — continued

Point groups $\mathscr{G}$ (group order g)			Classes of symmetry operations[a]
Irreducible representations[b] $\Gamma^{(\gamma)}$			Character[c] $\chi_j^{(\gamma)}$
…	B_{2u} 0 0 —	…	1 −1 1 −1 −1 1 −1 1 −1 1 1 −1
E_1 2 2 IR Ra	E_{1g} 0 2 Ra	E' 0 2 IR Ra	2 1 −1 −2 0 0 2 1 −1 −2 0 0
…	E_{1u} 2 0 IR	E'' 2 0 Ra	2 1 −1 −2 0 0 −2 −1 1 2 0 0
E_2 0 0 Ra	E_{2g} 0 0 Ra	…	2 −1 −1 2 0 0 2 −1 −1 2 0 0
…	E_{2u} 0 0 —	…	2 −1 −1 2 0 0 −2 1 1 −2 0 0
$\mathscr{D}_6$ ($g = 12$)			**E** $2C_6$ $2C_3$ C_2^1 $3C_2'$ $3C_2''$
	$\mathscr{C}_{6v}$ ($g = 12$)		**E** $2C_6$ $2C_3$ C_2^1 $3\sigma_v$ $3\sigma_d$
A_1 0 0 Ra	A_1 1 0 IR Ra	…	1 1 1 1 1 1
A_2 1 1 IR	A_2 0 1 —	…	1 1 1 1 −1 −1
B_1 0 0 —	B_1 0 0 —	…	1 −1 1 −1 1 −1
B_2 0 0 —	B_2 0 0 —	…	1 −1 1 −1 −1 1
E_1 2 2 IR Ra	E_1 2 2 IR Ra	…	2 1 −1 −2 0 0
E_2 0 0 Ra	E_2 0 0 Ra	…	2 −1 −1 2 0 0
	$\mathscr{D}_{6d}$ ($g = 24$)		**E** $2S_{12}$ $2C_6$ $2S_4$ $2C_3$ $2S_{12}^5$ C_2 $6C_2$ $6\sigma_d$
…	A_1 0 0 Ra	…	1 1 1 1 1 1 1 1 1
…	A_2 0 1 —	…	1 1 1 1 1 1 1 −1 −1
…	B_1 0 0 —	…	1 −1 1 −1 1 −1 1 1 −1
…	B_2 1 0 IR	…	1 −1 1 −1 1 −1 1 −1 1
…	E_1 2 0 IR	…	2 $\sqrt{3}$ 1 0 −1 $-\sqrt{3}$ −2 0 0

Table XXXIII — continued

Point groups $\mathscr{G}$ (group order g)			Classes of symmetry operations[a]
Irreducible representations[b] $\Gamma^{(\gamma)}$			Character[c] $\chi_j^{(\gamma)}$
…	E_2 0 0 Ra	…	2 1 —1 —2 —1 1 2 0 0
…	E_3 0 0 —	…	2 0 —2 0 2 0 —2 0 0
…	E_4 0 0 —	…	2 —1 —1 2 —1 —1 2 0 0
…	E_4 0 2 Ra	…	2 $-\sqrt{3}$ 1 0 —1 $\sqrt{3}$ —2 0 0
$\mathscr{T}$ ($g = 12$)			**E** $4C_3^1/4C_3^2$ $3C_2$
	$\mathscr{T}_h$ ($g = 24$)		**E** $4C_3^1/4C_3^2$ $3C_2$ **i** $4S_6^5/4S_6^1$ $3\sigma_d$
		$\mathscr{T}_d$ ($g = 24$)	**E** $8C_3$ $3C_2$ $6S_4$ $6\sigma_d$
A 0 0 Ra	A_g 0 0 Ra	A_1 0 0 Ra	1 1 1 1 1 1 1 1
…	A_u 0 0 —	A_2 0 0 —	1 1 1 —1 —1 —1 —1 —1
E 0 0 Ra	E_g 0 0 Ra	E 0 0 Ra	2 —1 2 2 —1 2 0 0
…	E_u 0 0 —	…	2 —1 2 —2 1 —2 0 0
F 3 3 IR Ra	F_g 0 3 Ra	F_1 0 3 —	3 0 —1 3 0 —1 1 —1
…	F_u 3 0 IR	F_2 3 0 IR Ra	3 0 —1 —3 0 1 —1 1
$\mathscr{O}$ ($g = 24$)			**E** $8C_3$ $3C_2$ $6C_4$ $6C_2'$
	$\mathscr{O}_h$ ($g = 48$)		**E** $8C_3$ $3C_2$ $6C_4$ $6C_2'$ **i** $8S_6$ $3\sigma_h$ $6S_4$ $6\sigma_d$
		$\mathscr{T}_d$ ($g = 24$)	**E** $8C_3$ $3C_2$ $6S_4$ $6\sigma_d$
A_1 0 0 Ra	A_{1g} 0 0 Ra	A_1 0 0 Ra	1 1 1 1 1 1 1 1 1 1
…	A_{1u} 0 0 —	…	1 1 1 1 1 —1 —1 —1 —1 —1

Table XXXIII — continued

Point groups $\mathscr{G}$ (group order g)			Classes of symmetry operations[a]									
Irreducible representations[b] $\Gamma^{(\gamma)}$			Character[c] $\chi_j^{(\gamma)}$									
A_2 0 0 —	A_{2g} 0 0 —	A_2 0 0 —	1	1	1	−1	−1	1	1	1	−1	−1
…	A_{2u} 0 0 —	…	1	1	1	−1	−1	−1	−1	−1	1	1
E 0 0 Ra	E_g 0 0 Ra	E 0 0 Ra	2	−1	2	0	0	2	−1	2	0	0
…	E_u 0 0 —	…	2	−1	2	0	0	−2	1	−2	0	0
F_1 3 3 IR	F_{1g} 0 3 —	F_1 0 3 —	3	0	−1	1	−1	3	0	−1	1	−1
…	F_{1u} 3 0 IR	…	3	0	−1	1	−1	−3	0	1	−1	1
F_2 0 0 —	F_{2g} 0 0 Ra	F_2 3 0 IR Ra	3	0	−1	−1	1	3	0	−1	−1	1
…	F_{2u} 0 0 —	…	3	0	−1	−1	1	−3	0	1	1	−1
$\mathscr{I}$ (g = 60)			**E**	**12C$_5$**	**12C$_5^2$**	**20C$_3$**	**15C$_2$**					
	$\mathscr{I}_h$ (g = 120)		**E**	**12C$_5$**	**12C$_5^2$**	**20C$_3$**	**15C$_2$**	**i**	**12S$_{10}^3$**	**12S$_{10}$**	**20S$_6$**	**15σ_h**
A 0 0 Ra	A_g 0 0 Ra	…	1	1	1	1	1	1	1	1	1	1
…	A_u 0 0 —	…	1	1	1	1	1	−1	−1	−1	−1	−1
F_1 3 3 IR	F_{1g} 0 3 —	…	3	p	m	0	−1	3	p	m	0	−1
…	F_{1u} 3 0 IR	…	3	p	m	0	−1	−3	−p	−m	0	1
F_2 0 0 —	F_{2g} 0 0 —	…	3	m	p	0	−1	3	m	p	0	−1
…	F_{2u} 0 0 —	…	3	m	p	0	−1	3	−m	−p	0	1
G 0 0 —	G_g 0 0 —	…	4	−1	−1	1	0	4	−1	−1	1	0
…	G_u 0 0 —	…	4	−1	−1	1	0	−4	1	1	−1	0
H 0 0 Ra	H_g 0 0 Ra	…	5	0	0	−1	0	5	0	0	−1	0
…	H_u 0 0 —	…	5	0	0	−1	0	−5	0	0	1	0

Table XXXIII — continued

Point groups $\mathscr{G}$ (group order g)									Classes of symmetry operations[a]							
Irreducible representations[b] $\Gamma^{(\gamma)}$									Character[c] $\chi_j^{(\gamma)}$							
$\mathscr{D}_{\infty h}$									E	$2C_\infty^\varphi$		$\infty\sigma_v$	i	$2S_\infty^\varphi$		∞C_2
				$\mathscr{C}_{\infty v}$					E	$2C_\infty^\varphi$		$\infty\sigma_v$				
Σ_g^+	0	0	Ra	$A_1 \equiv \Sigma^+$	1	0	IR	Ra	1	1	...	1	1	1	...	1
Σ_u^+	1	0	IR	...					1	1	...	1	—1	—1	...	—1
Σ_g^-	0	1	—	$A_2 \equiv \Sigma^-$	0	1	—		1	1	...	—1	1	1	...	—1
Σ_u^-	0	0	—	...					1	1	...	—1	—1	—1	...	1
Π_g	0	2	Ra	$E_1 \equiv \Pi$	2	2	IR	Ra	2	2 . cos φ	...	0	2	—2 . cos φ	...	0
Π_u	2	0	IR	...					2	2 . cos φ	...	0	—2	2 . cos φ	...	0
Δ_g	0	0	Ra	$E_2 \equiv \Delta$	0	0	Ra		2	2 . cos 2φ	...	0	2	2 . cos 2φ	...	0
Δ_u	0	0	—	...					2	2 . cos 2φ	...	0	—2	—2 . cos 2φ	...	0
Φ_g	0	0	—	$E_3 \equiv \Phi$	0	0	—		2	2 . cos 3φ	...	0	2	—2 . cos 3φ	...	0
Φ_u	0	0	—	...					2	2 . cos 3φ	...	0	—2	2 . cos 3φ	...	0
etc.				etc.								etc.				

[a] Classes of symmetry operations are given as symbols of a typical operation, preceded by an integral coefficient specifying the class order g_j ($g_j = 1$ is not given explicitly). If two different classes of operations have the same character values in this simplified notation, they are given at the head of a single column, separated by a slanted line. The sum of the integral coefficients is then taken as the g_j value (e.g. in group $\mathscr{T}$, column $4C_3^1/4C_3^2$, $g_2 = 8$ is taken for the calculation); [b] in the individual columns, the following data are successively given: the symbol of irreducible representation $\Gamma^{(\gamma)}$, the number of translational modes $n_{tr}^{(\gamma)}$, the number of rotational modes $n_{rot}^{(\gamma)}$ and the activity (IR — active in the infrared spectrum, Ra — active in the Raman spectrum); [c] in order to simplify the notation, the following symbols are used: $m = 2 . \cos 72° = (\sqrt{5} - 1)/2$; $p = -2 . \cos 144° = (\sqrt{5} + 1)/2$.

representations $\chi_R^{(\gamma)}$. Hence the notation of the tables of characters can be substantially abbreviated. At the heads of the individual columns, the operation representing the particular class is given, preceded by the value specifying the number of operations in the given class.

Once the individual normal vibrations are successfully assigned to the particular irreducible representations, the goal described in the beginning of Section 4.1, i.e. classification of vibrations according to their behaviour under the effect of symmetry operations, has also been attained. In spectroscopy the individual irreducible representations are usually termed types of symmetry or the races of normal vibrations and are denoted by conventional symbols. Types of symmetry for non-degenerate normal vibrations are denoted by letters A and B. Symbol A describes vibrations symmetrical with respect to all symmetry operations according to the n-fold rotational axis (or the rotation-reflection axis with groups $\mathscr{S}_n$), which represent the determining symmetry element (see Section 4.1.2); symbol B denotes all other non-degenerate vibrations. If there exist a greater number of irreducible representations satisfying these conditions, they are differentiated using various subscripts. Subscripts $_1$ and $_2$ differentiate between vibrations symmetrical and antisymmetrical with respect to rotation around a secondary two-fold axis (with groups $\mathscr{D}$), or with respect to reflection through plane σ_v (with groups $\mathscr{C}_{nv}$). For groups containing inversion operation $\mathbf{i}$, subscripts $_g$ or $_u$ are added to denote the type of symmetry with respect to the inversion operation. Subscript $_g$ (gerade) denotes symmetrical vibrations and subscript $_u$ (ungerade) antisymmetrical vibrations. This classification is especially important, since normal vibrations symmetrical with respect to $\mathbf{i}$ are active only in Raman spectra while those antisymmetrical with respect to $\mathbf{i}$ are active only in infared spectra. Infrared and Raman spectra are complementary in this case.

Finally, if a group contains symmetry operation $\boldsymbol{\sigma}_h$, vibrations symmetrical and antisymmetrical with respect to this symmetry operation are differentiated by adding one or two primes to the symbol of the irreducible representation, respectively (e.g. A' and A'').

The types of symmetry of doubly and triply degenerate vibrations are denoted by symbols E and F, respectively. The significance of numerical symbols employed for their differentiation is somewhat more complicated, but subscripts $_g$, $_u$, and superscripts ′ and ″ are the same as for non-degenerate vibrations.*)

*) With groups $\mathscr{I}$ and $\mathscr{I}_h$, irreducible representations denoted by symbols G and H, corresponding to quadruply and quintuply degenerate vibrations, can be encountered.

The rigorous significance of all the symbols used is best understood after a thorough study of the tables of the characters of irreducible representations.

4.1.4 The Number of Normal Vibrations for a Given Type of Symmetry and Their Activity

In the beginning of the analysis of vibrational spectra, the number of normal vibrations and their relation to a certain type of symmetry must first be determined and it must be found whether they will be active in the infrared and Raman spectra.

It follows from group theory that a certain irreducible representation, $\Gamma^{(\gamma)}$, appears $n^{(\gamma)}$-times in the notation of a certain fully reduced representation, where

$$n^{(\gamma)} = (1/g) \sum_{\mathbf{R}} \chi_{\mathbf{R}}^{(\gamma)} \chi_{\mathbf{R}}. \tag{4.1}$$

In Eq. (4.1) g denotes the group order, $\chi_{\mathbf{R}}^{(\gamma)}$ is the character of irreducible representation $\Gamma^{(\gamma)}$ for symmetry operation **R** and $\chi_{\mathbf{R}}$ is the character of the fully reduced representation. The summation is then performed over all symmetry operations **R** of the given group.

If the fact that the characters of representations are the same for all symmetry operations belonging to a single class is employed, Eq. (4.1) changes into

$$n^{(\gamma)} = (1/g) \sum_{\mathrm{j}} g_{\mathrm{j}} \chi_{\mathrm{j}}^{(\gamma)} \chi_{\mathrm{j}}, \tag{4.2}$$

where g is again the group order, g_{j} is the number of operations in the j-th class, $\chi_{\gamma}^{(\mathrm{j})}$ is the character of irreducible representation $\Gamma^{(\gamma)}$ for the j-th class of symmetry operations and χ_{j} is the value of the character of the representation for any symmetry operation **R** belonging in the j-th class. The value of χ_{j}, i.e. the sum of all diagonal elements in the whole transformation matrix, is constant over the process of representation reduction; therefore it can be derived from an arbitrary initial, i.e. non-reduced, representation. This renders knowledge of the normal coordinates unnecessary.

From the point of view of universality, the description of vibrations in terms of changes in the cartesian coordinates is most advantageous; the number of these coordinates is $3N$ for a molecule containing N atoms. However, certain combinations of deviations also describe translational movement of the molecule as a whole in space and other describe its rotation. Hence some of the irreducible representations correspond to translational and others to rotational modes. In order to obtain the number of vibrational coordinates proper, the improper coordinates (translational and rotational) must be excluded. The number of normal vibrations belonging to symmetry type $\Gamma^{(\gamma)}$, i.e. $n_{vib}^{(\gamma)}$, equals

$$n_{vib}^{(\gamma)} = n^{(\gamma)} - (n_{tr}^{(\gamma)} + n_{rot}^{(\gamma)})/\chi_E^{(\gamma)} = \\ = (1/g) \sum_j [g_j \chi_j^{(\gamma)} \chi_j] - (n_{tr}^{(\gamma)} + n_{rot}^{(\gamma)})/\chi_E^{(\gamma)} . \tag{4.3}$$

The values of $n_{tr}^{(\gamma)}$ (the number of translational modes) and $n_{rot}^{(\gamma)}$ (the number of rotational modes) can be found in Table XXXIII. The value of character $\chi_E^{(\gamma)}$ corresponding to the identity operation for the given irreducible representation equals the numerical value of the degeneration of this symmetry type; as $n_{tr}^{(\gamma)}$ and $n_{vib}^{(\gamma)}$ are related to the number of irreducible representation, the number of translations and rotations for the degenerate states must be divided by this value.

As character χ_R is formed by the sum of the diagonal elements of the transformational matrix, the changes in the coordinates of only those atoms that do not change their position on the selected symmetry operation can contribute to its value. It is usually said, not quite accurately, of these atoms that "they lie on the corresponding symmetry element". In identity operation **E** no atom changes its position; hence all atoms in the molecule contribute to χ_E. In operation $\mathbf{C}_n^k$ only atoms lying on axis C_n do not change their positions and in operation $\mathbf{S}_n^k$ only the atom located in the intercept of axis S_n with the imaginary reflection plane does not move.*) In reflection operation $\boldsymbol{\sigma}$, all atoms lying in plane σ do not change their positions and in inversion **i** the atom located at the centre of symmetry i remains unchanged. Operations **E**, $\boldsymbol{\sigma}$ and **i** can be converted into operations $\mathbf{C}_1^1$, $\mathbf{S}_1^1$ and $\mathbf{S}_2^1$; therefore it is sufficient to give the general relationships for the contributions $X(\mathbf{R})$ for operations of the $\mathbf{C}_n^k$-type,

$$X(\mathbf{C}_n^k) = 1 + 2 \,.\, \cos(2\pi k/n) \tag{4.4}$$

and of the $\mathbf{S}_n^k$-type,

$$X(\mathbf{S}_n^k) = -1 + 2 \,.\, \cos(2\pi k/n). \tag{4.5}$$

The values of contributions $X(\mathbf{R})$ for all common types of symmetry operations are given in Table XXXIV.

The number of vibrations corresponding to the individual types of symmetry is then calculated as follows:

---

*) This statement holds only for odd k; for even k, $\mathbf{S}_n^k$ is identical with operation $\mathbf{C}_n^k$ (see Table XXX).

1. One of the symmetry operations **R** is chosen from the j-th class of symmetry operations (according to the heads of the individual columns of the table of the characters of representations, XXXIII);

Table XXXIV

Contributions $X(\mathbf{R})$ of the atoms that do not change their positions to the character of the representation for various symmetry operations

Symmetry operation **R**		Contribution value[a] $X(\mathbf{R})$									
$\mathbf{C}_n^k$	n	k									
		1	2	3	4	5	6	7	8	9	10
	1	3	3	3	3	3	3	3	3	3	3
	2	—1	3	—1	3	—1	3	—1	3	—1	3
	3	0	0	3	0	0	3	0	0	3	0
	4	1	—1	1	3	1	—1	1	3	1	—1
	5	$1+m$	$1-p$	$1-p$	$1+m$	3	$1+m$	$1-p$	$1-p$	$1+m$	3
	6	2	0	—1	0	2	3	2	0	—1	0
$\mathbf{S}_n^k$	n	k									
		1	2	3	4	5	6	7	8	9	10
	1	1	3	1	3	1	3	1	3	1	3
	2	—3	3	—3	3	—3	3	—3	3	—3	3
	3	—2	0	1	0	—2	3	—2	0	1	0
	4	—1	—1	—1	3	—1	—1	—1	3	—1	—1
	5	$-1+m$	$1-p$	$-1-p$	$1+m$	—1	$1+m$	$-1-p$	$1-p$	$-1+m$	3
	6	0	0	—3	0	0	3	0	0	—3	0
	8	$\sqrt{2}-1$	1	$-\sqrt{2}-1$	—1	$-\sqrt{2}-1$	1	$\sqrt{2}-1$	3	$\sqrt{2}-1$	1
$\mathbf{E} \equiv \mathbf{C}_1^1$		3									
$\mathbf{i} \equiv \mathbf{S}_2^1$		—3									
$\sigma \equiv \mathbf{S}_1^1$		1									

[a] In order to simplify the transcription, the notation $m = 2 \cdot \cos 72° = (\sqrt{5}-1)/2$; $p = -2 \cdot \cos 144° = (\sqrt{5}+1)/2$ is employed.

2. The number of atoms of the molecule in the equilibrium configuration lying on the corresponding symmetry element and not changing their positions during this operation is found. If this number is denoted as $n(\mathbf{R})$, then the character of the representation of the j-th class is given by

$$\chi_j = n(\mathbf{R}) \,.\, X(\mathbf{R}); \tag{4.6}$$

the values of contributions $X(\mathbf{R})$ are either found from Table XXXIV or calculated from formula (4.4) or (4.5);

3. Steps 1. and 2. are gradually repeated for all symmetry classes of the given symmetry point group;

4. The calculated values of χ_j and the values of g, g_j, $\chi_j^{(\gamma)}$, $n_{tr}^{(\gamma)}$ and $n_{rot}^{(\gamma)}$ taken from Table XXXIII are then employed to calculate the number of normal vibrations corresponding to the individual vibration types with respect to symmetry $n_{vib}^{(\gamma)}$, using Eq. (4.3).

This procedure will be demonstrated on the methanesulphonyl chloride molecule. This molecule has a single symmetry element, reflection plane σ_h, and hence belongs to point group $\mathscr{C}_s$, which consists of operations $\{\mathbf{E}, \boldsymbol{\sigma}_h\}$ divided into two classes containing one operation each. With operation $\mathbf{E}$ from the first class, no atom of the eight present in the molecule changes its place; thus $n(\mathbf{E}) = 8$. The value, $X(\mathbf{E}) = 3$, is taken from Table XXXIV; hence χ_1 equals $8 \times 3 = 24$. In the reflection operation the positions of only four atoms (Cl, S, C and one of the three hydrogen atoms of the methyl group) do not change; consequently, $n(\boldsymbol{\sigma}_h) = 4$. As $X(\boldsymbol{\sigma}_h) = 1$ according to Table XXXIV, χ_2 equals $4 \times 1 = 4$. The further quantities required are found in the table of characters (Table XXXIII) and are substituted into Eq. (4.3) for the individual irreducible representations (lines A' and A''), finding that eleven normal vibrations of type A' (totally symmetrical) correspond to the methanesulphonyl chloride molecule,

$$n_{vib}(A') = (1/2)\,.\,(1\,.\,1\,.\,24 + 1.1\,.\,4) - 2 - 1 = 14 - 2 - 1 = 11,$$

(two translational modes and one rotational mode have been subtracted), as well as seven normal vibrations of type A'' (antisymmetrical with respect to σ_h),

$$n_{vib}(A'') = (1/2)\,.\,(1\,.\,1\,.\,24 + 1\,.\,(-1)\,.\,4) - 1 - 2 = 10 - 1 - 2 = 7,$$

(one translational and two rotational modes have been subtracted). This is symbolically written as $11A' + 7A''$. The vibrational amplitudes of the atomic nuclei in the individual vibrations are depicted in Fig. 57. It can also be found from Table XXXIII that both types of vibration are active both in infrared and Raman spectra.

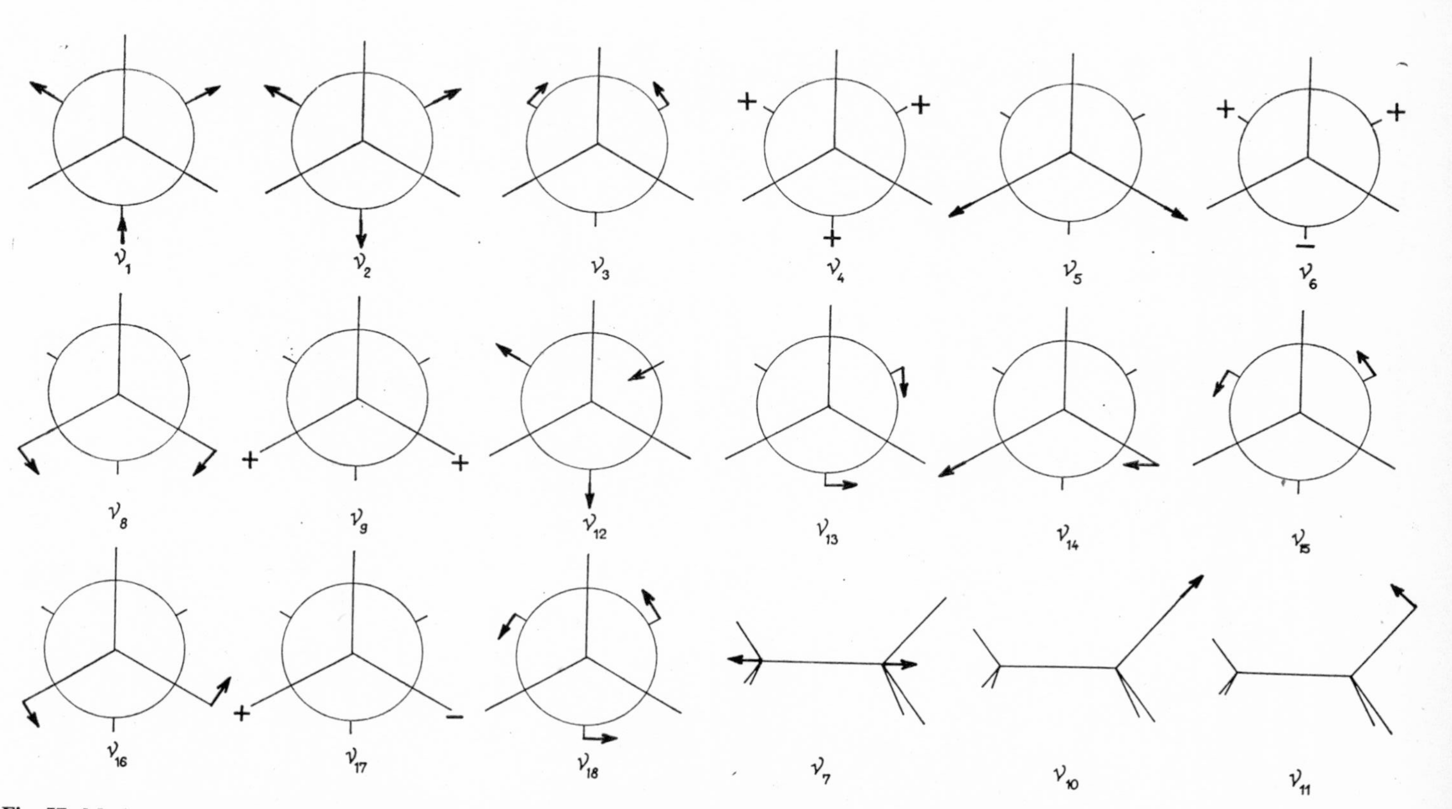

Fig. 57. Modes representing normal vibrations of the nuclei in the methanesulphonyl chloride molecule (the vibrations are denoted in agreement with the data in Table XXIX). Except for vibrations ν_7, ν_{10} and ν_{11}, the modes are represented on the model of the molecule in the Newman projection with the S and C atoms overlapping (the S atom is closer to the spectator).

The whole operation will further be demonstrated on the highly symmetrical methane molecule, CH_4, and on the molecules of its isotopomers. Methane CH_4 and its fully deuterated derivative CD_4 belong to symmetry point group $\mathcal{T}_d$, in which the symmetry operations fall into five classes. The first class contains only the identity operation $\mathbf{E}$ ($g_1 = 1$); $\chi_1 = 15$, as $n(\mathbf{E}) = 5$ and $X(\mathbf{E}) = 3$. The second class consists of the rotations around four three-fold axes C_3 passing through the C–H bonds; the total number of the operations of this class is eight ($g_2 = 8$). One of these axes is quite arbitrarily chosen and an arbitrary operation connected with it is selected (of course, the operation must belong to the class of operations studied), e.g. $\mathbf{C}_3^1$. In this operation, two atoms remain in their original places, the central carbon atom and the hydrogen atom located on the selected axis, C_3. Hence, $n(\mathbf{C}_3^1) = 2$; as $X(\mathbf{C}_3^1) = 0$, $\chi_2 = 0$. A further class is composed of rotations $\mathbf{C}_2^1$ around three two-fold axes halving the HCH valence angles ($g_3 = 3$). Any of these operations leaves only the position of the central atom unchanged; $\chi_3 = 1 \,.\, (-1) = -1$. The fourth class is formed by three $\mathbf{S}_4^1$ operations and three $\mathbf{S}_4^3$ operations ($g_4 = 6$), according to the three four-fold rotation-reflection axes (see Fig. 54); here again only the carbon atom remains unshifted, $\chi_4 = 1 \,.\, (-1) = -1$. The last class comprises reflection in planes σ_v, defined by groups of three HCH atoms (this class has altogether six symmetry elements and six operations, i.e. $g_5 = 6$), whose positions are unchanged during the symmetry operation; $\chi_5 = 3 \times 1 = 3$.

On substituting into Eq. (4.3), the following numbers of vibrations are obtained for the individual symmetry types:

$$n_{vib}(A_1) = (1/24) \,.\, [(1 \,.\, 1 \,.\, 15 + 8 \,.\, 1 \,.\, 0 + 3 \,.\, 1 \,.\, (-1) + \\ + 6 \,.\, 1 \,.\, (-1) + 6 \,.\, 1 \,.\, 3)] - (0 + 0)/1 = 1;$$

$$n_{vib}(A_2) = (1/24) \,.\, [1 \,.\, 1 \,.\, 15 + 8 \,.\, 1 \,.\, 0 + 3 \,.\, 1 \,.\, (-1) + \\ + 6 \,.\, (-1) \,.\, (-1) + 6 \,.\, (-1) \,.\, 3] - (0 + 0)/1 = 0;$$

$$n_{vib}(E) = (1/24) \,.\, [1 \,.\, 2 \,.\, 15 + 8 \,.\, (-1) \,.\, 0 + 3 \,.\, 2 \,.\, (-1) + \\ + 6 \,.\, 0 \,.\, (-1) + 6 \,.\, 0 \,.\, 3] - (0 + 0)/2 = 1;$$

(the following line in the table of characters is omitted, as it has no significance in group $\mathcal{T}_d$);

$$n_{vib}(F_1) = (1/24) \,.\, [1 \,.\, 3 \,.\, 15 + 8 \,.\, 0 \,.\, 0 \,.\, +3 \,.\, (-1) \,.\, (-1) + \\ + 6 \,.\, 1 \,.\, (-1) + 6 \,.\, (-1) \,.\, 3] - (0 + 3)/3 = 0$$

and finally,

$$n_{vib}(F_2) = (1/24) . [1 . 3 . 15 + 8 . 0 . 0 + 3 . (-1) . (-1) + + 6 . (-1) . (-1) + 6 . 1 . 3] - (3 + 0)/3 = 2.$$

It follows from Table XXXIII that vibrations of types A_1, E and F_2 are active in Raman spectra, while only vibrations of type F_2 are active in infrared spectra. As vibrations of types E and F are doubly and triply degenerate, respectively, the notation $A_1 + E + 2F_2$ represents nine normal vibrations which is in agreement with the value, $3N - 6$, for the calculation of the number of vibrational degrees of freedom for non-linear molecules.

Using the above procedure it can be derived analogously that the vibrations of monodeuterio- and trideuteriomethane, CH_3D and CD_3H, whose molecules belong to point group $\mathscr{C}_{3v}$, can be classified as $3A_1 + 3E$. All the vibrations are active in both infrared and Raman spectra. Dideuteriomethane CH_2D_2, whose molecule belongs to symmetry point group $\mathscr{C}_{2v}$, does not exhibit degeneration of the vibrational levels and its nine normal vibrations can be described as $4A_1 + A_2 + 2B_1 + 2B_2$. All these vibrations appear in both the infrared and Raman spectra, except for the vibration of type A_2, which is inactive in the infrared spectrum.

Data on vibration activity in infrared and Raman spectra are given in Table XXXIII, which represents a condensed form of the selection rules. It should be pointed out that the data on vibration activities hold exactly only for the gaseous state. Owing to intermolecular interactions in the liquid state and in crystals the symmetry of molecules may be decreased (even more pronounced changes may occur in crystal lattices) and apparent deviations from the selection rules may be encountered. Deviations from the selection rules may also be caused by other effects; e.g. vibration ν_2, which is not allowed by the selection rules, may appear in the infrared spectra of gaseous methane, due to the Coriolis interaction.

4.1.5 Symmetry of Higher Vibrational States

Knowledge of the symmetry of higher vibrational states of molecules is important for determination of the activity of higher harmonics and combination transitions in spectra.

The symmetry of a vibrational state for which two or more vibrational quantum numbers equal unity is described by an irreducible representation termed the product of the irreducible representations of the simple vibrational states. If e.g. the methanesulphonyl chloride molecule is in the state with both SO_2 stretching vibrations excited (symmetrical and antisymmetrical*), i.e. $v_5 = v_{14} = 1$ (for the other i, $v_i = 0$), its symmetry is described as the product, $A' \times A''$. Vibrational states with quantum numbers larger than unity are then described by a power of the irreducible representation (e.g. state $v_5 = 2$, $v_i = 0$ for $i \neq 5$, has symmetry $(A')^2$). It should be emphasized that the product of two identical irreducible representations is not generally equal to the square of this irreducible representation, with degenerate vibrations.

For calculation of more complex vibrational states, the associative and commutative laws can be used. For example, the state of methanesulphonyl chloride described by non-zero vibrational quantum numbers $v_5 = 1$, $v_{10} = 2$, $v_{17} = 1$ has symmetry $A' \times (A')^2 \times A''$.

Each product or power of irreducible representations can be converted into another irreducible representation (with non-degenerate vibrations) or into a combination of several irreducible representations (with degenerate vibrations). The calculation rules are relatively simple when non-degenerate vibrations alone or degenerate modes are combined with non-degenerate; they can be written as

$$A \times A = A^2 = B \times B = B^2 = A, \tag{4.7}$$

$$A \times B = B, \tag{4.8}$$

$$A \times E = B \times E = E \tag{4.9}$$

and

$$A \times F = B \times F = F. \tag{4.10}$$

Similar rules can be defined for subscripts added to symbols $\Gamma \in \{A, B\}$:

$$\Gamma_1 \times \Gamma_1 = (\Gamma_1)^2 = \Gamma_2 \times \Gamma_2 = (\Gamma_2)^2 = \Gamma_1, \tag{4.11}$$

$$\Gamma_1 \times \Gamma_2 = \Gamma_2, \tag{4.12}$$

$$\Gamma_g \times \Gamma_g = (\Gamma_g)^2 = \Gamma_u \times \Gamma_u = (\Gamma_u)^2 = \Gamma_g, \tag{4.13}$$

$$\Gamma_g \times \Gamma_u = \Gamma_u, \qquad (4.15)$$

$$\Gamma' \times \Gamma' = (\Gamma')^2 = \Gamma'' \times \Gamma'' = (\Gamma'')^2 = \Gamma', \qquad (4.15)$$

$$\Gamma' \times \Gamma'' = \Gamma''. \tag{4.16}$$

Groups $\mathscr{V}$ and $\mathscr{V}_h$ are exceptions to rule (4.12); the cyclic rule,

$$B_1 \times B_2 = B_3, \tag{4.17}$$

$$B_2 \times B_3 = B_1 \tag{4.18}$$

---

*) The subscripts employed for numbering the methanesulphonyl chloride vibrations are taken from Table XXIX.

and

$$B_3 \times B_1 = B_2 \tag{4.19}$$

holds in this case.

For the products of indexed, doubly degenerate vibrations E_1 and E_2 with types A and B it holds that

$$A \times E_1 = B \times E_2 = E_1 \tag{4.20}$$

and

$$A \times E_2 = B \times E_1 = E_2 . \tag{4.21}$$

If the above rules are applied to the examples of the methanesulphonyl chloride combination frequencies mentioned in the beginning of this section, it is found that symmetry $\nu_5 + \nu_{14}$ is A'', $2\nu_5$ is A' and $\nu_5 + 2\nu_{10} + \nu_{14}$ is A''.

The rules for multiplication and raising to a power of irreducible representations for mutual combinations of degenerate states are substantially more complicated and different for various groups. For this reason they will not be given in detail here; instead, a general procedure for determining the type of a higher vibrational state will be given. The procedure also applies to the simple relationships, (4.7) – (4.21) given above.

To determine the number and the types of irreducible representations describing the product of two vibration types, Eq. (4.2) can in principle be employed, provided that the character of the representation χ_j for the j-th class of symmetry operations in it is replaced by the product of the characters of the appropriate irreducible representations of the corresponding classes, i.e.

$$\chi_j = \chi_j^{(\gamma_1)} \cdot \chi_j^{(\gamma_2)} , \tag{4.22}$$

if two different vibrations are combined. If the type of symmetry is to be expressed for vibrational transition $2 \leftarrow 0$, a more complex formula must be employed for calculation of χ_j,

$$\chi_j = 1/2\{[\chi_{(\mathbf{R})}^{(\gamma)}]^2 + \chi_{(\mathbf{R}^2)}^{(\gamma)}\}; \tag{4.23}$$

Here $\mathbf{R}$ is the symmetry operation belonging to the j-th class of operations and $\mathbf{R}^2$ the operation obtained by repeating operation $\mathbf{R}$ twice in succession. It must be pointed out that operation $\mathbf{R}^2$ need no longer belong to the j-th class of operations. Formula (4.23) can be generalized for even higher harmonics.

The reader can verify that e.g. the product of irreducible representations E and F_2 for group $\mathcal{T}_d$ can be expressed as $F_1 + F_2$; the following procedure will then be observed: Types E (3rd line) and F_2 (6th line) are found for group $\mathscr{T}_d$ in the table of characters; the characters in these lines are gradually multiplied for the individual symmetry classes. The values, $\chi_1 = 2\times 3 = 6$; $\chi_2 = (-1)\times 0 = 0$: $\chi_3 = 2\times(-1) = -2$; $\chi_4 = 0\times(-1) = 0$ and $\chi_5 = 0\times 1 = 0$, are obtained. These values are gradually substituted into formula (4.2) for all symmetry types of group $\mathscr{T}_d$:

$$n(A_1) = (1/24)\,.\,(1\,.\,1\,.\,6 + 8\,.\,1\,.\,0 + 3\,.\,1\,.\,(-2) + 6\,.\,1\,.\,0 + 6\,.\,1.0) = 0,$$

$$n(A_2) = (1/24).(1\,.\,1\,.\,6{+}8\,.\,1\,.\,0{+}3\,.\,1\,.\,(-2){+}6\,.\,(-1)\,.\,0{+}6\,.\,(-1)\,.\,0) = = 0,$$

$$n(E) = (1/24)\,.\,(1\,.\,2\,.\,6 + 8\,.\,(-1)\,.\,0 + 3\,.\,2\,.\,(-2) + 6\,.\,0\,.\,0{+}6\,.\,0\,.\,0) = = 0,$$

$$n(F_1) = (1.24).(1\,.\,3\,.\,6{+}8\,.\,0\,.\,0{+}3.(-1).(-2){+}6\,.\,1\,.\,0{+}6.(-1)\,.\,0) = = 1,$$

$$n(F_2) = (1/24)\,.\,(1\,.\,3\,.\,6{+}8\,.\,0\,.\,0{+}3\,.\,(-1)\,.\,(-2){+}6\,.\,(-1)\,.\,0{+}6.1.0) = = 1.$$

It should be pointed out that the translational and rotational modes are not subtracted here.

The use of formula (4.23) will be demonstrated on the calculation of the symmetry of the first overtone of a type F_2 vibration, i.e. of the square, $(F_2)^2$, in the same point group $\mathscr{T}_d$. As $\mathbf{E}^2$ is again $\mathbf{E}$, then

$$\chi_1 = (1/2)\{[\chi^{(F_2)}_{(\mathbf{E})}]^2 + \chi^{(F_2)}_{(\mathbf{E})}\} = (1/2)\,.\,\{3^2 + 3\} = 6;$$

$(\mathbf{C}_3^1)^2$ represents operation $\mathbf{C}_3^2$, i.e.,

$$\chi_2 = (1/2)\{[\chi^{(F_2)}_{(\mathbf{C}_3^1)}]^2 + \chi^{(F_2)}_{(\mathbf{C}_3^2)}\} = (1/2)\,.\,\{0^2 + 0\} = 0;$$

however, with the third class of operations $(\mathbf{C}_2^1)^2$ equals $\mathbf{E}$, hence

$$\chi_3 = (1/2)\{[\chi^{(F_2)}_{(\mathbf{C}_2^1)}]^2 + \chi^{(F_2)}_{(\mathbf{E})}\} = (1/2)\,.\,\{(-1)^2 + 3\} = 2;$$

further it is found that $(\mathbf{S}_4^1)^2 = \mathbf{C}_2^1$, from which it follows that

$$\chi_4 = (1/2)\{[\chi^{(F_2)}_{(\mathbf{S}_4^1)}]^2 + \chi^{(F_2)}_{(\mathbf{C}_2^1)}\} = (1/2)\,.\,\{1^2 + (-1)\} = 0$$

and finally it holds for the last class that $(\boldsymbol{\sigma}_d)^2 = \mathbf{E}$ and

$$\chi_5 = (1/2)\{[\chi^{(F_2)}_{(\sigma_d)}]^2 + \chi^{(F_2)}_{(\mathbf{E})}\} = (1/2)\,.\,\{(-1)^2 + 3\} = 2.$$

The values of χ_j thus obtained are substituted into Eq. (4.2), giving $(F_2)^2 = A_1 + E + F_2$. On the contrary, the calculation of product $F_2\times F_2$ yields the result, $A_1 + E + F_1 + F_2$: hence $(F_2)^2 \neq F_2\times F_2$.

4.2 Classification of Molecular Vibrations According to Their Localization

Although in general all atomic nuclei in a given molecule participate in any normal vibration of this molecule, certain smaller parts, especially with larger molecules, make a predominant contribution to many vibrations. The internal coordinates, internuclear distances and valence angles, are useful parameters in this connection and with their help it is easy to correlate vibrational spectra with molecular structural elements.

The local character of vibrations is also manifest in the nomenclature of normal vibrational types. Vibrations in which only (or predominantly) the internuclear distances are varied are called stretching vibrations and are denoted by ν (not to be confused with the frequency with dimension s^{-1}!). The methanesulphonyl chloride molecule exhibits a number of stretching vibrations of C–S, S–Cl, etc. bonds, which are denoted as ν(C–S), ν(S–Cl), etc.

Vibrations in which chiefly the magnitude of the valence angles varies are called deformation vibrations. In a great majority of situations this name is insufficient or ambiguous and hence deformation vibrations must further be specified.

Most types of deformation vibrations can be explained on the methanesulphonyl chloride molecule. The basic deformation vibration type involves variations in the C\S/Cl valence angle. As all the three atoms forming this angle are located in the molecular plane, this vibration is in-plane; it is unambiguously described by the deformation vibration symbol, δ(CSCl).

Description of deformation vibrations in systems with a tetrahedrally located central atom, e.g. with methane derivatives is much more complicated. In methanesulphonyl chloride, this problem can be explained on deformation vibrations in the atomic arrangement CSO_2Cl.

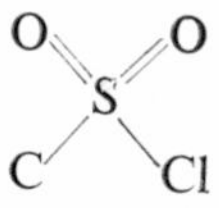

O   O
  S
C   Cl

If the deformation coordinate CSCl is omitted and the redundancy is taken into consideration, four deformation coordinates remain. The latter then correspond to four deformation vibrations which have been given special names for the particular type of molecule (or groups $-SO_2-$, $-CH_2-$, etc.), namely, "scissoring", "wagging" (symmetrical vibrations), "rocking" and "twisting" (antisymmetrical vibrations). These vibrations are depicted in Fig. 57 (ν_8, ν_9, ν_{16} and ν_{17}).

Another interesting deformation vibration is encountered with the methanesulphonyl chloride molecule, the torsion vibration, in which certain parts of the molecule are turned with respect to the other parts). With methanesulphonyl chloride, the methyl group is turned around the C–S bond with respect to the SO_2Cl group. The torsion coordinate actually describes the change in a four-particle system of the type,

$$\mathrm{H{\diagdown}C{-}S{\diagup}Cl} \; ;$$

more precisely, the angle between planes HCS and CSCl is changed. Torsion vibrations are denoted by letter τ with a precise specification of the turning parts. The torsion vibration is closely related to internal rotation in compound molecules and its wavenumber is related to the internal rotation barrier.

With planar molecules, out-of-plane deformation vibrations may further be encountered. As the methanesulphonyl chloride molecule is not planar, another molecule must be used for explanation of this vibration, e.g. phosgene. Of the six

$$\mathrm{O{=}C}\begin{matrix}\diagup\mathrm{Cl}\\ \diagdown\mathrm{Cl}\end{matrix}$$

vibrations of the phosgene molecule, all the atoms of which lie in a plane owing to the *sp*^{2}-hybridization of the carbon atom, three can be described as valence vibrations (those of the C=O bond and Cl–C–Cl group in

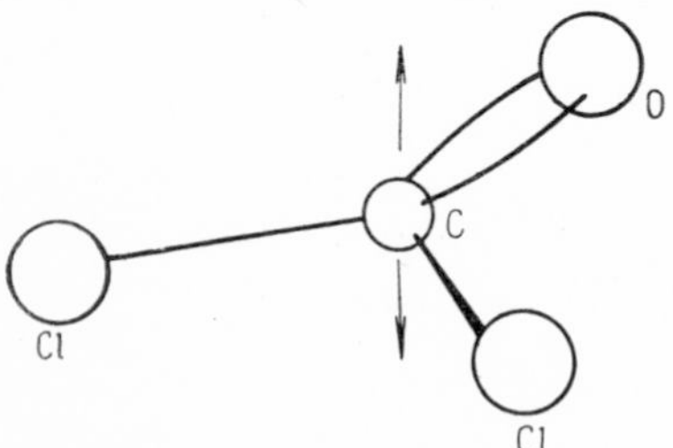

Fig. 58. Representation of an out-of-plane vibration of the planar phosgene molecule. Atom C deviates from the equilibrium position above and below the plane defined by the chlorine atoms and the oxygen atom.

and out of phase) and three as deformation vibrations. Only two deformation coordinates are, however, in the molecular plane (angles ClCCl and ClCO), as the third is redundant; the last deformation coordinate is therefore out of the plane. This coordinate corresponds to a vibration in which the C atom of the phosgene molecule periodically deviates from the equilibrium planar configuration (Fig. 58) given by *sp*^{2} hybridization.

Finally, the vibration type termed the puckering vibration must be mentioned. This is actually an out-of-plane deformation vibration of cyclic compounds, in which the ring atoms are not in plane in the equilibrium configuration. The cyclobutane molecule, in which the deviation of one carbon atom from the plane of the other three amounts to cca 35°, can be used as an example. In the puckering vibration, the out-of-plane carbon vibrates through the planar configuration (in which the molecule overcomes the maximum of the potential energy barrier) into the equivalent, mirror, out-of-plane configuration.

Trivial names are sometimes employed for vibrations in which the movements of vibrating atomic nuclei are analogous to some common movements. For example, the totally symmetrical vibration of atomic nuclei forming planar rings is called a "breathing" vibration, as the ring size periodically increases and decreases. Some deformation vibrations, e.g. the out-of-plane vibration of the hydrogens on the benzene ring, are called "umbrella" vibrations for their characteristic appearance, etc.

The local character of molecular vibrations is also manifested in other ways; the details of these manifestations will, however, be dealt with elsewhere, chiefly in Chapter 5, in the section devoted to the characteristic features of molecular vibrations.

4.3 Assignment of Spectral Bands

In Section 4.3, the individual procedures for band assignment in vibrational spectra will be discussed. Statements on relationships between the nature of molecular vibrations and the occurrence of certain bands in the vibrational spectra are made on the basis of particular indications whose general nature is not too difficult to explain.

4.3.1 Band Parameter Specificity

It has been found from experience that some bonds and groups can be identified according to the specific values of the parameters of the corresponding bands, be it the wavenumber, intensity, or an unusual band width or special shape. If this method could be employed generally, the problem of structural diagnostics would be completely and unambiguously solved. Unfortunately, this is not possible and the presence of certain groups or bonds in molecules can be assumed on the basis of the spectral band parameters only in a limited number of cases.

It has been found from the analysis of spectra that band wavenumbers and intensities are distributed statistically in each spectrum. Hence the occurrence can be judged in terms of the frequency; of course, bands whose parameters are exceptional, i.e. are substantially larger or smaller than the parameters with the highest frequency, have a greater importance in the assignment.

The vibrations of some bonds or groups can be identified according to the values of the wavenumbers of the corresponding infrared bands or Raman lines (the information obtained from the infrared spectrum is equivalent to that obtained from the Raman spectrum). These are chiefly the stretching vibrations of bonds and groups containing hydrogen atoms, e.g. O–H, N–H, C–H, Si–H, bonds, etc., or groups $-CH_3$, $-NH_3^+$, –COOH, etc., triple (C≡C, C≡N, etc.) and double (C=C, C=O, etc.) bonds and groups with cumulated double bonds (C=C=C, N=C=S, etc.), all of which, almost withou texception, appear between 4000 and 1500 cm^{-1}. The vibrations of these bonds are usually also characteristic and therefore each appears in a narrow wavenumber interval. The low probability of errors in assignment makes these bands exceptionally suitable for structural-diagnostic purposes.

The wavenumber stability of the vibrations of certain bonds can be explained on the basis of the Hook's law, which has the following form for the wavenumber of the vibration of a biatomic molecule

$$\tilde{\nu} = 1/(2\pi c)\sqrt{K/\mu}; \tag{4.24}$$

the wavenumber increases with increasing value of force constant K and decreasing value of the reduced mass of the nuclei of the atoms in the biatomic molecule, μ; the latter is defined as

$$\mu = (m_1 \,.\, m_2)/(m_1 + m_2). \tag{4.25}$$

Symbol c in Eq. (4.24) is the velocity of light in vacuo and m_1 and m_2 in Eq. (4.25) are the atomic masses in the biatomic molecule. The Hook's law is valid (of course, approximately) not only for biatomic molecules, but also for characteristically vibrating bonds (see Section 5.3.1.1). If the force constant for the given bond does not change very much from molecule to molecule and if the degree of characteristic nature of the vibration of this bond is roughly the same in the group of molecules to be compared, then the wavenumber of the vibration of this group lies within a certain limited interval.

The high wavenumbers of the X–H bond vibrations are caused by the low reduced mass μ and the high wavenumber of triple and double bonds by high force constant K. Bands appearing in "crowded" regions are difficult to assign. Such a region for organic molecules e.g. lies around 3000 cm^{-1}, where the absorption of all C–H bonds present in the molecule takes place (provided that these bonds do not substantially differ in their properties), and especially the so-called fingerprint region located between 1500 and 500 cm^{-1}. The assignment of bands in these regions on the basis of their wavenumbers alone is not, in principle, excluded, but a great risk of an error is involved. Further information is thus required for correct assignment. Band assignment in the fingerprint region is especially complicated with organic compounds, as atomic vibrations in single C–C bonds and single carbon bonds with heteroatoms (O, N) and further deformation vibrations of HCH, CCH, COH and other angles occur here. The vibrational bands of inorganic polyatomic ions also occur in the fingerprint region; however, their spectra are usually not as complex as those of polyatomic organic molecules common in chemical laboratories.

It is difficult and sometimes even impossible to assign bands of not very characteristic vibrations. With some groups or rings, many (or even all) of the atoms of the group contribute to most of the normal vibrations and structural changes thus cause considerable changes in the vibrational spectra. The number of contributing atoms and the contribution of the vibrations of their nuclei to normal vibrations are different in each case, the bands of the resultant normal vibrations are scattered over a wide spectral region and the wavenumbers of these bands exhibit no correlation rules. For example, the vibrations of all atomic nuclei in the cyclobutane molecule and in the molecules of its derivatives are strongly coupled; for this reason the spectra of cyclobutane and of its derivatives are so different in their band maximum wavenumbers that it has been impossible to detect the presence of the four-membered cyclobutane ring in them, using either a single band or a greater number of bands.

Some groups with strongly delocalized electrons, e.g. the phenyl-group or the peptide bond in secondary amides, are manifested by a system of bands with constant positions. This is the effect of group-characteristic vibrations; the vibrations of the nuclei of the atoms at the group are exceptionally strongly coupled and the effect of the connected parts of the molecule (e.g. residues R^1 and R^2 in the molecules of secondary amides, R^1CONHR^2) is small. This type of group is easy to recognize according to the frequencies of the bands of group-characteristic vibrations; the probability of correct determination of the presence of vibrationally-coupled groups and bonds increases with their size, so that the assignment of the bands for groups with a larger number of atoms is usually more reliable than for small groups.

Spectral band assignment is almost always based not only on the band wavenumbers, but also on the band intensities. Regarding the normal distribution of the vibrational spectral band intensities, bands with extreme intensities, especially very strong bands will be more valuable. Information on the band intensity gains importance when bands located in a "crowded" wavenumber region are to be assigned.

Information on the intensity of a band in an infrared spectrum and that on the intensity of a Raman line belonging to the same normal vibration are qualitatively different. This difference stems from the different mechanisms of absorption and scattering of radiation, which were discussed in detail in Section 2.2. But just this difference is another and important means for assignment of spectral bands.

The intensities of the bands of characteristic vibrations are, similar to the band wavenumbers, constant values to a considerable degree. This means that data on the intensities obtained with one molecule can be applied to other molecules with a high degree of reliability; however, a necessary condition is that the characteristic nature of the vibration in these molecules be preserved.

There are no direct relationships between the band wavenumbers and intensities. The wavenumbers of normal vibrations are determined by the elements of the force constant matrix and their intensities by the electro-optical parameters of molecules. For example, the position of the band of the C≡N triple bond stretching vibration in alkyl nitriles is predominantly determined by force constant K(CN) for the C≡N bond; however, a very considerable contribution to the intensity of this band is made by the electro-optical parameter connected with the C–C bond adjacent to the nitrile group. Hence substitution in the α-position with respect to the nitrile group gives rise to somewhat different trends in the wavenumbers and intensities of the ν(C≡N) band.

Sometimes information gained from band half-widths can also be employed in the band assignment. For example, stretching vibrations of O–H bonds in alcohols, N–H bonds in amines or C–H bonds in monosubstituted acetylene derivatives exhibit very broad bands in their spectra, when these bonds form an inter- or intramolecular hydrogen bond. The molecules of these compounds are always very strongly associated in the liquid state and the bands of the bonded forms then may attain half-widths of hundreds of cm^{-1} (especially with alcohols).

On the other hand, the vibrations of some groups give rise to extremely narrow bands, e.g. those of aryls, especially polycondensed ones.

Data on peculiarities in band shapes can also exceptionally be applied in band assignment. Carboxylic acids, RCOOH, RNH_3^+ ions, isothiocyanates, etc., can be quoted as examples; the vibrations of COOH, NH_3^+ and N=C=S groups appear in the infrared spectra of these compounds as multiplets with characteristic distribution of the component wavenumbers and intensities.

The bands of some vibrations are often doubled. Doubling (or even tripling) of bands may be a consequence of the rotational isomerism of molecules (this often occurs with the ν(C=O) band of ketone derivatives, esters, etc.) or it may be connected with the occurrence of hot bands (these are mostly located at lower wavenumbers close to the band of the fundamental vibration). Splitting into a doublet can also be caused by Fermi resonance or another resonance interaction. Fermi resonance, occuring often in the molecules of chemical compounds, results from the interaction of two levels which are close together on the energy scale and have the same symmetry; one is the fundamental level and the other must belong to a harmonic or a combination frequency. In the interaction the levels "repulse" one another and both bands have similar intensities. The result of the Fermi interaction is the Fermi doublet; a typical example of this resonance is exhibited by aldehydes. The level bound to the stretching vibrations of the C–H bond in the CHO group of aldehydes interacts with the second harmonic level, 2δ(CHO), derived from the fundamental frequency of the deformation vibration of the CHO group (2×1400 cm^{-1}). The result of the interaction is a Fermi doublet with branches around 2830 and 2730 cm^{-1} for various aldehydes. It is important for the Fermi resonance that the vibrations connected with the two interacting levels be localized in the same part of the molecule.

4.3.2 Isotope Substitution

Effects accompanying isotope substitution in molecules have quite exceptional importance in vibrational spectroscopy. Molecules differing only in their isotopic atom contents – isotopomers – exhibit virtually identical force fields and the differences in the band wavenumbers (and also small differences in their intensities) are caused only by the differences in the atomic masses. Isotope substitution is employed in experiments aimed at determining the nature of molecular vibrations, in structurally diagnostic and analytical procedures and in monitoring the kinetics of reactions and exchanges. Isotope substitution is indispensable in calculations of molecular force constants, as only the wavenumbers of a greater number of isotopomers yield a sufficient volume of experimental data for the solution of the secular equation.*)

The discussion of isotope substitutions will be limited here to situations when they are used in determining certain, exactly defined places in molecules and thus in verifying the correctness of the band assignement in vibrational spectra and evaluating the degree of localization of vibrations.

The manifestations of isotope substitution are treated in all general monographs on vibrational spectra; this topic is, however, especially thoroughly discussed in ref.[1].

4.3.2.1 Isotopes

Only certain isotopes are encountered frequently in chemical practice; the principal data on the most important of them are given in Table XXXV.

Deuterium, employed for isotopic replacement of hydrogen, is exceptionally important. The relative difference in the masses of hydrogen and deuterium is the largest of all pairs of isotopic atoms; the replacement of hydrogen by deuterium appears in the spectra in band shifts two or three orders of magnitude larger than the precision of the wavenumber determination, absolutely as much as 900 cm^{-1}. Of course, such large isotopic shifts are not always encountered. Deuterium is a stable isotope which can be used in work without any special safety precautions. The basic material for deuterium substitution is deuterium oxide, which is at present relatively readily available and cheap. Laboratory preparation of deuterium-labelled compounds is mostly simple and thus belongs among common laboratory procedures.

Substitutions with ^{15}N and ^{18}O isotopes are further met in spectroscopic practice. Both these isotopes are also stable, but very expensive and thus work with them is less common. It would be ideal to have isotopically

---

*) In a quadratic approximation, the number of force constants is $K_{ij} = n(n + 1)/2$, where $n = 3N - 6$ or $3N - 5$, for non-linear or linear molecules, respectively.

1 Pinchas S., Laulicht I.: Infrared Spectra of Labelled Compounds. Academic Press, New York 1971.

Table XXXV

Isotopes important in vibrational spectroscopy, their natural abundance and their effect on the vibrational frequency

Isotope X	Abundance[a]	$\tilde{\nu}_{X-H}/\tilde{\nu}_{X-D}$ [b]	$\tilde{\nu}_{X-H}/\tilde{\nu}_{X-D}$ [b]	Isotope M	Abundance[a]	$\tilde{\nu}_{X-H}/\tilde{\nu}_{X-D}$ [b]	$\tilde{\nu}_{X-D}/\tilde{\nu}_{X-H}$ [b]
^{1}H	99.98	1.155	0.866	^{28}Si	92.16	1.390	0.719
^{2}H (D)	0.02	1.225	0.816	^{30}Si	3.13	1.392	0.718
^{10}B	18.83	1.354	0.738	^{32}S	95.06	1.393	0.718
^{11}B	81.17	1.359	0.736	^{33}S	0.74	1.394	0.717
^{12}C	98.9	1.361	0.734	^{34}S	4.18	1.394	0.717
^{13}C	1.1	1.366	0.732	^{36}S	0.02	.395	0.716
^{14}N	99.62	1.370	0.730	^{35}Cl	75.43	1.395	0.717
^{15}N	0.38	1.372	0.729	^{37}Cl	24.57	1.396	0.718
^{16}O	99.76	1.374	0.728	^{79}Br	50.51	1.405	0.711
^{17}O	0.04	1.376	0.726	^{81}Br	49.49	1.405	0.711
^{18}O	0.20	1.378	0.725				

[a] Natural abundance of the isotope; [b] the ratio of the wavenumbers of stretching vibrations of the X—H and X—D bonds, determined from Eq. (4.28) in both variants, X—H/X—D and X—D/X—H.

pure samples at disposal for the study of isotopomer spectra; this state is only approached with expensive isotopes and the spectra of isotopic mixtures with various degrees of enrichment are usually followed.

Other isotopes, e.g. ^{13}C, ^{10}B, ^{11}B, ^{35}Cl and ^{37}Cl, are encountered much more rarely, usually at natural concentrations.

4.3.2.2 Isotopic Effects

As the force fields of isotopomer molecules are identical with a very high degree of precision, it is possible, according to Teller and Redlich, to express the wavenumbers for vibrations of any pair of structurally related isotopomers by means of the product rule,

$$\left[\prod_{j=1}^{n} \frac{\omega_{j,\,iso}}{\omega_j}\right]^2 = \left[\prod_{j=1}^{n}\left(\frac{m_j}{m_{j,\,iso}}\right)^{\alpha_j}\right]\cdot\left(\frac{M_{iso}}{M}\right)^{t}\cdot\left(\frac{I_{x,\,iso}}{I_x}\right)^{\delta x}\cdot\left(\frac{I_{y,\,iso}}{I_y}\right)^{\delta y}\cdot\left(\frac{I_{z,\,iso}}{I_z}\right)^{\delta z}, \qquad (4.26)$$

where ω_j denote the normal frequencies. The calculation is always carried out simultaneously for all the vibrations of a certain irreducible represent-tion (their number equals n). m_j are the masses of the individual atoms in groups of identical atoms*) and exponents α_j are the numbers of vibrations, including the improper ones, through which these groups of atoms contribute to the appropriate types according to the symmetry. M is the molecular mass and t the number of translational degrees of freedom contributing to the given representation. I_x, I_y and I_z are the main moments of inertia along axes x, y and z and exponents $^{\delta x}$, $^{\delta y}$ and $^{\delta z}$ attain values 0 or 1 according to whether the rotation around the appropriate fundamental axis contributes to the given representation or not. Subscript $_{iso}$ refers to the isotopically labelled compound. Degenerate vibrations are calculated only once, without respect to whether they belong to proper or improper vibrations.

The product rule is rather complicated and is applicable chiefly to calculations requiring high precision. Simplified versions and procedures have been proposed for less exact applications; they permit more rapid assessment of the vibration wavenumber for an isotopic molecule using a lower number of input parameters. The Decius and Wilson addition rule is one of them; when applied to the water molecule it can be written in the form,

$$\Sigma\omega_j^2(H_2O) = \Sigma\omega_j^2(D_2O) = 2\,\Sigma\omega_j^2(HDO), \tag{4.27}$$

where ω_j represents the wavenumber of the j-th vibration of the H_2O. D_2O or HDO molecule.

The rules for frequency shifts are considerably simplified if the isotopically labelled parts of molecules vibrate characteristically. If the bonds involved can be satisfactorily approximated by vibrations in biatomic molecules, the Hook's law (4.24) can be used for the estimation of isotopic frequencies,

$$\tilde{\nu}^i = \tilde{\nu}(\mu/\mu^i)^{1/2}; \tag{4.28}$$

where $\tilde{\nu}$ and $\tilde{\nu}^i$ are the wavenumbers and μ and μ^i the reduced masses of the atoms, (4.25), in the biatomic approximation. The relationship

$$\tilde{\nu}^i = \tilde{\nu}\{[(m\,.\,M)/(m + M)]/[(m_{iso}\,.\,M)/(m_{iso} + M)]\}^{1/2}, \tag{4.29}$$

where m and m_{iso} are the masses of the isotopically replaced atom and M is

---

*) These are atoms which are converted into indistinguishable configurations during the appropriate symmetry operations.

the mass of the other atom in the bond, can then be employed for the calculation.

The isotopic shifts in the wavenumbers of X–H bonds on replacement of hydrogen by deuterium (i.e. a change of X–H to X–D) can be estimated from the expression

$$\tilde{\nu}_{X-D} = \tilde{\nu}_{X-H}[(M + 2)/(2M + 2)]^{1/2} \quad (4.30)$$

where $\tilde{\nu}_{X-D}$ and $\tilde{\nu}_{X-H}$ are the approximate wavenumbers and M is the mass of atom X in the X–H and X–D bonds. The ratios, $\tilde{\nu}_{X-D}/\tilde{\nu}_{X-H}$ and $\tilde{\nu}_{X-H}/\tilde{\nu}_{X-D}$, for various bonds are given in Table XXXV.

The wavenumbers of some characteristic vibrations of the methanol molecule and of its deuterated derivatives are given in Table XXXVI. The data contained in the table show that the vibrating parts of the molecule are

Table XXXVI

The wavenumbers of the characteristic vibration bands of the methanol molecule and its isotopomers[a] (cm^{-1})

Assignment[b]	CH_3OH	CH_3OD	CD_3OH	CD_3OD
Stretching O—H or O—D	3,687	2,720	3,690	2,725
Stretching C—O	1,034	1,041	988	983
Stretching CH_3 or CD_3	2,973	2,965	2,235	2,228
Stretching CH_3 or CD_3	2,845	2,840	2,077	2,080

[a] Wavenumber values from the infrared spectrum of gaseous samples; [b] alternative assignment with isotopically labelled molecules.

Table XXXVII

The wavenumbers of the fundamental frequency bands of gaseous HCN, DCN and TCN (cm^{-1})

Vibration	Assignment[a]	HCN	DCN	TCN
ν_1	Stretching C—H, C—D and C–T	3,312	2,693	2,460
ν_2	Deformation (degenerate)	712	569	513
ν_3	Stretching C≡N	2,096	1,928	1,724

[a] Because of the mixed vibration character (see the isotope shifts) the description of the individual modes is not quite precise.

independent to a considerable degree; isotopic substitution in the methyl group has almost no effect on the O–H group vibrations and vice versa. On the contrary, all vibrations of the HCN molecule exhibit substantial changes in their wavenumbers on isotopic substitution of deuterium or tritium for the H atom (Table XXXVII). This wavenumber dependence is caused by coupling of the vibrations of all atoms in the HCN molecule in all vibrational modes (the vibration described as $\nu(C\equiv N)$ is not quite characteristic, as the H atom also contributes to the appropriate normal vibration, etc.). These examples show how isotopic substitutions can be utilized in estimating whether molecular vibrations are characteristic or coupled.

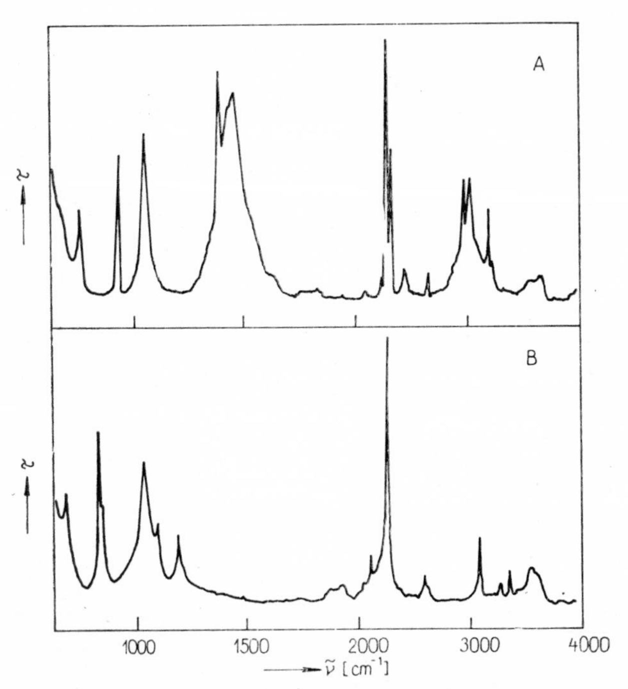

Fig. 59. Infrared spectra of liquid samples of acetonitrile CH_3CN (A) and its trideuterio derivative CD_3CN (B).

Extensive changes may sometimes occur in the spectra on substitution of hydrogen atoms by deuterium; these changes may be related to deep alterations in the nature of the vibrational modes accompanying these isotopic substitutions. For example, in molecules containing only hydrogen atoms in addition to heavy atoms, hydrogen atom vibrations participate in some normal vibrations; on isotopic substitution by deuterium, the deuterium atom vibrations need not take part in the same group vibrations as the hydrogen atom nuclei and consequently the character of the vibrational modes is quite different. A normal vibration of, for example, a molecule containing hydrogen atoms may also be involved in the Fermi resonance,

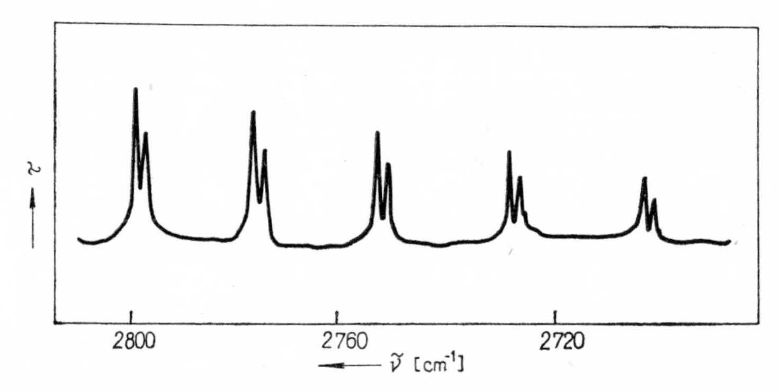

Fig. 60. Fine isotopic splitting found in the rotational-vibrational spectrum of the hydrogen chloride molecule (the fundamental frequency). In all doublets, the lower-intensity band corresponds to the $H^{37}Cl$ molecule and that with the higher intensity to the $H^{35}Cl$ molecule (the natural abundance of the isotopes is about 1 ^{37}Cl : 3 ^{35}Cl).

while the deuterated molecule does not exhibit this resonance. Level coupling as well as resonance always require mutual approaching of the energy levels, which may be considerably impeded by isotopic substitution.

An example of the changes taking place on substitution of hydrogen atoms by deuterium is given in Fig. 59, depicting the spectra of CH_3CN and CD_3CN. The fine splitting caused by the chlorine isotopes is shown in Fig. 60.

4.3.3 Appearance of Rotational-Vibrational Spectra

The rotational-vibrational bands in the spectra of gaseous samples of compounds contain very valuable information on the molecular geometry, yielding precise data about intermolecular distances and valence angles, as well as information useful in the assignment of spectral bands. It will be shown that the shapes of rotational-vibrational bands are closely related to the structure and symmetry of molecules; it is very advantageous that this information can even be obtained from spectra whose fine structure of the rotational-vibrational bands is not perceptible because of insufficient resolution.*)

The energy of the rotational-vibrational levels, E_{rv}, for a molecule can, to a first approximation, be described by the relationship,

$$E_{rv}/hc = G(v_1, v_2, \dots) + F_{(v)}(J, K), \tag{4.31}$$

where $G(v_1, v_2, \dots)$ is the vibrational energy (v_i are the vibrational quantum numbers), $F_{(v)}(J, K)$ is the rotational energy, h is the Planck constant and c is the velocity of light in a vacuum. J and K are the rotational quantum numbers; the rotational term is always related to the higher vibrational state (denoted by the vibrational quantum number v in $F_{(v)}$). Section 4.3.3 will deal only with the rotational part of Eq. (4.31).

An arbitrary system of cartesian coordinates, x, y, z, is first chosen, with the origin located at the centre of gravity of the molecule. Then the molecular rotation around an arbitrary rotation axis can be described by the three components of the vector of the angular rotation rate, $\omega_x, \omega_y, \omega_z$, in the directions of the appropriate axes of the coordinate system. The kinetic energy of the rotational movement can be expressed in this representation by the formula,

$$E_{kin} = (1/2) \, . \, (I_{xx}\omega_x^2 + I_{yy}\omega_y^2 + I_{zz}\omega_z^2 - 2I_{xy}\omega_x\omega_y - 2I_{yz}\omega_y\omega_z - 2I_{zx}\omega_z\omega_x), \tag{4.32}$$

where quantities

$$I_{\alpha\alpha} = \Sigma m_i r_{i,\alpha}^2, \qquad \alpha \in \{x, y, z\} \tag{4.33}$$

are termed the moments of inertia with respect to axis α; m_i is the mass of the i-th atom and $r_{i,\alpha}$ its distance from axis α. The summation is performed over all atoms in the molecule. The quantities,

$$I_{\alpha\beta} = \sum_i m_i r_{i,\alpha} r_{i,\beta}, \qquad \alpha, \beta \in \{x, y, z\}, \qquad \alpha \neq \beta \tag{4.34}$$

are termed the deviation moments.

A cartesian coordinate system can be found for each molecule such that all the deviation moments will equal zero; the relationships for the kinetic energy will then simplify to give

$$E_{rot} = (1/2) \, . \, (I_A\omega_A^2 + I_B\omega_B^2 + I_C\omega_C^2). \tag{4.35}$$

Quantities I_A, I_B and I_C are termed the principal moments of inertia of the molecule with respect to axes A, B and C, respectively. If the moments are

---

*) On the contrary, data on molecular geometry can only be obtained from perfectly resolved rotational-vibrational bands.

ordered according to their magnitude, so that $I_A \leqq I_B \leqq I_C$, then all molecules can be classified into three types

1. spherical ($I_A = I_B = I_C$),
2. symmetrical ($I_A = I_B \neq I_C$ or $I_A \neq I_B = I_C$) and
3. asymmetrical ($I_A \neq I_B \neq I_C \neq I_A$).

This classification is closely related to the overall molecular symmetry. If the point group contains two or more three- or more-fold rotational axes of symmetry (i.e. groups of the $\mathscr{T}$, $\mathscr{O}$ and $\mathscr{I}$ classes), the molecule belongs to the spherical rotator type. If it contains only a single three- or more-fold rotational axis, the molecule is a symmetrical rotator (e.g. groups $\mathscr{C}_{3v}$, $\mathscr{D}_{6d}$, etc.). A special case in this category are linear molecules (groups $\mathscr{C}_{\infty v}$ and $\mathscr{D}_{\infty h}$), for which $I_A = 0$. Other molecules belong among asymmetrical rotators. It may, of course, happen that a molecule, which should exhibit a lower rotator symmetry according to the point group of symmetry, has identical values of two or all three principal moments of inertia.

For molecules belonging among spherical rotators, the rotation term, $F_{(v_J)}$, from Eq. (4.31), attains the values

$$F_{(v)} = B_{(v)}J(J + 1); \tag{4.36}$$

where J is the rotational quantum number. The rotational levels of this type of molecule are, similar to linear molecules, unambiguously determined by this single rotational quantum number.

With molecules of the symmetrical rotator type, two quantum numbers, J and K, are required for unambiguous determination of the rotational level. The expressions for the rotational level energies are somewhat different for molecules of the prolate top symmetrical rotator type, whose principal moments of inertia are described by the relationship, $I_A < I_B = I_C$ (e.g. the methyl chloride molecule),

$$F_{(v)} = B_{(v)}J(J + 1) + (A_{(v)} - B_{(v)})\,K^2, \tag{4.37}$$

and for molecules of oblate top symmetrical rotators, with the moments of inertia, $I_A = I_B < I_C$ (e.g. the benzene molecule),

$$F_{(v)} = B_{(v)}J(J + 1) + (C_{(v)} - B_{(v)})\,K^2. \tag{4.38}$$

The energy levels of an asymmetrical rotator cannot be described by such simple relationships. Quantities $A_{(v)}$, $B_{(v)}$ and $C_{(v)}$ in Eq. (4.36) – (4.38) are the rotation constants, defined by the relationships,

$$A_{(v)} = h/(8\pi^2 c I_A), \tag{4.39}$$
$$B_{(v)} = h/(8\pi^2 c I_B), \tag{4.40}$$
$$C_{(v)} = h/(8\pi^2 c I_C). \tag{4.41}$$

The position of the principal axes of inertia in the molecule is related to the orientation of the symmetry elements in this molecule. The orientation is quite arbitrary in a spherical rotator; with symmetrical rotators, the orientation of the axis with a different value of the moment of inertia is identical with that of the three-fold (or more-fold) axis of symmetry. With asymmetrical rotators, the principal axes of inertia lie on the two-fold axes of symmetry or are perpendicular to the planes of symmetry.

So far, rotational interaction with vibration has been completely neglected, which cannot, of course, be done in rigorous interpretations. Here it is sufficient to point out that this interaction is represented by the terms including the coefficient of centrifugal distortion, $D_{(v)}$, e.g. in the expression for the spherical rotator,

$$F_{(v)} = B_{(v)}J(J + 1) - D_{(v)}J^2(J + 1)^2. \tag{4.42}$$

The "rigid rotator" approximation neglects the distortion term.

The appearance of rotational-vibrational spectra is determined by the rotational selection rules. Rotational quantum number J usually varies by one unit, $\Delta J = \pm 1$, or does not change at all, $\Delta J = 0$. Components with $\Delta J = +1$ are concentrated in the R-branch of the rotational-vibrational bands, those with $\Delta J = -1$ in the P-branch and those with $\Delta J = 0$ in the Q-branch. With biatomic molecules, in which the electric dipole moment along the axis changes during the vibration, the transitions among the rotational levels are governed by the selection rules for purely rotational transitions, $\Delta J = \pm 1$. Therefore, only rotational-vibrational bands with R- and P-branches and without the Q-branch appear in the infrared spectra of all heteronuclear biatomic molecules. The NO molecule is exceptional, but the activity of the Q-branch here is caused by angular momentum due to the presence of an unpaired electron. A similar situation occurs with polyatomic linear molecules, for which rotational-vibrational transitions in which only the electric dipole moment along the molecular axis is changed, are involved. The Q-branch appears with vibrations in which the dipole moment perpendicular to the molecular axis changes. Consequently, rotational-vibrational bands without the Q-branch are termed parallel and those with the Q-branch perpendicular.

The same selection rules hold for molecules of the spherical rotator type and for linear molecules. The selection rules for symmetrical rotator molecules determine the conditions for both quantum numbers simultaneously; there are also differences in the band appearance for transitions in which the electric dipole moment varies perpendicularly or in parallel with respect to the rotation axes. The selection rules for parallel bands can be written in the form

$$\Delta K = 0, \qquad \Delta J = \pm 1 \qquad \text{for} \qquad K = 0 \tag{4.43}$$

or

$$\Delta K = 0, \qquad \Delta J = \pm 1; 0 \qquad \text{for} \qquad K \neq 0 \tag{4.44}$$

and for perpendicular bands in the form

$$\Delta K = \pm k, \qquad \Delta J = \pm 1; 0. \tag{4.45}$$

Although an analytical expression of the energies for asymmetrical rotators analogous to Eqs. (4.36)–(4.38) is impossible, the shapes of rotational-vibrational bands for these molecules have been theoretically studied in a number of papers.[1,2] In order to simplify the problem, only

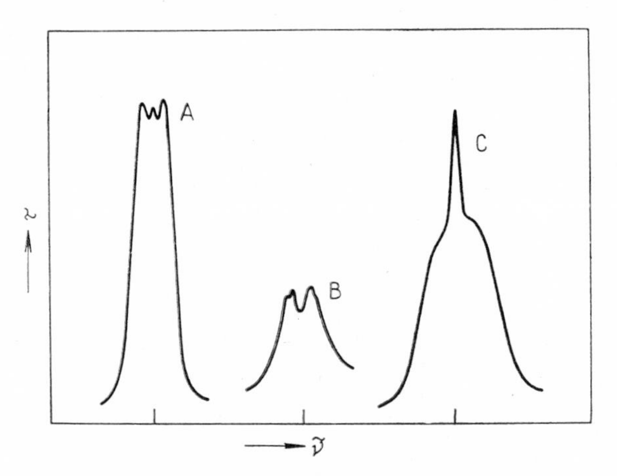

Fig. 61. Envelopes of unresolved rotational-vibrational bands in the infrared spectrum of *p*-fluoroanisol vapours, types A, B and C.

those cases when the orientation of the electric dipole moment or of its change is identical with one or more of the principal axes of inertia have been studied. In each of the three possible cases for the given ratio of the principal moments of inertia ($I_A < I_B < I_C$), a characteristic band shape is attained (see Fig. 61), denoted by letters A, B and C depending on which

[1] Badger R. M., Zumwalt L. R.: The Band Envelopes of Unsymmetrical Rotator Molecules I. Calculation of the Theoretical Envelopes. *J. Chem. Phys. 6*, 711 (1938).

[2] Ueda T., Shimanouchi T.: Band Envelope of Asymmetrical Top Molecules. *J. Mol. Spectry 28*, 350 (1968).

axis of inertia the change in the dipole moment is parallel to. All three branches occur with bands of the A-type (parallel with the *A*-axis), the Q-branch is suppressed in bands of the B-type and all three branches are perceptible with type C, but the intensity of the Q-branch is very high compared with branches P and R.

Theoretical (calculated) spectra for forty different ratios of $I_A : I_B : I_C$ and for three different spectrometer spectral slit widths (the parameter specifying the instrument resolution) were published in ref.[2]. This catalogue

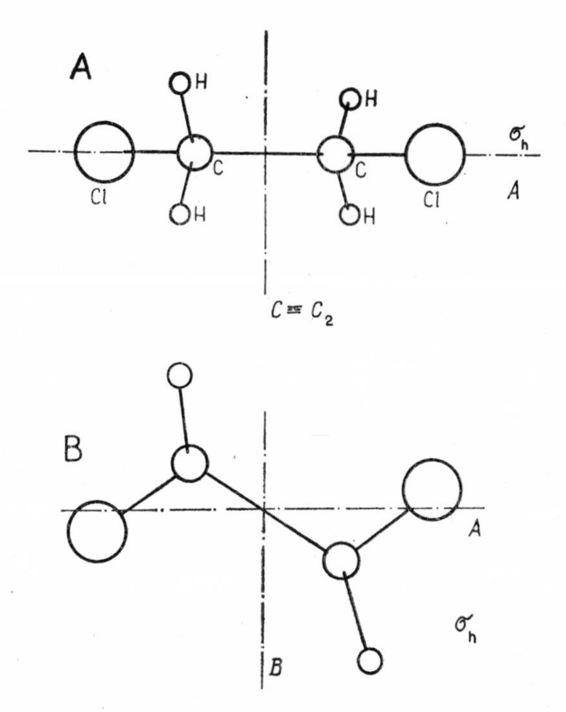

Fig. 62. Representation of the anti-periplanar conformation of the 1,2-dichloroethane molecule in the ground plan (A) and front view (B). A, B and C are the principal axes of inertia of the mol ecule, C_2 and σ_h are the symmetry elements of the molecule.

of spectra enables an a priori assessment of the band shapes for any asymmetric rotator. If the dipole moment or its change is not oriented in parallel with some of the principal molecular axes of inertia, the band profile can be roughly approximated by a linear combination of the bands of types A, B and C.

The 1,2-dichloroethane molecule in the trans conformation can serve as an example (Fig. 62); on the basis of its symmetry it belongs to point group $\mathscr{C}_{2h}$. Rotation axis of symmetry C_2 is identical with the rotation axis belonging to the largest moment of inertia (axis C) and rotation axes A and B lie in plane of symmetry σ_h. According to the selection rules (see Table XXXIII), only vibrations belonging to classes A_u and B_u are allowed in the infared spectra for this point group. As vibrations of type A_u are symmetrical with respect to rotational operation $\mathbf{C}_2^1$, vector $\partial \boldsymbol{\mu}/\partial q_i$ must be parallel with the rotation axis with the largest moment of inertia and the band profiles of class A_u must thus belong to pure type C. On the contrary, vibrations B_u are antisymmetrical with respect to $\mathbf{C}_2^1$ and symmetrical with respect to $\boldsymbol{\sigma}_h$. Hence vector $\partial \boldsymbol{\mu}/\partial q_i$ must lie in plane σ_h, but may be oriented differently with respect to axes A and B for each vibration. Therefore, bands of class B_u can be expressed by various linear combinations of bands of types A and B, the content of type A varying between 0 and 100 %.

4.3.4 Polarization Effects

The measurement of polarized infrared radiation is important chiefly with anisotropic media of single crystals or oriented films. As the problems of solid samples are not of prime interest in this book, they will not be discussed further.

Information about polarization of the scattered radiation in the Raman effect, which was treated in detail in Section 3.3.5.2, is much more interesting. Data concerning the polarization of Raman lines are very valuable in the assignment of normal vibrations. Totally symmetrical vibrations may attain values of the depolarization factor approaching zero (e.g. the totally symmetrical vibration of the tetrachloromethane molecule); antisymmetrical vibrations are depolarized. Polarization effects are disturbed by mutual interaction of molecules; with many molecules the totally symmetrical vibrations should actually exhibit zero depolarization factors (e.g. the CCl_4 molecule) and non-zero values stem from intermolecular interactions in the liquid state or in solution.

The Raman spectrum of methanesulphonyl chloride measured with $\parallel$ and $\perp$ polarizers is given in Fig. 63. Vibrations of type A', which should be polarized, exhibit larger intensity changes than depolarized vibrations of type A''.

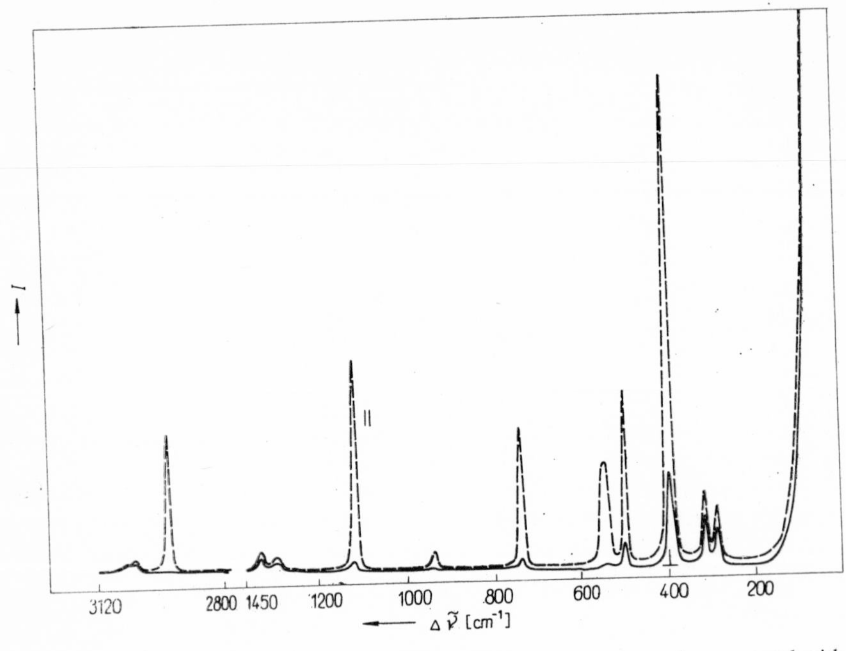

Fig. 63. Raman spectrum of a methanesulphonyl chloride liquid sample, measured with ⊥ and || polarizers.

4.3.5 Solvent Effects

Band shifts and changes in intensities caused by the effect of solvents are very specific; while the bands of certain molecular vibrations are very sensitive to the effect of solvents, others are almost insensitive. This sensitivity of some bands toward the solvent effect can also be utilized in band assignment.

For example, the carbonyl group stretching vibration band could be distinguished in the spectra of 4-pyridone and its N-methylderivative; this band is shifted to the region of unusually low wavenumbers, owing to strong delocalization of the electrons in these molecules. Only one band changed its parameters significantly when the spectra of the two compounds were measured in a series of solvents differing in their polarity; this band was then assigned to the stretching vibration of the C=O bond in 4-pyridone and its N-methylderivative, as sensitivity toward solvents could be assumed for no other vibration appearing in the region around 1500 cm^{-1}.

Using the solvent effect, the existence of the bands of the O–H, N–H and other bonds can very easily be verified. On adding basic solvents to solutions of compounds containing these bonds in e.g. tetrachloromethane, hydrogen-bonded complexes are formed, the intensity of the bands of the non-associated form decreases and the bands of the complexes appear.

The method employed for assignment of the bands of the C–O bond vibration in ethers and alcohols can also be given here. Solutions of these compounds in e.g. tetrachloromethane are saturated with gaseous hydrogen chloride; the oxonium salts are formed, the band intensities corresponding to the C–O bond vibration decrease and the bands corresponding to the vibrations of C and O^+ atomic nuclei in the $C-O^+$ bond appear.

The use of solvent effects for identification of the bands of various conformers is described in Section 5.2.5.2.

4.3.6 Calculation of Thermodynamic Quantities

The correctness of fundamental frequency band assignment in spectra can be very reliably verified by calculating the thermodynamic quantities for the studied compound. Naturally, this calculation is also carried out to obtain data on the magnitude and course of the thermodynamic functions; however, here we are interested in the former application, based on matching the calculated and experimental values of the thermodynamic functions.

Thermodynamic quantities are determined using partition function Q, which can be generally written as

$$Q = \sum_{i} \exp(-\varepsilon_i/kT); \tag{4.46}$$

The summation is carried out over all the energy states of the molecule, ε_i. It is advantageous to define the partition function as the sum of the energy levels,

$$Q = \sum_{j} g_j \exp(-\varepsilon_j/kT), \tag{4.47}$$

where g_j is the statistical weight, i.e. the level degeneracy.

To a good approximation the partition function can be written in the form,

$$Q = Q_{tr} \cdot Q_{int}, \tag{4.48}$$

where Q_{tr} is the translational part of the partition function and Q_{int} is the part including the electronic, vibrational and rotational levels. Hence,

$$Q_{int} = Q_{tr} \cdot Q_{el} \cdot Q_{vibr} \cdot Q_{rot}. \tag{4.49}$$

The translational part of the partition function can be expressed by

$$Q_{tr} = \sum_{tr} g_{tr} \exp(-\varepsilon_{tr}/kT); \tag{4.50}$$

Eq. (4.50) can be replaced with sufficient accuracy by the expression,

$$Q_{tr} = V[(2\pi \, . \, M \, . \, k \, . \, T)/(h^2 \, . \, N_A)]^{3/2}, \tag{4.51}$$

where V is the volume for which Q_{tr} is determined, M is the relative molecular mass, k is the Boltzmann constant, T is the absolute temperature, h is the Planck constant and N is Avogadro's constant. On substitution of the constant values in the SI unit system, the relationship,

$$Q_{tr} = 5.9423707 \, . \, 10^{30} \, . \, V \, . \, M^{3/2} \, . \, T^{3/2} \tag{4.52}$$

is obtained, from which the translational part of the partition function is readily determined.

Because of the very low probability of the population of higher electronic states in molecules through collisions (the electronic state separations are rather large), the electronic part of the partition function is close to unity at laboratory temperatures and can often be completely neglected. The remaining contributions, Q_{vibr} and Q_{rot}, are then determined from the relationships,

$$Q_{vibr} = \sum_{vibr} g_{vibr} \exp(-\varepsilon_{vibr}/kT) \tag{4.53}$$

and

$$Q_{rot} = \sum_{rot} g_{rot} \exp(-\varepsilon_{rot}/kT), \tag{4.54}$$

into which the appropriate vibrational or rotational levels, ε_{vibr} or ε_{rot}, are substituted. The problem is readily defined for vibrational levels if the fundamental frequency wavenumbers (i.e. transitions from the ground to the first excited state) are employed for the calculation. The calculations are usually carried out in the harmonic approximation, i.e. without correcting the wavenumbers for anharmonic character; this approximation is satisfactory for laboratory temperatures and molecules containing mostly heavier atoms. More rigorous calculations of the thermodynamic quantities, especially for higher temperatures and compounds whose molecules contain larger numbers of hydrogen atoms require, however, the use of an anharmonic model (vibrations of bonds involving a hydrogen atom are anharmonic to a high degree).

The function is formulated as a product in order to avoid summation over the individual rotational-vibrational levels. The rotational partition function can then be constructed independently, which is very advantageous. The procedures for its determination are somewhat different for molecules of spherical, symmetrical or asymmetrical rotators. The calculation is the simplest for biatomic and linear polyatomic molecules, for which the solid rotator approximation leads to the expression

$$Q_{\mathrm{rot}} = \sum_{J} (2J + 1) \exp[-BJ(J + 1)\, hc/(kT)]. \tag{4.55}$$

The individual thermodynamic functions are then calculated from partition function Q using Eqs (4.56)–(4.60). For example, for internal energy E^0,

$$E^0 = E_0^0 + RT^2[\mathrm{d}(\ln Q)/\mathrm{d}T], \tag{4.56}$$

for enthalpy H^0

$$H^0 = E_0^0 + RT + RT^2[\mathrm{d}(\ln Q)/\mathrm{d}t], \tag{4.57}$$

for entropy S^0

$$S^0 = R(1 - \ln N_A) + RT[\mathrm{d}(\ln Q)/\mathrm{d}T] + R \ln Q, \tag{4.58}$$

for free energy F^0

$$F^0 = E_0^0 + RT \ln N_A - RT \ln Q, \tag{4.59}$$

for heat capacity at constant pressure C_p^0

$$C_p^0 = R + R(\mathrm{d}/\mathrm{d}T)[T^2\, \mathrm{d}(\ln Q)/\mathrm{d}T] \tag{4.60}$$

etc.; E_0^0 is the zero-point energy, N_A is Avogadro's constant, etc., cf. Table of Symbols on p. 379.

The problems in the calculation of the thermodynamic quantities from vibrational and rotational-vibrational spectra are discussed in most general monographs cited at the end of Chapter 2. A very detailed treatment is given in ref.,[1] especially in the first volume.

4.4 Accuracy and Completeness of the Analysis of Vibrational Spectra

Many difficulties are encountered in analyses of spectra and in band assignment, which make these operations complicated. One of the serious problems is simultaneous occurrence of bands of fundamental frequencies, higher harmonics, combination frequencies and hot bands in infrared and Raman spectra. A spectrum containing only fundamental frequencies would be ideal; Raman spectra approach this ideal more closely than infrared spectra, since the fundamental frequencies are most pronounced in them for probability reasons.

---

[1] Gurvich L. V. et al.: Termodynamicheskaya Svoistva Individualnykh Veshchestv. Vol. I and II. Izdatelstvo Akad. Nauk USSR, Moscow 1962.

Fundamental frequency bands, provided that they are allowed by the selection rules, predominantly determine the structure of infrared and Raman spectra. Bands of higher harmonic and combination frequencies are usually weaker but must be taken into account in practice. Moreover, an increase in the intensity of the bands of higher harmonics and combination frequencies due to Fermi resonance must also be considered; this occurs especially frequently with certain compound types and in certain spectral regions (e.g. with hydrocarbons around 3000 cm^{-1} and with aldehydes around 2800 cm^{-1}).

The higher harmonic and combination frequency bands can be suppressed employing certain measuring techniques; for example, the layer (cell) thickness or the solution concentration may be selected so that only the strongest bands in the spectrum appear. When measuring on a thicker layer or a more concentrated solution, weaker bands also become visible in the spectrum and thus rather strong absorption can appear in regions where initially only a linear background could be observed (cf. the CH_3SO_2Cl spectrum in a 0.1 cm cell).

It sometimes happens that the band of some normal vibration almost does not appear in the spectrum, owing to limitations following from the selection rules. The assignment of such a band is then problematic, as the reliability of distinguishing a weak band among other weak bands is poor. It would be advantageous if the band wavenumbers for higher harmonics and combination frequencies could be calculated using the values for the fundamental frequencies. However, it is mostly difficult because these frequencies are not integral sums, difference or multiples of the fundamental frequencies.

Anharmonicity of biatomic molecular vibrations can be shown using the expression for the vibrational term $G(v)$,

$$G(v) = \tilde{\nu}(v + 1/2) - \tilde{\nu}x(v + 1/2)^2 + \tilde{\nu}y(v + 1/2)^3 - \ldots, \qquad (4.61)$$

$\tilde{\nu}$ being the vibrational frequency wavenumber, v the quantum number and $x, y, \ldots$ anharmonic constants. Anharmonic constant y is much smaller than x and thus all terms beyond the quadratic one are disregarded in Eq. (4.61). Anharmonic constant x is calculated from the relationships,

$$x = (\tilde{\nu}_{2\leftarrow 0} - 2\tilde{\nu}_{1\leftarrow 0})/2, \qquad (4.62)$$

$$x = (\tilde{\nu}_{3\leftarrow 0} - 3\tilde{\nu}_{1\leftarrow 0})/6, \text{ etc.} \qquad (4.63)$$

where $\tilde{\nu}_{2\leftarrow 0}$, $\tilde{\nu}_{3\leftarrow 0}$, etc. are the wavenumbers of the second, third, etc., harmonic frequency, respectively.

An example of the distortion of the harmonic frequency values for selected compounds due to anharmonicity is given in Table XXXVIII. As the anharmonic constant can be reasonably estimated from Eqs. (4.62) and (4.63) for biatomic molecules (and significant characteristic vibrations), the experimental values obtained for them can be readily corrected for anharmonicity. This is generally a complex problem with polyatomic molecules; sometimes the fact that even the bond vibration anharmonicity may be characteristic can be utilized to advantage.[1]

Table XXXVIII

Harmonic frequencies of the C—H bond vibration of the HCN and $HCCl_3$ molecules

Harmonic frequency	Wavenumber $\tilde{\nu}$ (cm^{-1})			
	HCN	$\Delta\tilde{\nu}$[a]	$HCCl_3$	$\Delta\tilde{\nu}$[a]
First (fundamental)	3,312		3,019	
		3,210		2,881
Second (the first overtone)	6,522		5,900	
		3,105		2,800
Third	9,627		8,700	
		3,009		2,615
Fourth	12,636[b]		11,315	
		2,916		2,545
Fifth	15,552[b]		13,860[b]	
		—		2,440
Sixth	—		16,300[b]	

[a] Simple differences in the wavenumbers of two successive harmonic frequencies;
[b] the bands already lie in the visible spectral region.

Anharmonic constants are more frequently negative than positive: the wavenumbers are increased by corrections for anharmonicity. Some anharmonic constants also have a positive sign, but their contributions are not very significant, because of low constant values.

With polyatomic molecules, the number of higher harmonic and especially combination frequencies increases to such dimensions that they form the so-called quasicontinuum of the spectrum. In addition to binary combinations ($\nu_i \pm \nu_j$), ternary ($\nu_i \pm \nu_j \pm \nu_k$ or $2\nu_i \pm \nu_j$), quaternary and

[1] Papoušek D.: *Collect. Czech. Chem. Commun. 29*, 2277 (1964).

higher combinations occur. Although the intensity of the higher harmonic and the combination frequency bands decreases with increasing order for probability reasons and many limitations are imposed by the selection rules, a large number of such bands can still appear in the spectra of polyatomic molecules. Large numbers of weak bands cannot usually be resolved; the bands begin to merge and thus form the spectral background. As the frequencies are distributed non-uniformly in the spectra, a non-linear spectral background can appear in the regions of their random concentration.

Higher harmonic or combination frequencies can sometimes be utilized for verification of the correctness of the fundamental frequency assignment, for example, in the analysis of the spectra of benzene derivatives in the region of the out-of-plane deformation vibrations of the benzene ring hydrogens.

According to the selection rules, some of these bands have very low intensities (e.g. in molecules with the $\mathscr{C}_{2v}$ symmetry some vibrations belong to the A_2 symmetry class and thus their assignment is often ambiguous). However, the higher harmonic and combination frequencies derived from these fundamental frequencies are relatively intense and can readily be detected and analyzed. They occur between $2000-1500\ cm^{-1}$ in infrared spectra; analysis of this region then permits evaluation of the correctness of the fundamental frequency bands in the region between $1000-700\ cm^{-1}$. A similar situation exists with compounds of the RNH_3^+X-type; under certain circumstances an "indicator" band appears in their spectra around $2100\ cm^{-1}$, corresponding to a combination frequency with participation of torsion (especially when the RNH_3^+ ion has $\mathscr{C}_{3v}$ local symmetry). Using the indicator band, the frequency of the torsion vibration in the RNH_3^+ ion can be determined.

Structural-diagnostic applications of vibrational spectroscopy are very valuable and relatively easy. Here answers are usually sought to simple questions (e.g. whether the double bond is saturated after reduction of an unsaturated ketone, or the ketogroup or both). The parameters (wave-numbers, intensities, half-widths, shapes) of easily recognizable bands of characteristically vibrating bonds and groups are followed. Simultaneously, it is borne in mind that the presence of certain groups and bonds cannot be reliably detected using vibrational spectra.

For other applications, complete analysis of vibrational spectra must be carried out (e.g. for the calculation of force constants). Analyses of this type are rather demanding; it must be taken into consideration that the spectral band assignment is not always equally easy (and reliable) and that complete analysis of the spectra of more complex molecules can only be more or less approached. This fact can best be verified on continuously repeated analyses of the spectra of virtually all compounds, with a series of new corrections and additions.

In structural-diagnostic applications, either only infrared or only Raman spectra can specifically be used and wavenumbers, intensities, etc. studied, depending on the nature of the job. In complete spectral analyses, the broadest possible experimental material must naturally be available. Both infrared and Raman spectra are evaluated and not only the common band parameters, but also differences among them, the results of polarization measurements, envelopes of infrared spectra of gaseous samples and the results of solvent and temperature effects are studied. The correctness of the assignment of the fundamental frequency bands is often further checked by calculation of force constants or thermodynamic functions.

CHAPTER 5

Chemical Applications of Vibrational Spectroscopy

Chapter 5 summarizes the rules for application of infrared and Raman spectroscopy in chemical practice. There are very many possibilities of utilizing these spectra and therefore it is advantageous to order this information in more general application schemes. In many places in Chapter 5 the reader will find references to computer programs serving for numerical solution of the given problems.

5.1 Characterization and Identification of Chemical Compounds

The infrared and Raman spectra represent "special" constants for molecules, useful for the characterization and identification of chemical compounds. The usefulness of vibrational spectra for these purposes follows from their molecular basis, i.e. from the fact that the number of vibrational states, the separation between them, the transition activities, etc., are inherent properties of molecules and depend on the number and kind of bonded atoms, on the nature of the chemical bonds and on the spatial structure and symmetry of the molecules considered.

The identification of chemical compounds concerns the kind of molecules; information is obtained on whether or not a certain sample consists of the same kind of molecules as another sample. In general, any property, e.g. physical, that carries a sufficient amount of characterizing information can be employed for this purpose. However, the methods of vibrational spectroscopy have proven successful compared with other methods; the information contained in them is kind-specific and its amount is sufficient for selection from large sets of compounds (with magnitudes of the order of 10^5).

Finally it should be added that vibrational spectra are easy and relatively cheap to measure. They are well suited for computer handling; many authors dealt with the problem of automation of computer identification procedures, using collections of spectra stored in the computer memory.[1,2]

5.1.1 Characterization of Individual Compounds

With any newly encountered compound, whether it has been isolated from a natural material or prepared synthetically or its history is unknown, it is first ascertained whether it belongs among known (and described) compounds or is a new compound.

The identification process is based on comparing the test compound with a set of known compounds, using data termed "characterization indices". In ascertaining the compound identity, data (characterization indices) concerning the complete set of known compounds should generally be available. It is impossible and, after all, not really necessary to gather information on such an immense scale. Using certain criteria, groups of compounds which may be called "pre-sorted" sets can always be assembled, in which the studied compound can be sought with greater probability than in other sets. Pre-sorted sets involve compounds of similar chemical structures (steroids, aromatic compounds), of similar origin (alkaloids, atmospheric pollutants) or with similar properties or technological use (detergents, solvents); other aspects can also be used. Larger sets are usually assembled when the great practical importance of the compounds serves as the selection criterion.

As the characterization indices, physical constant values are most frequently employed, as they can be obtained by objective measurement. These are e.g. the temperatures of some phase transitions (the melting and boiling points), further physical properties depending on the state of aggregation and chiefly spectra. The same characterization indices are employed within each pre-sorted set; differences may occur among various sets.*)

---

1 Thomas L. C.: A New Chemical Structural Code for Data Storage and Retrieval in Molecular Spectroscopy. Heyden, London 1970.

2 Rann C. S.: Automatic Sorting of Infrared Spectra. *Anal. Chem. 44*, 1669 (1972).

*) For example, physical properties typical for liquids (boiling points, density) will be advantageous for a set including predominantly liquid compounds (solvents), but they need not be suitable for characterization of compounds in a set with predominantly crystalline compounds.

When composing a characterization model for compounds in a certain pre-sorted set, the greatest possible differentiability among the individual compounds in the set should be sought. If the characterization indices do not contain a sufficient amount of information, the statements concerning the identity cannot be unambiguous and the effectiveness of the whole process is limited. It has been found from experience that even infrared spectra cannot be employed as the only characterization index in large sets, in spite of their complexity; however, it is sufficient to combine them with further indices and the identification effectiveness increases to the required degree.

The presence of impurities unfavourably affects the degree of unambiguity of identification. In view of the difficulties in defining purity, here all compounds will be considered as pure for the purposes of characterization, provided that their infrared and/or Raman spectra are not disturbed by the spectra of impurities more than within the experimental error of the spectrum measurement. The same formal definition of purity will be assumed in employing other physical constants.

The usefulness of vibrational spectroscopy for the characterization of compounds depends on many circumstances, e.g. on the state of aggregation. The complex rotational-vibrational spectra of gaseous compounds are excellently suited to this purpose. Especially compounds that are gaseous under laboratory conditions (temperature, pressure) can be unambiguously characterized by their infrared spectra. There are very few compounds belonging in this category and their spectra are substantially different. The only drawback is the limitation of the infrared activity with homonuclear biatomic molecules (H_2, N_2, O_2, etc.); the vibrations of these molecules can be monitored in the Raman spectra, but these are not commonly measured for purposes of characterization of gaseous compounds.

Vibrational spectra are also relatively well suited for characterization of compounds that are liquid under laboratory conditions (temperature, pressure). The complex character of the liquid state sometimes has an adverse effect, especially with associating liquids; the medium effect can be eliminated only by rigorous standardization of the sample preparation procedure and the spectrum measurement, as spectra obtained under different experimental conditions may differ. The number of compounds that are liquid under laboratory conditions is substantially larger than the number of gaseous compounds and the purely vibrational spectra of liquids contain less characterizing information than the rotational-vibrational spectra of gases; therefore, the characterization effectiveness of the vibrational spectra decreases considerably on transfer from gases to liquids.

Vibrational spectra can, of course, also be employed for characterization of substances that are crystalline under laboratory conditions (temperature, pressure). The dependence of the spectrum appearance on the sample preparation and spectrum measuring conditions is even more pronounced here than with liquids. Complications may arise not only owing to the ability of molecules in crystalline samples to associate (or to form ion-pairs, complexes, etc.), but also due to polymorphism. The spectra of crystalline samples measured using the nujol or KBr technique may differ from those of melts, solutions or films or even can be different themselves. Polymorphous changes may also occur during sample preparation, depending on the method used for KBr pellet preparation, the duration of milling, the pressure magnitude, etc. Hence compounds sensitive in this way may be characterized only with utmost care. With crystalline compounds, the standardization is more difficult than with liquids, although standardization is required more.

The ability of vibrational spectra to characterize crystal unit cell structures can sometimes be utilized advantageously, as the individual crystalline forms of many compounds can be characterized. Thus, if it is known that the test compound is polymorphous and if the spectra of the polymorphous forms are available, the sample crystalline form can then be determined from the measured spectrum. Unfortunately, knowledge of the crystalline forms is generally inadequate, especially with organic compounds, and it may often happen that the identification procedure fails only because two different crystalline forms of a compound are compared without realizing it. The spectra of solutions of all polymorphous forms of a compound are naturally identical, as polymorphism is connected with differences at the crystal unit cell level and not at the molecular structure level.

The application of vibrational spectra for characterization is also inherently limited, chiefly in connection with determination of the structure of compounds. The applicability decreases with increasing complexity of molecules. With small molecules (with five to ten atoms), any structural change is relatively large and is consequently accompanied by a large change in the spectrum. The situation is different with large molecules, where an analogous structural change need not be marked. The limitation of the characterization effectiveness of vibrational spectra connected with the size of the molecule can be demonstrated on the spectra of two pairs of n-alkanes. While the spectra of liquid samples of n-octane and n-nonane (Fig. 64) reflect the difference in the presence of one $-CH_2-$group, no corresponding structural change appears in the spectra of liquid samples of n-pentadecane and n-hexadecane; a difference by one

$-CH_2-$group is too small with the latter pair of compounds to be manifested in the spectrum.

Naturally, not all spectra of similar compounds with large molecules are similar. The similarity of the spectra of higher n-alkanes are related

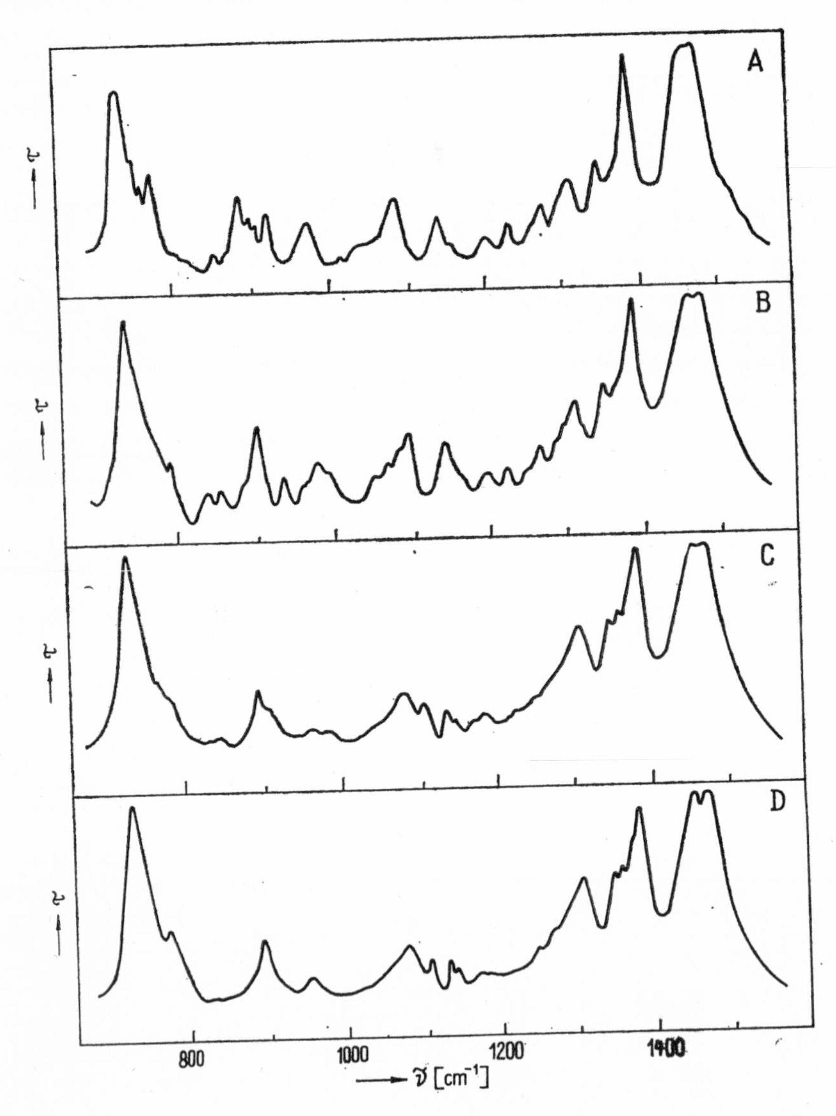

Fig. 64. Infrared spectra of n-octane (A), n-nonane (B), n-pentadecane (C) and n-hexadecane (D) in liquid samples.

rather to the group character of the spectra of n-alkanes in the liquid state. This feature (i.e. the group character) is encountered with all compounds whose molecules are composed of repeating units, e.g. polymers (with n-alkanes, the repeating unit is the CH_2 group, with polymers, the monomere unit). The vibrational spectra of these compounds are then characteristic of the basic units, rather than of the individual molecules. For example, the spectra of the styrene dimer and trimer differ only a little and the difference between the spectra of the trimer and tetramer is even smaller, etc.

Another limitation of the application of vibrational spectra follows from the polarity of molecules. For example, molecules with large permanent electric dipole moments exhibit especially intense infrared bands of all

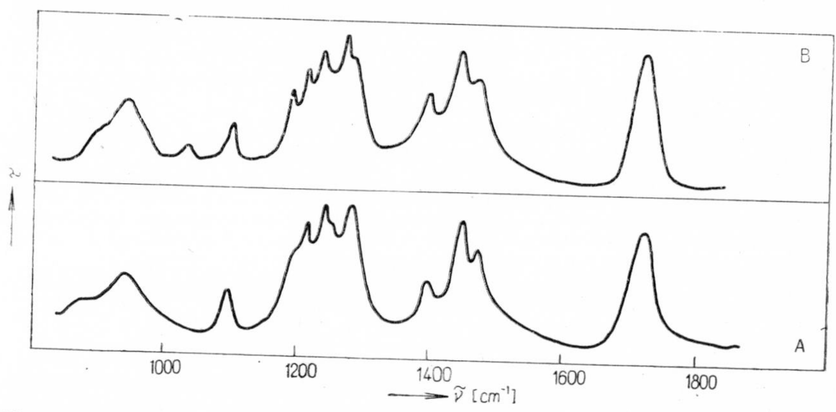

Fig. 65. Infrared spectra of carboxylic acid liquid sample: A — valeric, B — capronic acid.

normal vibrations in which the polar parts, bonds and groups participate; e.g. the bands of the $-COOH$ group vibrations are pronounced in the infrared spectra of carboxylic acids, the intensity of these bands sometimes being so high that they actually determine the shape of the whole spectrum. This statement can be demonstrated on the spectra of liquid samples of valeric acid, $CH_3(CH_2)_3COOH$, and caproic acid, $CH_3(CH_2)_4COOH$ (Fig. 65). The two acids from a homologous series differ by one $-CH_2-$group and should therefore exhibit at least the same differences as n-octane and n-nonane (Fig. 64). However, this is not so, as the structural difference is in the hydrocarbon part of the molecule, whose effect in the infrared spectrum is suppressed by the strong absorption of the carboxyl. Differences in the structure of the hydrocarbon chain of compounds, e.g. carboxylic acids, could readily be monitored in the Raman spectra, in which the effects of the polar parts of molecules are suppressed by the scattering on the hydrocarbon skeleton. However, only very few collections of Raman spectra are now available, so that identification combining the two vibrational spectroscopic methods still remains to be developed.

The suitability of vibrational spectra for the characterization of unstable, reactive compounds must be evaluated with exceptional care. Application of an unsuitable experimental technique, insufficient care in the sample preparation or other circumstances can cause that the spectrum obtained does not correspond to the original compound but to a mixture with decomposition products in various ratios.

The whole infrared spectrum is suited for compound characterization, but the part called the fingerprint region is especially suitable. This region extends from 1500 cm^{-1} to 700 cm^{-1} (if cells with NaCl windows are employed) or lower. This region is very complex, especially with organic compounds; a large number of variously intense and overlapping bands give rise to a complicated outline, rich in characterization information. The vibration bands for inorganic polyatomic ions are also located in this region.

The Raman spectra of aqueous solutions of inorganic ionic compounds can be obtained. With sufficient dilution, bands characterizing quite unambiguously polyatomic cations or anions appear in the spectra of salts of strong acids and bases. The appearance of the spectra of salts that form ion-pairs is somewhat more complicated.

5.1.2 Compound Identification

In the identification of individual compounds using vibrational spectra, identity of the test compound spectrum with the spectrum of a known compound is sought. On the basis of agreement between the spectra, identity of the two compounds is assumed; disagreement between the spectra (provided that gross errors are excluded) indicates that the compounds are different. However, a final statement concerning identity is rarely based only on agreement between the spectra. Usually it is also based on comparison of the values of further characterization indices and is often supplemented by chemical proofs.

The test and reference compounds are sometimes compared directly. However, no laboratory owns a sufficiently large collection of compounds to be able to carry out direct comparisons as a rule. Therefore data used for compound characterization are summarized in collections of physical constants and spectra, of which there is a great number on the contemporary book market.

There are specialized collections containing compounds selected according to certain rules. In addition to these monothematic collections there also exist universal collections listing the data on the substances most frequently required in chemical laboratories. The universal collections are more voluminous, the largest ones containing data on up to 100,000 compounds. These collections are usually supplemented periodically, the supplements of universal collections appearing regularly similarly to journals. For information on current collections, see the end of Chapter 5.

Owing to the great progress in infared spectroscopy after World War II, a large collection of infrared spectra is available for identification purposes. Raman spectra could also be successfully used in many cases, but their application is limited by the low number of spectra so far published. Moreover, the Raman spectra of the same compound obtained on various types of instruments can differ substantially. For this reason Raman spectra are now used rather as supplements to infared spectra.

5.1.3 Compound Purity

Infrared (and sometimes also Raman) spectroscopy can be utilized to determine the purity of compounds. The spectra of impure compounds are usually more complex than those of pure compounds, as they contain bands of impurities in addition to the bands of the main component. On the basis of the nature and degree of the spectrum perturbation, the character and the amount of the admixture (impurity) can even be found.

In monitoring the compound purity, the sample spectra are measured in cells somewhat thicker than usual, as even weak bands, through which e.g. a small amount of an admixture can be manifested, should be apparent. If the bands of impurities lie at the perceptibility limit, they are monitored using the techniques described for very weak bands (e.g. the photometric scale is expanded). The most modern spectrometers directly coupled with a computer are advantageous for these applications, as weak impurity bands can be filtered off from the noise using standard procedures.

If the bands of impurities are perceptible in the spectrum, differential spectroscopy can be used advantageously. A sample of the impure compound is placed in the measuring cell and a pure compound sample is placed in the reference cell (it is very advantageous to employ a variable thickness reference cell so that the absorbance of main component absorption is compensated).

The spectrometer then directly records the spectrum of the impurity(ies). The recording is, of course, useless in the regions of 100 % absorption by the main component (because of the poor signal-to-noise ratio), but the spectrum can be recorded in other regions. From the thus-obtained spectrum conclusions can be drawn concerning the impurity nature and under optimum conditions the impurity can even be identified. If there are several impurities, the conclusion is no longer as straightforward.

Infrared spectroscopy has quite exceptional importance in following the isotopic purity of labelled compounds. This technique is most frequently employed for control of the degree of hydrogen replacement by deuterium, but can also be applied to other isotope pairs. A number of procedures have been described for the determination of natural concentrations of isotopes, e.g. deuterium oxide in water, ^{13}C, etc. The application of infrared spectroscopy is indispensable in monitoring exchanges in molecules of organic and inorganic compounds, in the determination of contamination of labelled samples, etc.

Vibrational spectroscopic methods have often been used for monitoring the amount of moisture in liquids (solvents) and the presence of water of crystallization (or coordinated water) or of residues of other compounds in crystals. The stretching vibration bands of water are very intense (the very polar water molecule strongly absorbs infrared radiation) and make the detection of water in samples relatively easy. Water molecules often participate in bonding in crystalline hydrates (e.g. of inorganic salts), often form strong hydrogen bonds with the basic atoms in molecules and hence the stretching vibration bands of water may appear at a considerable distance from the absorption regions of non-associated water molecules (i.e. water vapour). These bands are often shifted as far as to the region around 3000 cm^{-1}, where they merge with the bands of other vibrations. The bands of water can then be detected by deuteration of the sample; a deuterated compound can be prepared from the hydrate by exchange reaction with deuterium oxide; the spectrum then exhibits the bands of D_2O of crystallization in regions completely different from those of H_2O absorption in the hydrate.

The spectrum of an impure compound often does not contain the individual bands of the impurity, but exhibits minima which are not as deep as those in the spectrum of the pure compound. This phenomenon can indicate pollution with several impurities, especially those whose spectra are not very different (e.g. pollution with compounds from a homologous series).

5.1.4 Identification of Components in Mixtures and in Equilibrium Systems

As has already been stated, infrared and Raman spectra are suitable for determining the existence of forms in which molecules of pure compounds may occur in equilibrium and non-equilibrium mixtures.

For example, in mixtures of some imperfectly dried aldehydes the water bands need not appear; however the bands of hydroxyl groups of a hydrate formed by addition

$$RCHO + H_2O \rightleftarrows RCH(OH)_2$$

from the two components can be present. The water bands also do not appear in the spectra of strong acid monohydrates; only the hydronium ion H_3O^+ exists in these systems owing to the specific conditions. In the spectrum of the catalyst for the Friedel–Crafts synthesis prepared from stoichiometric amounts of NaCl and $AlCl_3$, the bands of anion $AlCl_4^-$ formed by addition of the two salts were found. The nitronium ion, NO_2^+, was spectroscopically detected in the nitration mixture, acetylium ion CH_3CO^+ in acetylation mixtures, etc.

This is very valuable information about the components of compound mixtures obtained without disturbing the mixture composition and without isolation of the components of interest.

Very valuable data are also obtained from vibrational spectra concerning the distribution of isotopomers in systems with rapid exchange. For example, infrared spectra enable monitoring of the deuterium distribution in ammonia. The equilibrium,

$$NH_3 \rightleftharpoons NH_2D \rightleftharpoons NHD_2 \rightleftharpoons ND_3,$$

among the four possible isotopomers, each of which yields separate bands in the spectrum, is established due to rapid exchange reactions. The infrared spectrum can thus yield unique information about the instantaneous composition of this mixture.

Vibrational spectra are also employed for the determination of the isotopomer contents in systems whose separation is impossible for practical or economical reasons. For example, the contents of isotopomers with various amounts of ^{35}Cl and ^{37}Cl (i.e. $^{35}Cl_4C$, $^{35}Cl_3C\,^{37}Cl$, $^{35}Cl_2C\,^{37}Cl_2$, $^{35}ClC^{37}Cl_3$ and $C^{37}Cl_4$) can be determined from the intensities of the components of the vibration multiplet ν_1 for tetrachloromethane in the Raman spectrum.

A very valuable property of vibrational spectra is their applicability to the identification of short-lived species.*) An example is the suitability of infrared spectroscopy for unambiguous identification of very weak complexes, whose components are probably held together by van der Waals forces alone (e.g. the phenol complexes with n-hexane). The reasons why vibrational spectra are suited for these purposes have been discussed in Chapter 2. Vibrational spectra cannot, of course, be used for identification of electronically excited species, as their life-times are too short. However, works have been published in which vibrational spectra were employed for identification and structure determination on species cooled to very low temperatures.

Modern chemistry and chemical technology require analyses of especially complicated mixtures more and more frequently. These are the detection and determination of individual components of natural mixtures, such as oil and etheric oils, or of mixtures produced in chemical reactions, e.g. cracking, pyrolysis or combustion. Mixtures with very large numbers of components cannot be analyzed without separation; therefore, separation methods are usually combined with methods suitable for identification of the separated products.

Gas chromatographic methods are now almost exclusively used for separations in analyses of complex mixtures. The mixture components are separated in the chromatographic column and eluted. The amounts separated in analytical chromatographs are very small, unfortunately smaller than the amounts required for standard measurement of infrared spectra; moreover, the separated components are eluted at intervals shorter than the time required for obtaining spectra. Despite many attempts to couple gas chromatography with infrared spectroscopy, good results have not been obtained; hence mass spectrometry has been placed in "tandem" with gas chromatography, forming a so far unsurpassed analytical method for complex mixtures on the present instrumental basis. Both the coupled methods operate under their particular optimum conditions, while the application of infrared spectroscopy required the use of the method under extreme conditions. This fact does not adversely affect the usefulness of infrared spectroscopy for characterization and identification of compounds when coupling with gas chromatogrphy is not required, when a sufficient

---

*) The detection ability of infrared and Raman spectroscopy extends to short-lived species, provided that their life-time is in a suitable relationship to the frequency of their vibrations, as the latter is expressed in s^{-1}.

sample amount is available and the time for obtaining the spectrum is not limited.

Infrared spectroscopy is often employed for identification of gaseous components in analyses of very complex mixtures, e.g. the CO, CO_2, NH_3, CH_4 and HCN molecules, nitrogen oxides, etc.; however, the components of liquid and solid phases are identified by gas chromatography combined with mass spectrometry.

5.2 Quantitative Analysis

Infrared spectroscopy is also suited for quantitative determination of components in mixtures and systems. It is not an ideal method; compared with other methods it is subject to many limitations and difficulties and to a relatively large experimental error. However, there exist chemical fields where it occupies the dominant position, following chiefly from the fact that infrared spectra enable monitoring of some compounds and their forms not detectable by other analytical methods. Infrared spectroscopy is used frequently for quantitative analytical purpose, but Raman spectroscopy has also been used.

It seems that new prospects in quantitative infrared analysis stem from direct coupling of infrared spectrometers with computers, such as with the new Perkin–Elmer models (models 281, 283 and 580). This new generation instrument enables attainment of a substantially better photometric precision and hence of better precision of quantitative concentration determinations. The improvement in the photometric precision and better signal separation from the noise then permit a decrease in the sample size, thus complying better with the requirements of modern chemistry characterized by a frequent change in the experimental technique from a macro- to a microscale; moreover, it is then possible to simultaneously determine components present in large concentrations in mixtures and minor components. Practice will show how successful quantitative infrared analysis on these instruments will actually be.

The most unpleasant circumstance in quantitative infrared analysis stems from the complicated relationship between the compound concentrations and the absorption band intensities; the proportionality constant (the molar absorption coefficient) is different for each band in the spectrum of a single compound and naturally differs with the analytical bands of various compounds in mixtures. In no case can quantitative analyses be carried out without knowing the values of these coefficients for the analytical bands considered.

5.2.1 Infrared Spectra of Mixtures of Compounds

Quantitative infrared analysis is based on the assumption that the spectra of mixtures are linear superpositions of the spectra of all the components of the mixture. This means practically that the spectra of components mixed in arbitrary ratios are identical with those of the pure components.

This assumption actually cannot be met exactly and it can only be approached in the spectra of mixtures of some compounds. Good additivity of the spectra was found with mixtures of non-polar liquids, in dilute solutions of non-polar liquids, etc. Mixtures of aliphatic hydrocarbons are a good example. Their molecules are mutually bound only by weak intermolecular forces of the van der Waals type. Moreover, in a binary mixture of hydrocarbons A and B the molecular interactions, A ... A, B ... B and A ... B will be qualitatively and quantitatively comparable, which ensures good miscibility of the components of the mixture. However, non-polar liquids form only a minute fraction of the total number of chemical compounds, for which the assumption of perfect additivity of the spectra is no longer met so well.

If the assumption of spectral additivity is satisfied, the number of components in a mixture determinable by infrared spectroscopy is limited only by the number of suitable "analytical" bands and by the requirement that the contents of the individual components to be determined are not comparable with the error of the method (which is usually around 5 %). Analyses of ten-component mixtures can be considered.

In systems of polar compounds, considerable deviations from spectral additivity are encountered, especially with compounds readily forming associates and complexes. Even these mixtures can be analyzed, but the determination of the components must be based on deeper understanding of the interaction mechanism, as simple linear combination of the spectra is no longer sufficient.

The selection of the so-called analytical bands is connected with the question of whether the quantitative analysis of a mixture can be carried out or not. These are spectral bands typical of only a single component of a mixture; only weak absorption by the other components of the mixture can occur in the analytical band region, but not strong absorption bands. Any number of analytical bands can be formed for each component of the mixture, but at least one is required. As will be seen in Chapter 5.2.4, the transmittance or absorbance value must be known at one more point, which is usually located in a region where none of the test components strongly absorbs. In principle it is not necessary to measure the absorbance or transmittance exactly at the analytical band maxima; it is only important that these values are measured at the same wavenumber in all spectra.

5.2.2 The Lambert-Beer Law and its Validity

The Lambert – Beer law,

$$D(\tilde{\nu}) = \varepsilon_X(\tilde{\nu}) \,.\, c_X \,.\, d, \tag{5.1}$$

which forms the basis of quantitative infrared spectral analysis, was derived in Chapter 3.3.3.4. In the derivation it was assumed that monochromatic radiation with wavenumber $\tilde{\nu}$ can encounter only a single kind of molecule, X, along the whole optical path length in the absorbing medium, d.

The first assumption, monochromaticity of the absorbed radiation, is relatively well satisfied with modern grating spectrometers and represents no limitation from the point of view of practical applications. However, the validity of the law is complicated by many effects whose magnitude and direction follow from the analysis of the individual quantities on the right-hand side of Eq. (5.1).

With increasing concentration c_X, the assumption of mutual independence of molecules, from which constancy of quantity $\varepsilon_X(\tilde{\nu})$ follows, ceases to be valid. If quantity $\varepsilon_X(\tilde{\nu})$ is considered as a proportionality constant, then the Lambert – Beer law should describe a linear relationship between activity a_X and the absorbance at wavenumber $\tilde{\nu}$ and should correctly be written in the form

$$D(\tilde{\nu}) = \varepsilon_X(\tilde{\nu}) \,.\, a_X \,.\, d = \varepsilon_X(\tilde{\nu}) \,.\, f_X \,.\, c_X \,.\, d, \tag{5.2}$$

where f_X is the activity coefficient. Activity a_X is replaced by the analytical concentration, c_X, chiefly because it is simply not known in systems containing a larger number of compounds. As has already been mentioned in Chapter 3.3.3.4, the additivity of the absorbance contributions from the individual substances follows from the derivation procedure; hence the Lambert – Beer law can be written for a mixture with n components in the form

$$D(\tilde{\nu}) = d \sum_{i=1}^{n} \varepsilon_i(\tilde{\nu})\, a_i. \tag{5.3}$$

If the molecules in the system analyzed interact only to a negligible degree and if the interactions of all kinds of molecules making up the system are similar, then the activity coefficients, f_X, approach unity and replacement of the activity by the concentration is justified. However, if strong interactions occur among the molecules in the system or associates are formed, the processes being concentration-dependent, then the activity coefficients differ from unity and the relationship between the absorbance and the concentration is no longer linear. This non-linearity does not prevent quantitative determinations in some simple systems; however, it must be considered in complex systems. Procedures for mathematical correction for this non-linearity have been developed (introduction of quadratic or cubic terms).

The validity of the Lambert–Beer law is usually limited to certain concentration regions. Associating molecules are usually present in the monomeric form at low concentrations; at higher concentrations, association becomes important and the dependence departs from linearity. However, the concentration regions in which polar molecules associate are often difficult to assess; hence the D vs. c_X relationship is examined over the whole concentration range in which the relationship is followed.

Non-constancy of cell thickness d also contributes to errors of the method. Intense infrared radiation sources are mostly employed in modern spectrometers and consequently their cell rooms are heated to temperatures

Table XXXIX

Changes in the cell thickness (in %) on heating by 10° or 20 °C relative to its absolute thickness d

d (cm)	Change by +10 °C	Change by +20 °C
$5 . 10^{-4}$	0.5	0.9
$2 . 10^{-4}$	1	2
$1 . 10^{-4}$	5	12
$5 . 10^{-5}$	13	28

of 35–50 °C. Infrared cells change their thickness when heated to various temperatures owing to expansion of the component parts; these changes are relatively larger when the cell is thinner. The magnitude of these errors, for a set of cells tempered in the cell room of a Perkin–Elmer model 621

infrared spectrometer under normal conditions is given in Table XXXIX. Cells attain a temperature equilibrium with the instrument within cca 20 – 30 min; over this time interval the cell thickness varies. A further error caused by cell heating that is not negligible is the change in the mixture component concentrations with changing temperature. This error is connected with temperature changes in the liquid density and is especially large when sample concentrations are measured in solutions of liquids with high temperature dilatation coefficients. Both errors are reflected in the value of D.

Further difficulties arise owing to differences in placing cells in the cell room. If the cell is not fixed well in the holder and its axis can form an angle of $\pm\alpha$ with the axis of the passing beam, then its effective thickness d may vary within a range of $d_s < d < d_s/\cos\alpha$, where d_s is the actual cell thickness. The more the cell axis deviates from the direction of the passing radiation, the larger is its effective thickness d and consequently also the measured absorbance value, D.

Certain complications are also caused by quantity $\varepsilon_X(\tilde{\nu})$, which is assumed to be a molecular constant for given wavenumber $\tilde{\nu}$. This assumption is actually satisfied only under exactly defined conditions. The molar absorption coefficient is a function of the temperature, pressure and the medium in which the test compound molecules are placed; hence the measuring conditions must be exactly specified. The use of ε_X values under conditions different from those under which they were determined brings about the danger of committing variously large errors, which may attain as much as tens of percent.

All the errors mentioned are reflected in the value of absorbance D as systematic or random errors, which are very difficult to eliminate.

5.2.3 Absorbance Measurement on Double-Beam Spectrometers

The transmittance $\tau(\tilde{\nu})$, on which the calculation of absorbance $D(\tilde{\nu})$ is based, is defined by Eq. (2.9) as the ratio of the spectral radiant powers before and after passage through the absorbing medium. However, double-beam instruments measure the ratio of the radiant powers in the reference and the measuring beams (see Eq. (3.99)), i.e. substantially different quantities. The spectral slit width, which is mostly sufficiently small in grating spectrometers, does not play an important role here; however, the fact that radiation beams are compared which, in the common experimental arrangement, passed through different cells, one containing the test substance and the other being a reference, is important. Hence, the apparent absorbance $D'(\tilde{\nu})$, defined by the relationship,

$$D'(\tilde{\nu}) = -\log \tau'(\tilde{\nu}) \tag{5.4}$$

is physically different from quantity $D(\tilde{\nu})$ appearing in Eqs (5.1)–(5.3), and consequently it cannot principally obey the Lambert–Beer law.

Let it be assumed, for the sake of simplification, that the spectral slit width is infinitely small.*) Eq. (3.99) can then be rearranged to give

$$\tau'(\tilde{\nu}) = \Phi_{\tilde{\nu}}^{(\mathrm{meas})}(\tilde{\nu})/\Phi_{\tilde{\nu}}^{(\mathrm{ref})}(\tilde{\nu}); \tag{5.5}$$

the Lambert–Beer law can be used for the calculation of radiant powers $\Phi^{(\mathrm{meas})}(\tilde{\nu})$ and $\Phi^{(\mathrm{ref})}(\tilde{\nu})$.

Let it further be assumed that the same spectral radiant power, $\Phi_{\tilde{\nu}}^{0}(\tilde{\nu})$, is incident on the two cells.**)

A cell filled with pure solvent***) is placed in the reference beam and a solution of n substances in the same solvent in the measuring beam.

The losses in the spectral radiant powers on the cell windows†) due to absorption, reflection, scattering, etc. are expressed collectively by function $Z(\tilde{\nu})$ $(0 \leqq Z(\tilde{\nu}) \leqq 1)$. It then holds for the reference beam that

$$\Phi_{\tilde{\nu}}^{(\mathrm{ref})}(\tilde{\nu}) = \Phi_{\tilde{\nu}}^{0}(\tilde{\nu}) \,.\, Z^{(\mathrm{ref})}(\tilde{\nu}) \,.\, \exp\left[-2{,}3029\varepsilon_{\mathrm{ref},0}(\tilde{\nu})\, a_{\mathrm{ref},0} d_{\mathrm{ref}}\right] \tag{5.6}$$

and for the measuring beam that

$$\Phi_{\tilde{\nu}}^{(\mathrm{meas})}(\tilde{\nu}) = \Phi_{\tilde{\nu}}^{0}(\tilde{\nu})\, Z^{(\mathrm{meas})}(\tilde{\nu}) \exp\left[-2{,}3029\left(\varepsilon_{\mathrm{solv}}(\tilde{\nu})\, a_{\mathrm{solv}} + \sum_{i=0}^{n} \varepsilon_i(\tilde{\nu})\, a_i\right) d_{\mathrm{meas}}\right]. \tag{5.7}$$

On substituting these values into Eq. (5.5),

$$\tau'(\tilde{\nu}) = \frac{Z^{(\mathrm{meas})}(\tilde{\nu})}{Z^{(\mathrm{ref})}(\tilde{\nu})} \times \exp\left[-2{,}3029\left(-\varepsilon_{\mathrm{solv},0}(\tilde{\nu})\, a_{\mathrm{solv},0} d_{\mathrm{ref}} + \varepsilon_{\mathrm{solv}}(\tilde{\nu})\, a_{\mathrm{solv}} d_{\mathrm{meas}} + d_{\mathrm{meas}} \sum_{i=0}^{n} \varepsilon_i(\tilde{\nu})\, a_i\right)\right] \tag{5.8}$$

---

*) The instrument function degenerates to the so-called delta function.

**) This basically corresponds to the requirement of correct adjustment of the transmittance scale to 100 % τ.

***) If no solvent is employed, it suffices to take $\varepsilon_{\mathrm{solv}}(\tilde{\nu}) = 0$.

†) Or those along the whole path length beyond the cells up to the point at which the measuring and the reference beam are joined again.

is obtained; taking the logarithms,

$$D'(\tilde{\nu}) = K(\tilde{\nu}) + \varepsilon_{solv}(\tilde{\nu})\, a_{solv} d_{meas} - \varepsilon_{solv,0}(\tilde{\nu})\, a_{solv,0} d_{ref} + + d_{meas} \sum_{i=0}^{n} \varepsilon_i(\tilde{\nu})\, a_i, \tag{5.9}$$

where

$$K(\tilde{\nu}) = \log\left[Z^{(ref)}(\tilde{\nu})/Z^{(meas)}(\tilde{\nu})\right]. \tag{5.10}$$

It can be seen that this expression differs from the simple form of the Lambert–Beer law in that three terms have been added. The first describes unequal losses in the cell windows (or outside the cells) and the other two the uncompensated effects of solvent absorption.

Provided that the optical path outside the cells is equally long in the two beams, then term $K(\tilde{\nu})$ is virtually independent of wavenumber $\tilde{\nu}$ and merely represents a vertical shift of the zero-line.

The effect of the solvent is somewhat more complicated. First and foremost it must be considered whether the molar absorption coefficients of the solvent alone and the solvent in the test solution, $\varepsilon_{solv,0}(\tilde{\nu})$ and $\varepsilon_{solv}(\tilde{\nu})$, respectively, can be considered to be identical. If this is so, then these two terms are eliminated when

$$a_{solv} \cdot d_{meas} = a_{solv,0} \cdot d_{ref}, \quad \text{resp.} \quad d_{meas}/d_{ref} = a_{solv,0}/a_{solv}. \tag{5.11}$$

If the ratio of the cell thickness cannot be maintained, then the magnitude of the contribution from the second and third terms in Eq. (5.9) is directly proportional to the magnitude of the solvent molar absorption coefficient $\varepsilon_{solv}(\tilde{\nu})$. This is another argument for the necessity of exact measurements being carried out in spectral regions in which the solvent does not absorb.

5.2.4 Analysis of Mixtures of Stable, Non-Reacting Compounds

It is assumed that the spectra of pure substances and of their mixture are measured in an experimental arrangement in which the second and third terms can be neglected in Eq. (5.9) and that function $K(\tilde{\nu})$ can be replaced by its mean value, $\bar{K}$, over the part of the spectrum that is required for the analysis. Then the absorbance for the *i*-th pure substance can be expressed by

$$D_i(\tilde{\nu}) = \bar{K}_i + \varepsilon_{i,0}(\tilde{\nu})\, a_{i,0} d_i; \tag{5.12}$$

for the mixture it then follows that

$$D_s(\tilde{\nu}) = \bar{K}_s + d_s \sum_{i=1}^{n} \varepsilon_i(\tilde{\nu})\, a_i. \tag{5.13}$$

Assuming that the molar absorption coefficient for substances measured alone, $\varepsilon_{i,0}(\tilde{\nu})$, and in the mixture, $\varepsilon_i(\tilde{\nu})$, are identical, then the equation,

$$D_s(\tilde{\nu})/d_s = \overline{K}_s/d_s + \sum_{i=1}^{n} \frac{D_i(\tilde{\nu}) - \overline{K}_i}{a_{i,0}d_i} a_i, \quad (5.14)$$

is obtained on simple algebraic rearrangement. In determinations that do not require a high precision it can be assumed that the mean values of K_i equal zero. In more precise determinations these coefficients must be determined from dependence (5.12) of $D_i(\tilde{\nu})$ on the activity (or concentration). Term $\overline{K}_s$ cannot usually be determined in this way and therefore it is mostly included among the unknown quantities. In order to solve this case, matrix $\mathbf{E}$ with dimensions $(n + 1) \times m \ (m \geqq n + 1)$ is constructed, with elements

$$\mathbf{E} = \begin{bmatrix} 1 & \varepsilon_1(\tilde{\nu}_1) & \varepsilon_2(\tilde{\nu}_1) & \dots & \varepsilon_n(\tilde{\nu}_1) \\ 1 & \varepsilon_1(\tilde{\nu}_2) & \varepsilon_2(\tilde{\nu}_2) & \dots & \varepsilon_n(\tilde{\nu}_2) \\ \vdots & & & & \vdots \\ 1 & \varepsilon_1(\tilde{\nu}_m) & \varepsilon_2(\tilde{\nu}_m) & \dots & e_n(\tilde{\nu}_m) \end{bmatrix}. \quad (5.15)$$

where

$$\varepsilon_i(\tilde{\nu}_j) = [D_i(\tilde{\nu}_j) - \overline{K}_i]/[a_{i,0} \cdot d_i] \quad (5.16)$$

and vector $\boldsymbol{D}_s$ of the experimental absorbance values in the mixture

$$\boldsymbol{D}_s = [D_s(\tilde{\nu}_1), D_s(\tilde{\nu}_2), \dots, D_s(\tilde{\nu}_m)], \quad (5.17)$$

where $\tilde{\nu}_1$ to $\tilde{\nu}_m$ are the wavenumbers of the selected analytical bands. If the vector of the unknown quantities is termed $\boldsymbol{a}$,

$$\boldsymbol{a} = [K_s/d_s, a_1, a_2, \dots, a_n], \quad (5.18)$$

then the set of the definition equations can be written in matrix form,

$$\mathbf{E} . \boldsymbol{a} = \boldsymbol{D}_s, \quad (5.19)$$

whose solution by the least squares method is (see Chapter 3.2.2.2)

$$\boldsymbol{a} = (\mathbf{E}^T\mathbf{E})^{-1} . \mathbf{E}^T . \boldsymbol{D}_s. \quad (5.20)$$

The individual points can be ascribed various statistical weights, as described in Chapter 3.2.2.2.2.

If zero value of $\overline{K}_s$ is not assumed and cannot be determined experimentally, then for the determination of the activities of n substances it is necessary to measure all spectra in at least $n + 1$ points. In the opposite case, when $\overline{K}_s$ is known, the first column in matrix $\mathbf{E}$ and the first element in vector $\boldsymbol{a}$ are omitted; then n experimental points in each spectrum, corresponding to the position of the analytical bands, are sufficient. It is, however, advantageous to employ more points (more analytical bands), as the estimates of the appropriate variances are then obtained simultaneously with the calculated activities of the individual components and the overall analytical precision increases.

If the spectra of the pure substances and mixtures are separated into individual bands, molar integrated intensities A (see Chapter 3.3.3.4) of the individual analytical bands can be advantageously employed for quantitative determinations instead of the molar absorption coefficients. The basis of the calculations discussed in this and the following sections is not changed. It can be seen that the integrated intensities are much more suitable than the molar absorption coefficients especially in monitoring of the temperature dependences during the determination of thermodynamical constants (see Chapter 5.2.5.2).

5.2.4.1 Practical Procedures

In quantitative analysis based on infrared spectra various procedures can be employed, differing either in the experimental technique or in the way the obtained spectra are handled. Before discussing the various procedures, the general rules that must always be observed will be treated.

The analysis of a mixture of compounds can be carried out only if all the components are known. The only exception is the determination of a single component in a mixture of unknown compounds; this determination is successful only when the analytical band of the test compound lies entirely outside the absorption by the other components in the mixture or when the content of the other components of the mixture is constant.

Knowledge of the components of a mixture is not the only condition for quantitative analysis; another important condition is independence of the occurrence of analytical bands. Each component of the mixture must give rise to an at least partially resolved analytical band in the spectrum. If the analytical bands are not sufficiently removed from intense bands of other components, the uncertainty of the analysis increases. Sometimes, the calculation can be facilitated by computer separation of the individual spectral bands.

If the basic conditions for the quantitative analysis procedure are satisfied, the operation must be evaluated from the point of view of the errors involved. The question of the signal-to-noise ratio is primarily evaluated; if the analytical bands are located in a region of almost complete absorption by the other components (the solvent in solutions, carbon dioxide from the air, etc.), all attempts at quantitative analysis are unsuccessful, as noise fluctuations become more pronounced in the spectrum and distort the information contained in the recording. For the reasons discussed in the previous section the analytical bands should be close together; (they are then measured in a single spectral region in which the instrument resolution is almost constant). However, sometimes the analytical bands cannot be selected close together because of the absorption of other components; the error which would be encountered due to the noise would exceed that stemming from the use of distant bands.

Another important condition for performing quantitative analyses is the verification of the validity of the Lambert–Beer law. The molar absorption coefficients, $\varepsilon_i(\tilde{\nu}_j)$, must be determined for virtually any analysis; these values have been recently given in some journals for various mixtures, but their use introduces certain errors into the analytical procedure, owing to different precision and accuracy of the absorbance values obtained on different instruments.

The molar absorption coefficient values must be determined for each substance at the wavenumbers of all analytical bands of the other components of the mixture considered. The least squares method (see Chapter 3.2.2.2) applied to Eq. (5.12) or (5.16) in the region of the dependence of the absorbance D on the concentration (or activity) is usually employed for precise determinations. For control purposes, a series of "synthetic" mixtures with exactly known compositions is mostly analyzed. From the comparison of these actual concentrations with the values calculated from Eq. (5.20), information about the systematic error of the determination of the individual components is obtained.

In view of the fact that the individual analytical bands may introduce various errors into the calculation procedure (e.g. due to unequal absolute values of the molar absorption coefficients of the analytical bands and other effects), each component of the mixture is usually determined with different precision.

The basic operation of quantitative analysis is the determination of the analytical band absorbances, using the procedure described in detail in Chapter 3.3.3.4.1. Prior to the measurement, the values of 0 % and 100 % transmission of the instrument are carefully adjusted without the cell inserted and the measurement itself is then performed in the absorbance scale or the values are transformed from the transmittance to the absorbance scale. The method for concentration calculation based on Eq. (5.20) is called the compensation method. It requires only knowledge of the values of D, the molar absorption coefficient, ε, and the cell thickness, d. The last value is determined interferometrically, after temperature equilibration in the cell room; the time required for attaining temperature equilibrium is different for instruments of different types. The wavenumbers of the interferometric maxima or minima are corrected for the systematic error of the spectrometer wavenumber scale and the thickness is calculated by the least squares method using the relationship,

$$d \,.\, \tilde{\nu}_i + k = i/2, \tag{5.21}$$

where $\tilde{\nu}_i$ are the wavenumbers of the interference band maxima, i is the maximum serial number and k is the calibration curve intercept, or from

$$d \,.\, \tilde{\nu}_i + k = i/4, \tag{5.22}$$

where $\tilde{\nu}_i$ are the wavenumbers of both maxima and minima.

If only a single component is to be determined in a mixture, the so-called base-line method is employed. Here it is assumed that the analytical band is superimposed on a constant background spectrum and its maximum intensity is determined with respect to a newly defined scale, in which 0 %τ is the same as in the previous case, but 100 % is located at the inter-

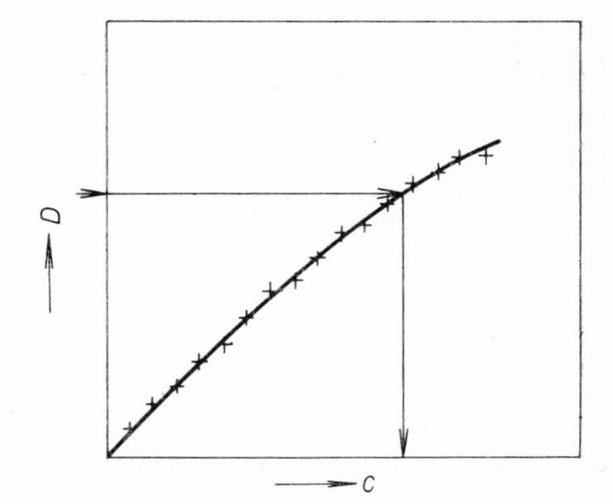

Fig. 66. An example of a calibration graph for quantitative determination of one component in a mixture. Concentration c is determined from experimental absorbance D by the procedure indicated by the arrows.

cept of the straight line constructed perpendicularly from the analytical band maximum to the base-line represented by a straight line connecting the two minima adjacent to the analytical band (Fig. 24). It is evident e.g. from the approximation of the base-line by a straight line, although it is actually a curve, that the method is subject to a certain error. However, in practice a calibration plot is employed, constructed from the absorbance dependence on the concentration of the test compound obtained with "synthetic" mixtures. The analysis itself is not performed by calculation, but the concentration is read from graphs for experimental absorbance values (Fig. 66). The method is very rapid and can even be employed when the absorbance concentration relationship is non-linear.

In practice, the so-called compensation method is also introduced in the quantitative analysis. Here the spectrum of the test solution containing two components (this is the most suitable situation) is compensated by the spectrum of one component of the mixture placed in the reference beam in a variable-thickness cell. A cell thickness is then sought for which the spectrum of the compound compensated is just eliminated (this procedure is chiefly used with solutions and the solvent spectrum is subtracted). The concentration is thus determined as a function of the cell thickness; it need not be added that the determination precision depends on the precision of the description of the variable-thickness cell parameters. One component can also be compensated for by varying its concentration in the reference cell (at a constant thickness). The component concentration in the reference cell is selected experimentally and hence the procedure is substantially less advantageous than the previous one.

5.2.4.2 Specific Problems of Analyses of Samples in Various States of Aggregation

So far, rules common for virtually all quantitative analyses without respect to the sample state of aggregation have been discussed. Now the specific features should be examined.

The analysis of gaseous samples has specific features, e.g. the use of gas cells with relatively long optical paths (usually 10 cm) with consequent low dependence of the error on this quantity.

Pressures are usually employed instead of concentrations in working with gases. The Lambert–Beer law is then modified to give

$$D(\tilde{\nu}) = \varepsilon'(\tilde{\nu}) \, . \, p \, . \, d, \qquad (5.23)$$

where p is the pressure*) at which absorbance D was determined. The dimension of the main unit of the molar absorption coefficient, ε', is $s^2 \, . \, kg^{-1}$, i.e. $m^{-1} \, . \, Pa^{-1}$. Outdated units can also be found in the literature, e.g. $cm \, . \, kp^{-1}$, $cm^{-1} . \, Torr^{-1}$ or $cm^{-1} \, . \, at^{-1}$.

---

*) For conversion of Anglo-American pressure units it holds that 1 psi = 6894.76 Pa or 1 in Hg = 3386.38 Pa.

The quantitative analysis of gaseous mixtures is subject to a relatively large error when the infrared spectrum of the gaseous mixture has a resolved rotational-vibrational structure. As has been emphasized in Chapter 3, the measurement of bands with a very small half-width (rotational-vibrational bands are very narrow, with a width of the order $10^{-2} - 10^{-3}$ cm^{-1}) on medium dispersion instruments is subject to a large error. Their shape and intensity are very distorted; the use of the integrated intensity.is somewhat more advantageous. The use of various instruments, accompanied by changes in the spectral slit width, or choice of different measuring conditions when using one instrument render comparison of the results quite useless. It is better when the rotational-vibrational bands no longer exhibit fine structure; then the precision of the method approaches that for liquids.

A specific feature of infrared spectroscopy is its use for determining absolute amounts of gases formed in reactions, e.g. carbon dioxide and monoxide, nitrogen oxides, hydrogen cyanide, etc. Infrared spectroscopy is often employed in combination with the gas chromatography-mass spectrometry tandem. The gases produced in the reaction are determined by infrared spectroscopy and liquid and solid products by gas chromatography combined with mass spectrometry.

Infrared spectroscopy can also be successfully employed for the determination of trace amounts of gaseous atmospheric pollutants. The spectra of air samples are usually measured in cells with very long optical paths (up to 50 or more metres). The detectable amounts of gases are of the order of 10^{-4} volume percent (1 ppm); principally, the higher the absorption coefficient of the analytical band of the gas, the lower the detectable amount. For example, nitrogen dioxide is determined using the antisymmetrical valence vibration band at 1626 cm^{-1} ($\varepsilon = 309$ l . mole^{-1} . cm^{-1}) down to 0.014 ppm, carbon monoxide using the band at 2160 cm^{-1} ($\varepsilon =$ $= 5.25$ l . mole^{-1} . cm^{-1}) down to 0.8 ppm, sulphur dioxide using the antisymmetrical valence vibration band at 1370 cm^{-1} ($\varepsilon = 152$ l . mole^{-1} . . cm^{-1}) down to 0.03 ppm, methane using the triply degenerate deformation vibration at 1300 cm^{-1} ($\varepsilon = 13.4$ l . mole^{-1} . cm^{-1}) down to 0.034 ppm etc.

Liquid mixtures are analyzed rather frequently and solutions even more often. Non-polar compound mixtures are mostly analyzed in the liquid phase. In analyses of liquids, the principles expounded in Chapter 5.2.3.1 are observed and the concentration of the components is given either absolutely, e.g. in g . l^{-1} or mole . l^{-1}, or relatively in volume or weight per cent (or exceptionally in molar or volume fractions). The dimensions of the molar absorption coefficient units depend on the way the concentration is expressed. The main molar absorption coefficient unit is $m^2 . mole^{-1}$; if concentrations and cell thickness are substituted into Eq. (5.1) in mole . l^{-1} and cm, respectively, then ε is expressed in practical units l . $mole^{-1}$. cm^{-1}.

Work with liquid samples is advantageous chiefly because of their easy manipulation and the greatest disadvantage is the use of very thin cells for spectrum measurements, whose thickness is subject to the largest errors due to imperfect thermostatting. This fact is not very important with hydrocarbons, e.g. n-alkanes or branched saturated aliphatic hydrocarbons, as the bands in the infrared spectra of these compounds mostly have low absorption coefficients; with other compounds, e.g. unsaturated hydrocarbons, the absorptivity of many bands increases.

In work with solutions the experimental conditions can be adjusted relatively easily, so that the measurement is performed under the optimum conditions for quantitative infrared analysis. The proper choice of the concentrations and the cell thickness virtually always enables work in a region most suitable from the point of view of the photometric precision. Cells about 0.01 cm thick are most frequently employed; the concentration is then adjusted to this thickness. Even mixtures of polar compounds can

Table XL

Dependence of the molar absorption coefficient value, ε, for the phenol ν (O—H) band on the solvent nature[a]

Solvent	ε[b]	Solvent	ε[b]
n-Hexane	190	Tetrachloromethane	205
Isooctane	190	Tetrachloroethylene	220
Cyclohexane	205	Carbon sulphide	140

[a] Obtained from the infrared spectra of very dilute solutions in order to prevent association of phenol itself; [b] dimension l . $mole^{-1}$ . . cm^{-1}.

be analyzed in solution, as the mutual interaction of their molecules can be more or less suppressed by dissolution in a non-polar solvent. In this case the solvent absorption in the test region must be determined before calibration and before each measurement by recording the solvent spectrum without compensation. The spectral background must further be known after compensation for the solvent; for this purpose, the spectrum is obtained over the test region with cells containing the pure solvent placed both in the measuring and in the reference beam. The recording obtained is the accessible approximation to the background of the mixture experimental spectrum. In the analyses themselves the measuring cell contains the solution and the reference cell the pure solvent for compensation. The use of variable-thickness reference cells has given good results. Even with small sample weights the relative loss in the solvent in the measuring cell is not negligible and the error thus incurred can be readily and accurately corrected for only by compensation with a variable-thickness cell. This cell type should actually also be employed for the spectral background measurement.

In measurements on solutions the values of the molar absorption coefficients, ε, determined in solutions in the same solvents must be employed. As shown in Chapter 5.2.2, the coefficient values differ considerably in dependence on the medium in which they were determined (Table XL).

As can be expected, the greatest difficulties are encountered in attempts to quantitatively analyze mixtures of solid compounds. The only exception are analyses of films of polymeric materials. The components (especially admixtures, plasticizers, antioxidants, etc.) are determined on direct passage. The only difficulties are connected with the determination of the thickness of these films and sometimes with interference effects. However, interference can usually be eliminated by pressing the film between two transparent plates (e.g. NaCl).

Problems with analyses of samples in nujol mulls and KBr pellets are connected with the fact that dispersed materials are involved. The reproducibility of the sample preparation is poor and the size of the dispersed particles causes errors owing to scattering and reflection phenomena. Therefore, internal standards are most frequently employed here, i.e. admixtures whose analytical bands do not interfere with those of the solid mixture components (cyanides, thiocyanates). The intensity of these bands in relation to their analytical concentration enables calculation of correction factors for the absorbances in the analytical bands of the test components.

The sample concentration is expressed somewhat unusually for KBr pellets, because the pellet thickness cannot generally be determined accurately. If the internal standard method is not employed, the overall amount of sample in the pellet (expressed as the weight in grams or the material amount in moles) is related to the pellet surface area, S, which is unambiguously given by the matrix shape. It is simultaneously assumed that the whole amount of KBr weighed with the sample is used for the pellet preparation. Eq. (5.1) is then converted into the form,

$$D(\tilde{\nu}) = \varepsilon_{KBr} \, . \, n_X/S = \varepsilon'_{KBr} \, . \, m_X/S, \tag{5.24}$$

where n_X is the amount of substance and m_X the mass of substance X in the pellet and S is the pellet surface area. The main unit of ε_{KBr} thus has the dimension $m^2 \, . \, mole^{-1}$, or, for ε'_{KBr}, $m^2 \, . \, kg^{-1}$.

The so-called mosaic effect, caused by particle aggregation, often appears in nujol mulls. Therefore the mulls are sometimes prepared using semisolid vaseline in place of liquid paraffin oil. Fresh mulls are usually well dispersed and aggregation occurs on ageing.

Further errors in the quantitative analyses of solid compound samples stem from the simultaneous effects of absorption and reflection in poorly prepared samples. The Christiansen effect results, manifested by distortion of the absorption bands. Samples whose bands exhibit the Christiansen effect must be analyzed carefully; the error incurred is impossible to eliminate simply.

5.2.4.3 Errors of Quantitative Analyses

Gross errors must, of course, be avoided in quantitative analyses. One of them is the occurrence of a component in the mixture, which was not originally considered. If this component does not absorb too strongly in the region of the analytical bands of the mixture components, its presence leads to errors proportional to its content in the mixture. However, it is critical if it absorbs or forms complexes with the mixture components.

Random (statistical) errors are introduced into analyses chiefly by imprecise determination of the cell thickness and the concentration.

The interferometric procedure of the cell thickness measurement (see Chapter 5.2.4.1) yields a certain "average" cell thickness which does not account for irregularities on the cell window surface. However, the factor given by non-rigid placement of the cell with respect to the direction of the beam passage through the cell is much more significant. It plays a role especially when the cells supplied with one instrument are used in another instrument and the differences in the dimensions are compensated for by using adaptors.

Errors in the determination of analytical concentration c_X can be classified in two large groups. The primary source of error is poor precision of solution preparation, i.e. weighing of the test substance and measurement of the solvent. As the measurement of volume is relatively less precise, it is recommended that the solvent also be weighed in analyses. If the amount of solvent is to be expressed in terms of volume (e.g. when the concentration is to be expressed in units of mole . l^{-1}), the solvent mass is recalculated to volume employing density tables (at given temperature).

The other source of error in concentration c_X stems from the fact that the measurement on the spectrometer is performed at a temperature different from that of solution preparation. In precise measurements the temperature of the solution in the cell must thus be determined, which is often difficult. The overall error in the determination of the concentration can then be expressed by the variance estimate, using the relationships described in Chapter 3.2.2.1.2.

Thus if the overall variance of the given cell thickness is $\sigma^2(d)$, the variance of the sample mass $\sigma^2(m)$, of the solvent mass $\sigma^2(m_s)$ and of the temperature measurement in the cell $\sigma^2(t)$, then the estimate of the overall variance of the molar absorption coefficient is given by

$$\sigma^2(\varepsilon) = \left[\sigma^2(D)m_s^2 + \sigma^2(m_s)\, D^2(\tilde{\nu}) + (\sigma^2(m)/m^2 + \sigma^2(d)/d^2 + \right.$$
$$\left. + \sigma^2(t)\left(\frac{d\varrho}{dt}\right)^2/\varrho^2)\,(D(\tilde{\nu})\, m_s)^2\right][M_r/(m\varrho d)]^2 \qquad (5.25)$$

where $D(\tilde{\nu})$ is the measured absorbance, M_r the relative molecular mass of the test compound, ϱ the solvent density at the given temperature and $d\varrho/dt$ the density temperature gradient.

The error introduced into analyses by stray radiation should finally be mentioned. On adjusting the spectrometer to a certain wavenumber, $\tilde{\nu}$, not only radiant power $\Phi_{mon}(\tilde{\nu})$ which passed through the monochromator as expected (and whose distribution is given by the instrument function), but also scattered radiation with overall radiant power Φ_{stray} with completely different wavenumbers is incident on the detector. The contribution of the stray radiation to the overall detector signal level is then expressed by the ratio, $\Phi_{stray}/(\Phi_{mon}(\tilde{\nu}) + \Phi_{stray})$. The effect of stray radiation can be included in function $Z(\tilde{\nu})$ or $K(\tilde{\nu})$ (see Chapter 5.2.3); it thus has systematic character.

Stray radiation primarily contains short-wave components, as the maximum of the blackbody spectral radiation (approximating the infrared source radiation) lies in the near infrared region. Thus the magnitude of the stray radiation in the region of wavenumber $\tilde{\nu}$ can be, to a first approximation, estimated using an experiment in which a plate made of a material with total absorption in this region but transparent for shorter-wave radiation is placed in the measuring beam. If the spectrometer does not indicate 0 % transmittance, short-wave stray radiation has passed through the filter; the experimental spectra must then be corrected for the measured transmittance value of the stray radiation. Optical material plates are suitable for placement in the beam: quartz (1 mm thick) below 2000 cm^{-1}, LiF (10 mm) below 1250 cm^{-1} and CaF_2 (3 mm) below 850 cm^{-1}. The effect of stray radiation is not as marked in grating spectrometers as in older prism instruments; there is practically no stray radiation in spectrometers with double pass monochromators.

5.2.5 Analysis of Equilibrium Systems

Quantitative analyses of mixtures of stable compounds, as described in Chapter 5.2.4, can principally also be performed on the basis of separational procedures. For example, a multicomponent mixture can be analyzed gas chromatographically; the areas under the chromatographic peaks of the separated components give the component amounts. It need not be emphasized that also other spectroscopic or non-spectroscopic methods are suited for quantitative analyses of mixtures of stable compounds and this severe competition limits the application of infrared quantitative analysis to really advantageous cases.

However, analyses of equilibrium mixtures form a field in which infrared spectroscopy does not suffer from such strong competition from other methods and in some cases is even unique. Various components, forms, species, etc. exist in equilibrium mixtures and their contents are primarily determined by the law of mass action; the components of equilibrium mixtures are in dynamic equilibrium, often have only short life-times and are converted into other species. Equilibrium mixture components cannot be separated; sometimes new equilibrium occurs after separation (e.g. with rotational isomers, provided that the isolated forms are not "conserved" by unusual external conditions, e.g. a very low temperature), in other cases the separation of the equilibrium mixture components destroys the possibility of following the conditions in the mixture, as quite different systems are formed.

Infrared spectroscopy is one of the very few methods enabling insight into the conditions in equilibrium mixtures. Methods whose reaction times are too long permit only monitoring of the average statistical behaviour of molecules in equilibrium systems; however, many, even short-lived, species can be monitored in the stationary state using infrared spectroscopy.

Infrared spectra permit qualitative identification of the individual components of equilibrium mixtures by means of their analytical bands. Quantitative analyses are subject to certain difficulties. The individual components of equilibrium mixtures cannot be isolated in the pure form and thus there are no direct methods of determining the molar absorption coefficients of the component analytical bands. As will be shown in the following sections, procedures are available for circumventing this obstacle.

5.2.5.1 Description of Equilibrium Systems by Thermodynamical Quantities

As equilibria are studied by infrared spectroscopy methods mostly under a constant pressure, three quantities are chiefly met in the description of equilibrium systems: enthalpy H, Gibbs free energy G and entropy S. The internal energy of the system, U, and free energy F appear less frequently. For better orientation in the following treatment, some basic definitions and relationships will be given here.

$$H = U + p \,.\, V, \tag{5.26}$$

where p is the pressure and V the volume of the system,

$$F = U - T \,.\, S, \tag{5.27}$$

where T is the absolute temperature of the system,

$$G = U + p \,.\, V - T \,.\, S = H - T \,.\, S = F + p \,.\, V. \tag{5.28}$$

The Gibbs free energy (the Gibbs function), G, of the given thermodynamic system equals the sum of the chemical potentials (molar free energies), μ_i, for the individual components, multiplied by the number of moles (material amount) of these components, thus

$$G = \sum_{i=1}^{n} n_i \,.\, \mu_i \,. \tag{5.29}$$

Chemical potential μ_i depends on the activity of the i-th component of the system according to the equation

$$\mu_i = \mu_i^0 + RT \,.\, \ln\,(a_i/a_i^0) = \mu_i^0 + RT \,.\, \ln \hat{a}_i, \tag{5.30}$$

where R is the universal gas constant (R = 8.314 33 J . K^{-1} $mole^{-1}$ = 1.9872 cal . K^{-1} $mole^{-1}$), μ_i^0 is the chemical potential of the i-th compound at standard activity a_i^0 and $\hat{a}_i$ is the dimensionless ratio of the given activity to the standard activity. It is suitable to select standard activity a_i^0

equal unity, especially in solutions of solids; then $\hat{a}_i$ numerically equals the substance activity in the solution, but it is actually a dimensionless number.

Let it be assumed that equilibrium among substances X_i is established in solution,

$$q_1X_1 + q_2X_2 + \dots q_nX_n \rightleftharpoons q_{n+1}X_{n+1} + \dots q_mX_m.$$

As the standard change of the Gibbs free energy for the given reaction, the difference,

$$\Delta G = \sum_{\text{products}} q_i \cdot \mu_i^0 - \sum_{\substack{\text{initial}\\ \text{substances}}} q_j \cdot \mu_j^0, \tag{5.31}$$

is defined, q_i and q_j being the stoichimetric coefficients of the reactants.*) It holds for the equilibrium state that

$$\sum_{\text{products}} q_i \cdot \mu_i = \sum_{\substack{\text{initial}\\ \text{substances}}} q_j \cdot \mu_j, \tag{5.32}$$

which corresponds to the condition that the change of the chemical potential is zero in the vicinity of equilibrium state. Substitution of Eq. (5.30) into Eq. (5.32) and rearrangement yields

$$\Delta G = \sum_{\substack{\text{initial}\\ \text{substances}}} q_j \,.\, RT \,.\, \ln \hat{a}_j - \sum_{\text{products}} q_i \,.\, RT \,.\, \ln \hat{a}_i \tag{5.33}$$

or

$$\Delta G = RT \,.\, \ln \Big[\Big(\prod_{\substack{\text{initial}\\ \text{substances}}} \hat{a}_j^{q_j}\Big)/\Big(\prod_{\text{products}} \hat{a}_i^{q_i}\Big)\Big] = -RT \,.\, \ln K, \tag{5.34}$$

quantity K being termed the equilibrium constant of the system. As values $\hat{a}_i$ and $\hat{a}_j$ are dimensionless, so is equilibrium constant K.

Values of ΔG are less frequently calculated for standard conditions defined as the state of the pure compound; then relative activities $\hat{a}_i$ in Eq. (5.34) must be replaced by the appropriate mole fractions, x_i (or the products of the mole fractions with the activity coefficients, $f_i x_i$). Hence it must always be specified to which standard state the Gibbs free energy change is related in the given reaction.

The enthalpy change, ΔH, for the given reaction is obtained from Eq. (5.28) on substituting the changes of the quantities of state, G and S, i.e.

$$\Delta G = \Delta H - T \,.\, \Delta S. \tag{5.35}$$

Assuming that ΔH is temperature-independent, it can be determined in

---

*) These are selected as the smallest integral numbers satisfying the given reaction scheme.

practice from the dependence of the Gibbs free energy, ΔG, on the temperature, mostly using the least squares method. If ΔG is known for two temperatures T_1 and T_2, the enthalpy change is directly calculated from the relationship,

$$\Delta H = [\Delta G(T_1)/T_1 - \Delta G(T_2)/T_2]/(T_1^{-1} - T_2^{-1}) = \\ = R \ln [K(T_2)/K(T_1)]/(T_1^{-1} - T_2^{-1}). \qquad (5.36)$$

5.2.5.2 Determination of the Rotational Isomer Contents

Compounds with hindered internal rotation around single bonds often form stable rotational isomers; samples of these compounds are actually equilibrium mixtures in which all the stable rotational isomers are present simultaneously. Their contents are primarily determined by the differences in their potential energies and by the temperature; they basically obey the statistical distribution laws discussed in Chapter 2.

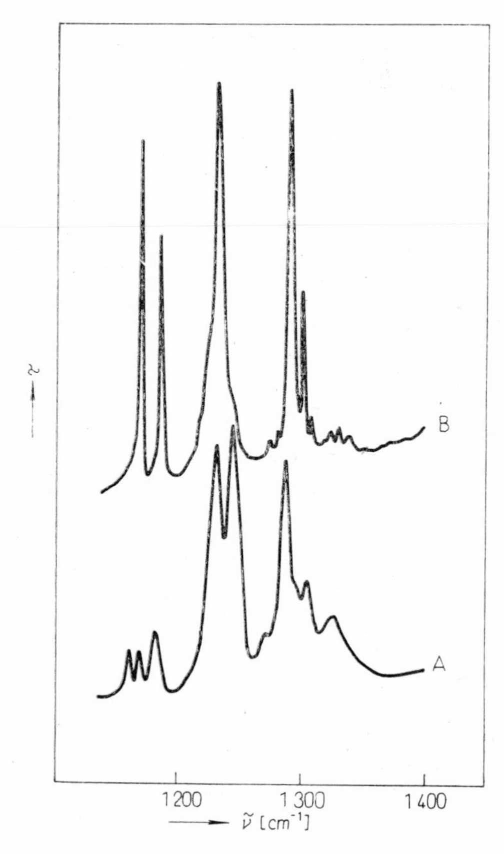

Fig. 67. The infrared absorption spectrum of *m*-iodoanisol: A — liquid sample measured at the temperature of the spectrometer cell space, B — crystalline sample cooled to the temperature of liquid nitrogen.

If a single hindered rotation centre is present in the molecule, not more than two or three rotational isomers are formed. The molecules of rotational isomers differ in their geometry and consequently also in their polarity and spectra. The spectrum of a mixture of rotational isomers consists of contributions from the spectra of the individual forms, in the ratio of their contents. If the analytical bands of all isomers present in the mixture are found in the spectrum, then the isomers can be identified on the basis of these bands and their relative contents determined from the band intensities The number of rotational isomers increases rapidly with an increasing number of hindered rotation centres in molecules; as has already been mentioned, infrared spectra are unsuitable for the study of such complicated mixtures.

The first step taken in the analysis of an equilibrium mixture of rotational isomers is the detection of the bands of the forms present. For this purpose, the compound sample is cooled to the temperature of liquid nitrogen, at which it usually crystallizes; only a single rotational isomer, generally the thermodynamically most stable, remains in the supercooled crystalline samples and the infrared spectrum becomes simpler (Fig. 67). It has been shown in some recent works that very high pressures of the order of 10^9 Pa (tens of kbars) can lead to enrichment of the rotational isomer mixture by thermodynamically less stable forms.

The low-temperature and high-pressure methods thus supplement one another and yield information on the appearance of the spectra of the individual forms present in samples of compounds. They also help in selecting analytical bands for the individual forms.

The qualitative detection of the existence of rotational isomers and the assignment of their spectral bands must further be supplemented by data on the polarity of the forms; for this reason the spectra of a compound forming several rotational isomers are measured in solutions in liquids with various relative permittivities.

In polar media (in solutions of liquids with large permittivities), rotational isomers with larger permanent electric dipole moments are populated to an increased degree and, on the other hand, less polar forms are more populated in non-polar solvents. A decrease or an increase in the contents of the individual forms is manifested in a decrease or an increase in the appropriate band intensities.

Quantitative analyses are usually based on the assumption that the molar absorption coefficients of the analytical bands of the rotational isomers present are identical. This is, of course, an erroneous assumption and can be justified only for rough estimation of the population of the individual forms or for monitoring changes in the populations. In more precise analyses, the molar absorption coefficients determined on model compounds are employed; e.g. in monitoring the contents of the axial and equatorial forms of cyclohexanol in cyclohexanol samples, the coefficients obtained from the spectra of *cis*- and *trans*-4-tert. butylcyclohexanol are employed. In these compounds the tert. butyl group is always in the equatorial position and therefore the hydroxyl is either axial (*cis*-form), or equatorial (*trans*-form).

Thus the molar absorption coefficient values, $\varepsilon_0(\tilde{\nu}_j)$, are obtained for the individual "pure" rotational isomers; the mole fraction, $x_{i,j}$, of the *j*-th rotational isomer in an equilibrium mixture of the *i*-th substance is then calculated from the relationship

$$x_{i,j} = \varepsilon_i(\tilde{\nu}_j)/\varepsilon_0(\tilde{\nu}_j), \tag{5.37}$$

where $\varepsilon_i(\tilde{\nu}_j)$ is the molar absorption coefficient related to the overall analytical concentration of the *i*-th substance.

Such model substances, i.e. those containing only a single rotational isomer, are not always available. In sufficiently large series of analogous substances, the population of the individual rotational isomers can be calculated by the least squares method, again assuming that the molar absorption coefficients for the analytical bands characterizing the individual isomers are transferable. This assumption is based on the fact that the sum of the mole fractions $x_{i,j}$ of all the rotational isomers (subscript *j*) for each substance (subscript *i*) must equal unity, i.e.,

$$\sum_{j=1}^{m} x_{i,j} = \sum_{j=1}^{m} \varepsilon_i(\tilde{\nu}_j)/\varepsilon_0(\tilde{\nu}_j) = 1; \qquad i \in \langle 1, n \rangle; \; n \geqq m. \tag{5.38}$$

The $\varepsilon_0(\tilde{\nu}_j)$ values then represent the molar absorption coefficients of the analytical bands of the individual rotational isomers in a hypothetical standard*) for the given rotational isomer. The reciprocal values of $\varepsilon_0(\tilde{\nu}_j)$ are obtained from Eqs. (5.38) using the least squares method described in Section 3.2.2.2.

The molar integrated intensities can also be employed for these purposes in place of the molar absorption coefficients.

To a limited extent, this procedure can also be applied to the study of changes in the composition of equilibrium mixtures of conformers (rotamers)

---

*) I.e. in a non-existant substance which would contain only the *j*-th isomer from all the possible rotational isomers.

of a single substance due to temperature changes, provided that substantial changes in the contents of the individual forms occur in the temperature range studied. Here it is assumed that the absorption coefficients, $\varepsilon_0(\tilde{\nu}_j)$, (or the molar integrated intensities) do not change substantially with changes in the temperature. Subscript i in Eq. (5.38) then corresponds to various temperatures, not to various compounds. However, the $\varepsilon_i(\tilde{\nu}_j)$ value also does not change very much with changing temperature, so that the solution of Eqs. (5.38) is subject to a large error.

If the enthalpy difference, ΔH, is to be determined from the temperature changes of the intensities of the rotational isomer analytical bands, then knowledge of the absolute ratio of the mole fractions of the individual rotational isomers or of their equilibrium concentrations (activities) is not required. It suffices to assume that the molar absorption coefficients or the more frequently employed integrated molar intensities are transferable from one temperature to another.

Equilibrium constant $K(T_1)$ at temperature T_1 for a pair of rotational isomers 1 and 2 is defined by the ratio,

$$K(T_1) = [A_1(T_1)/A_1^0(T_1)]/[A_2(T_1)/A_2^0(T_1)], \tag{5.39}$$

where A_1^0 are the integrated molar intensities of the standards. Consequently, the Gibbs free energy difference, $\Delta G(T_1)$, at temperature T_1 equals

$$\begin{aligned}\Delta G(T_1) &= -RT_1 \ln K(T_1) = \\ &= -RT_1 \ln [A_1(T_1)/A_2(T_1)] + RT_1 \ln [A_1^0(T_1)/A_2^0(T_1)].\end{aligned} \tag{5.40}$$

The $\Delta G(T_2)$ value is determined analogously at temperature T_2. Substitution into Eq. (5.36) for calculation of the enthalpy difference yields

$$\begin{aligned}\Delta H = R\,.\,\{&\ln [A_1(T_1)/A_2(T_1)] - \ln [A_1(T_2)/A_2(T_2)] - \\ &- \ln [A_1^0(T_1)/A_2^0(T_1)] + \ln [A_1^0(T_2)/A_2^0(T_2)]\}/(T_1^{-1} - T_2^{-1}).\end{aligned} \tag{5.41}$$

Provided that the assumption of constancy of the standard intensities, A_0, with varying temperature is satisfied,*) then the third and fourth terms in the braces in Eq. (5.41) compensate one another and the expression simplifies to give

$$\Delta H = R\,.\ln \{[A_1(T_1)\,.\,A_2(T_2)]/[A_1(T_2)\,.\,A_2(T_1)]\}/(T_1^{-1} - T_2^{-1}) \tag{5.42}$$

from which it is clear that the integrated molar intensities for the individual rotational isomer standards need not be known.

---

*) It even suffices to assume that the standard integrated intensity ratio, A_1^0/A_2^0, does not change with changing temperature.

5.2.5.3 Study of Equilibria in Mixtures

Infrared spectroscopy has found extensive use as an analytical method for monitoring the equilibrium composition of mixtures in which complexes are formed. The underlying principle is the fact that most complexes exhibit individual bands in infrared spectra, by means of which the complexes can be identified, and that the band intensities in equilibrium mixtures are proportional to the complex concentrations over a wide interval.

The equilibrium reaction giving rise to a complex can be formally written as

$$A-H + B \rightleftharpoons C;$$

A – H is the donor molecule, B is the acceptor molecule and C is the complex, whose structure is actually A – H ... B (this is an example of a hydrogen-bonded complex). Complex formation is a relatively rapid equilibrium reaction and the equilibrium mixture composition depends on the temperature. Equilibrium constant K_t can be expressed in terms of the law of mass action,

$$K_t = \hat{a}_C/(\hat{a}_{A-H} \cdot \hat{a}_B), \tag{5.43}$$

where $\hat{a}_B$, $\hat{a}_{A-H}$ and $\hat{a}_C$ are the relative real activities.*)

Employing the common procedure, the equilibrium constant is calculated from the analytical concentrations of components A – H and B and from the actual concentration of free component A – H in the equilibrium mixture, using the formula,

$$K_t = (1 - \alpha)/\alpha[\hat{c}_B^0 - \hat{c}_{A-H}^0 \cdot (1 - \alpha)], \tag{5.44}$$

where $\hat{c}_{A-H}^0$ and $\hat{c}_B^0$ are numerical (i.e. dimensionless) values of the analytical concentrations of components A – H and B in the mixture,**) respectively and quantity α is the degree of dissociation, given by,

$$\alpha = D/D_a. \tag{5.45}$$

Expression $(1 - \alpha)$ in Eq. (5.44) is proportional to the complex concentration. Quantity D in Eq. (5.45) is the experimental absorbance in the analytical band of free component A – H and D_a is the absorbance in the same band on condition that the concentration of component A – H equals the analytical concentration.

Quantities D_a and ε_a (according to the relationship, $D_a = \varepsilon_a \cdot \hat{c}_{A-H}^0 \cdot d$) are temperature – dependent and must therefore be determined over the

---

*) I.e. the activities divided by a standard (here unit) activity (see Chapter 5.2.5.1).

**) I.e. the concentrations calculated from the weights of the components added.

temperature interval in which the temperature–dependence of equilibrium constant K_t is monitored. Table XLI indicates the changes in the molar absorption coefficient of the free phenol band in a tetrachloromethane solution with changing temperature. For calculation of the thermodynamic quantities, the equilibrium constants are determined at a larger number of temperatures (at least 10 values over a range of at least 50 °C).

Table XLI

Temperature dependence of the molar absorption coefficient value[a], ε, for the phenol ν (O—H) band in a tetrachloromethane solution

Temperature	ε^b	Temperature	ε^b
5	225	35	200
15	218	42	193
25	209		

[a] Correction has been made for volume changes due to solvent expansion; [b] dimension $l \cdot mole^{-1} \cdot cm^{-1}$.

Thermodynamic functions ΔG and ΔH are determined from the equilibrium constants using Eqs. (5.34), (5.35) and (5.36).

The values of K_t, ΔG or ΔH are frequently employed for evaluation of the interaction energy of the molecules bound in the complex, the ΔS value reflecting the effect of steric factors during this process. Some experimental values of the thermodynamic quantities are given in Table XLII. The method is suitable for the study of equilibrium mixtures, whose components yield pronounced analytical bands in the infrared spectrum; it is relatively precise, but very tedious and requires precise experimental technique.

5.2.6 Determination of the Number of Groups in a Molecule

Infrared spectra have sometimes been used to determine the number of groups occurring repeatedly in the test compound molecule. Quantitative analyses of this type employ the characteristic vibration bands for these groups; these groups must then behave as independent oscillators. The

Table XLII

The values of the thermodynamic quantities for the complex formation with the organic bases[a]

Base	—Δ*F*		—Δ*H*		—Δ*S*		$\Delta\nu$(O—H)[b]
	(kJ . mol^{-1})	(kcal . mol^{-1})	(kJ . mol^{-1})	(kcal . mol^{-1})	(J.mol^{-1}.K^{-1})	(cal . mol^{-1} . K^{-1})	(cm^{-1})
Benzene	—3.18	—0.76	6.49	1.56	22.7	7.8	47
n-Butyraldehyde	3.94	0.94	19.55	4.67	52.3	12.5	180
Di-n-butyl-sulphide	0.96	0.23	14.23	3.40	44.4	10.6	256
Diethyl ether	5.40	1.29	22.65	5.41	57.4	13.8	277
Pyridine	9.63	2.30	24.07	5.85	49.8	11.9	—[c]
Di-n-butyl-sulphoxide	3.36	3.19	26.12	6.24	42.7	10.2	391
Quinuclidine	15.0	3.6	36.0	8.6	70.8	16.9	—[c]

[a] Determined using infrared quantitative analysis from the spectra in dilute CCl_4 solutions;
[b] the difference in the wavenumbers of the ν(O—H) bands of phenol in pure CCl_4 and of the complex;
[c] imprecise data.

molecules of classical carbon compounds can be analyzed, but these analyses are more frequently performed on compounds of other elements, e.g. silicon, where the assumption of independence of the vibrating groups is better satisfied.

n-Alkanes are typical examples of carbonaceous compounds with which some vibrations exhibit independence of the individual groups. The number of CH_2 groups increases in the homologous series of n-alkanes. The bands of stretching vibrations of the C–H bonds in these groups increase in proportion to the increase in the number of CH_2 groups; the band intensity for an antisymmetrical stretching vibration is given by

$$\varepsilon_a = 77n - 18 \tag{5.46}$$

and for a symmetrical stretching vibration by

$$\varepsilon_s = 46n - 64, \tag{5.47}$$

where n is the number of CH_2 groups (this holds for samples in CCl_4 or CS_2). Therefore, the value of n can be calculated, provided that ε_a or ε_s are determined experimentally.

The procedure is thus based on the knowledge of molar absorption coefficient ε for the analytical band of the test group; this value is determined from the spectra of model compounds with an accurately known number of test groups. The total absorption coefficient, ε_X, in Eq. (5.1) is replaced by the product of the known number of groups, n_X, in compound X with unit molar absorption coefficient ε_{gr}, which then equals

$$\varepsilon_{gr} = D_X/(n_X \cdot c_X \cdot d), \tag{5.48}$$

where d is the cell thickness, c_X is the analytical concentration of compound X and D_X is the analytical band absorbance. The same formula, rewritten as

$$n_Y = D_Y/(\varepsilon_{gr} \cdot c_Y \cdot d) \tag{5.49}$$

is then employed for the determination of the unknown number of groups, n_Y, in compound Y. The n_Y value is naturally rounded off to the closest integer.

The assumption that the vibrations are randomly degenerate and their molar absorption coefficients identical is never met sufficiently accurately and hence the n_X or n_Y values are always only approximate.

Stretching vibrations of C–H bonds of some types of hydrocarbons and the valence vibration of the O–H bonds of some polyhydroxy compounds (whose spectra must be measured in pyridine solutions*) are very

*) To limit the self association of their molecules.

well suited for these purposes; however, this is not a general rule. For example, with compounds containing heavier S or Si atoms, vibration bands from the fingerprint region can also be employed (e.g. the symmetrical deformation vibration band of the CH_3 group around 1250 cm^{-1} can be used for $SiCH_3$ groups).

A special application of this experimental technique is the determination of the position of the three-membered cyclopropane ring in the molecules of steroid or triterpenic compounds. It is then determined whether the number of CH_3 or CH_2 groups increased on opening the ring using hydrogen chloride, obtaining this information from the intensities of the appropriate bands in the infrared spectrum. The result can be verified by measuring the spectrum after addition of DCl to the initial compound.

5.2.7 Monitoring of Reaction Courses

Infrared spectroscopy has so far found use in "static" analyses rather than in the study of reaction kinetics. However, it has been used for the latter purpose in exceptional cases.

Discontinuous measurements can be carried out with sufficiently slow reactions. Samples are taken from the reaction mixture at certain time intervals and are (usually after certain modifications) normally analyzed quantitatively. In principle, these reactions could also be monitored continuously, provided that they could proceed directly in the cell; the analytical band wavenumber for the test component would be adjusted on the infrared spectrometer and the data required obtained from the intensity-time dependence. However, reaction mixtures are mostly too complicated for direct spectroscopic monitoring.

Much better possibilities are provided by gaseous state reactions or reactions in which one of the reactants or products is gaseous or has a high vapour pressure; the content of this component in the gaseous phase can then be monitored according to the rules given in Chapter 5.2.4.2.

Fast-response spectrometers have been manufactored recently; these instruments depict the spectrum on an oscilloscope screen, in contrast to the classical instruments. They employ a somewhat limited wavenumber range but a spectrum can be obtained within microseconds. Such instruments are suitable for the study of the kinetics of reactions with at least one gaseous component which absorbs infrared radiation.

5.2.8 Determination of Compound Acidity or Basicity

Infrared spectroscopy can also be used for determination of the acidity and basicity of compounds; mostly the Lewis acidity or basicity and sometimes also the Brönstedt acidity or basicity of very weak acids and bases are determined in non-aqueous media.

The determination of the acidity and basicity of large groups of substances is very advantageous and rather common. The data are relative, always related to a certain standard; e.g. the acidity of compounds is related to a certain base, such as pyridine, basicity is related to a standard acid, e.g. phenol, pyrrole, etc.

For orientative evaluation of the properties of acid-base compounds, it suffices to follow the differences in the $\nu(X-H)$ band wavenumbers for acids (both in determining the acidity, where the acid vibration bands are monitored directly, and the basicity, which is derived from the changes in the $\nu(X-H)$ band wavenumber for a standard acid free and conjugated with the base). The values of these differences,

$$\Delta\tilde{\nu} = \tilde{\nu}_{free} - \tilde{\nu}_{bound}, \qquad (5.50)$$

serve, to a first approximation, as a measure of the acidity (directly) and basicity (indirectly); an increasing $\Delta\tilde{\nu}$ value represents increasing acidity or basicity.

Determinations of this type are very simple; the wavenumber for the free acid is determined from the spectra of dilute acid solutions and the value of $\tilde{\nu}_{bound}$ from the spectra of the same solutions after addition of the base. A drawback of the procedure lies in that it is suitable only for very similar compounds; the $\Delta\tilde{\nu}$ value reflects the acid-base properties of the interacting components only partially and an important role is also played by the properties of the $X-H$ bond itself. Thus very good relative acid-base data are obtained for groups of compounds where X is, e.g., oxygen, i.e. with phenol derivatives, alcohols, etc., but these data cannot be correlated with the $\Delta\tilde{\nu}$ values of $N-H$ bond vibrations in the series of secondary amides, as the shifts for the $O-H$ and $N-H$ bonds depend on the different properties of the two bond types. Compounds with sterically hindered acidic or basic centres are also excluded from the series studied; if the steric hindrance does not completely prevent complex formation between the acid and the base, then it considerably distorts the $\Delta\tilde{\nu}$ values (the values are smaller than corresponds to the acidity or basicity).

The evaluation of the acidity or basicity using the ΔG value whose determination was described in detail in Chapter 5.2.5.3 is much more objective. The formation of a hydrogen-bonded complex between the acid and the base is considered as an equilibrium reaction; the thermodynamic properties of hydrogen-bond formation, which are used in evaluation of these equilibrium reactions, are determined from the temperature dependences of the equilibrium constants. Infrared spectroscopy is used here as a quantitative-analytical method for monitoring the concentrations of the equilibrium mixture components.

The method described can also be used for qualitative detection of intermolecular hydrogen bonds, such as in 1-alkines $RC{\equiv}C{-}H$ (that exhibit autoassociation but can also associate with other bases). It has been employed for the study of changes in the basicity of alkenes and alkines, which increases with an increasing number of alkylsubstituents, for following the basicity of diethylchalcogenides $(C_2H_5)_2X$ (X = O, S, Se and Te), which decreases from oxygen toward tellurium, etc. A significant contribution of the $(p \rightarrow d)$ π-conjugation of O and Si atoms was estimated from the decrease in the basicity in the series of compounds, $R_3C{-}O{-}CR_3$, $R_3C{-}O{-}SiR_3$ and $R_3Si{-}O{-}SiR_3$, etc.

Attempts have often been made to correlate the wavenumber shift values, $\Delta\tilde{\nu}$, with the ΔG values for the hydrogen bond formation in systems of conjugated acids and bases; the attempted linear correlation of the two quantities is termed the Badger–Bauer relationship. As already pointed out, this relationship cannot be expected to be generally valid, since the $\Delta\tilde{\nu}$ values do not fully represent the interaction between an acid and a base. The verification of the validity of this relationship published by some authors is fallacious, as the sets studied by them contained only very similar systems, for which the relationship can accidentally be satisfied.

5.3 Vibrational Spectra and Molecular Structure

The most practically important application of infrared and Raman spectroscopy is structural diagnostics and the determination of the structure of molecules. Both procedures are usually applied to newly prepared (or isolated) compounds, whose structure is for some reason unknown. Structural-diagnostic procedures can sometimes also be applied to known compounds, in order to verify the results obtained by the given identification techniques.

Structural diagnostics is usually considered to be a group of procedures in which the presence of certain bonds or groups in molecules is assumed on the basis of the information contained in the vibrational spectra. Structure determination is a basically similar, but more demanding task; it should result in the determination of the complete molecular structure, which practically means the determination of the constitution, configuration and conformation (the absolute configuration is not considered, because it is not manifested in vibrational spectra). Information contained in vibrational spectra is mostly insufficient for complete structure elucidation and therefore several methods are combined, vibrational spectra playing a greater or lesser role. At present, when not only efficient separation methods (chiefly gas chromatography), but also other spectroscopic methods (mass spectrometry, nuclear magnetic resonance, ultraviolet spectroscopy, photoelectron spectroscopy, ESCA) and non-spectroscopic methods (electron, X-ray and neutron diffraction, dieletric methods, etc.) are available, it is necessary to choose a suitable set of methods for each individual structural analyses. A proper choice permits the task to be completed in minimum time and at a reasonable cost. The structure of more complex compounds cannot be economically solved without using physical methods and vibrational spectroscopy; it took almost 150 years to determine the structure of the alkaloid strychnine molecule, after concentrated application of the most modern chemical and especially physical methods.

The importance of structural-diagnostic applications of vibrational spectroscopy and their use for the elucidation of molecular structures is documented by the exceptionally large number of books which have recently been published; they are listed in the References at the end of Chapter 5.

5.3.1 Structural Diagnostics

Structural diagnostics is based on the information following from the spectral manifestations of characteristic vibrations. This question was explained in detail in the beginning of Chapter 4; however, the problems of characteristic vibrations will still have to be discussed.

Characteristically vibrating groups and bonds are, in principle, independent of the residues of molecules to which they are connected or of which they form a part. They are manifested in spectra by constantly located bands with very typical intensities, widths and often also shapes. Characteristic vibrations have mostly been discovered empirically; data about them come from the knowledge that bands with very similar parameters appear in a narrow wavenumber interval in the spectra of compounds which are different but contain the same characteristically vibrating bond or group. The values of the wavenumbers, intensities and band halfwidth are more or less scattered around mean values; the more, the less characteristic is the vibration.

At present, data on vibration manifestations of almost all common chemical bond types and groups and often also on the degree of their characteristic nature are available; this question is discussed in Chapter 5.3.1.1.

Structural diagnostics is more frequently employed with organic compounds. Vibrational spectra yield information on the bonds and groups in the hydrocarbon parts of molecules (CH_3 or CH_2 groups, double and triple bonds, cyclopropane or benzene ring), but first of all about functional groups containing heteroatoms O, N, S, etc. The X–H bonds can be differentiated for virtually any atom X. However, structural diagnostics can also be met in inorganic chemistry, most frequently in determining the types of bonds, groups, molecules bonded as ligands or otherwise, ions, particles, etc.

5.3.1.1 The Characteristic Nature of Vibrations

So far, characteristic vibrations have been described only qualitatively or by means of their external manifestations. In view of the importance of their characteristic nature for such a wide application range, a somewhat more rigorous explanation of this phenomenon will be attempted.

For any normal vibration of a polyatomic molecule, the amplitudes of the individual vibration coordinates which contribute to it can be calculated. In these terms the relationship,

$$\sum_{i,j}^{3N-6} F_{ij} L_i^{(k)} L_j^{(k)} = \lambda_k, \tag{5.51}$$

can be written, where F_{ij} is the appropriate force constant and $L_j^{(k)}$ is the normalized amplitude of vibrational coordinate r_i in the k-th normal vibration. Quantity λ_k on the right-hand side of the equation is the vibrational frequency. If Eq. (5.51) is compared with the expression for the potential energy V,

$$\sum_{i,j}^{3N-6} F_{ij} r_i r_j = 2V, \tag{5.52}$$

it can be seen that they are formally identical. It can thus be stated that

Eq. (5.52) expresses the potential energy of the k-th normal vibration. As it is senseless to determine the concrete energy value for a vibration (because it depends e.g. on the initial state of the test system), only the relative amplitude values are considered. This is the reason for normalization, best by setting the potential energy equal to λ_k.

Eq. (5.51) is employed for determining the contributions from the individual terms, $F_{ij}L_i^{(k)}L_j^{(k)}$. For a completely characteristic vibration, which is related only to coordinate r_m in a polyatomic molecule, it holds that

$$F_{mm}L_m^{(k)}L_m^{(k)} = \lambda_k; \qquad (5.53)$$

the contributions from all other terms, for which $i, j \neq m$, are zero. With an incompletely characteristic vibration, the other coordinates, r_j, also play a role in addition to coordinate r_m; the relative participation of the individual vibration coordinates in the given vibration then determines the degree of its characteristic nature (if a certain coordinate significantly predominates) or of its mixed nature (if no contribution from the coordinates is decisive).

Table XLIII gives the distributions of the potential energy for vibrations of two similar molecules, chloropropinal, $ClC{\equiv}CCHO$, and bromopropinal, $BrC{\equiv}CCHO$. It can be seen from the table that the stretching and deformation coordinates contribute to the individual vibrations to different degrees. When the participation of a single coordinate predominates, the vibration can be considered characteristic (stretching vibrations of the C–H and C≡C bonds, out-of-plane deformation vibration of CHCO); the other vibrations are mixed. The description of characteristic vibrations is simple; it is always bound to the coordinate which plays the decisive role in the given vibration. The description (even qualitative) of vibrations to which a larger number of coordinates contribute is more complex; as can be seen, description based on the knowledge of the potential energy distribution is suitable here.

The characteristic nature of the vibration need not be related to a single coordinate. With characteristic vibrations of larger groups of atoms, the characteristic group frequencies involve constant participation of the individual coordinates.

5.3.1.1.1 Diagnostics of Bonds and Groups

Chemical practice offers a very extensive selection of tasks in which the presence of certain bonds and groups in molecules is diagnosed using vibrational spectroscopic methods. All diagnostic tasks are based on the

Table XLIII

Distribution of the potential energy of vibrations of the ClC≡CCHO and BrC≡CCHO molecules (in %)

Vibration	Chloroderivative		Bromoderivative	
	Wavenumber $\tilde{\nu}$ (cm^{-1})	Contributions[a]	Wavenumber $\tilde{\nu}$ (cm^{-1})	Contributions[a]
In-plane vibrations A'				
ν_1	2860	C—H(98)	2858	C—H(98)
ν_2	2220	C≡C(79)	2196	C≡C(82)
ν_3	1694	CCH(54) C=O(51)	1692	CCH(54) C=O(51)
ν_4	1387	CCH(45) C=O(42)	1386	CCH(45) C=O(43)
ν_5	1077	C—Cl(42) C—C(41)	1030	C—C(52)
ν_6	738	CCO(48) C—Cl(26)	691	CCO(47) C—Br(28)
ν_7	473	CCC(43) C—C(28)	395	CCC(49) C—Br(32) CCBr(17) C—C(17)
ν_8	312	CCO(39) CCCl(23) CCC(24) C—C(15)	290	CCO(34) CCBr(20) CCC(16) C—C(19) C—Br(15)
ν_9	114	CCCl(67) CCC(21)	105	CCBr(64) CCC(22)
Out-of-plane vibrations A''				
ν_{10}	945	CCHO(90)	(950)	CCHO(87)
ν_{11}	352	CCC(99) CCCl(41)	340	CCC(99) CCBr(15) CCHO(20)
ν_{12}	152	CCCl(71)	143	CCBr(68)

[a] The potential energy includes contributions from the vibrations described by the stretching coordinates of the C—H, C≡C, C—C, C=O and C—Hal bonds and the deformation coordinates of the CCH, CCO, CCC and CCHal angles; with the out-of-plane vibrations these are out-of-plane coordinates.

possibility of assigning bands or lines in infrared or Raman spectra to characteristic vibrations of bonds and groups. Specific manifestations of characteristically vibrating atomic nuclei underlie the statements and hypotheses concerning the presence of certain bonds and groups in molecules.

The regions of occurence of the characteristic vibration bands for certain bonds and groups are given in Table XLIV. The selection is limited to cases for which the prediction is reliable; of course, a much larger number of bonds and groups can be diagnosed with a lower degree of reliability, as the reader can find in the specialized literature. The selection of diagnosable bonds and groups does not and actually cannot correspond to the real requirements of chemical practice; it should be borne in mind that many bonds and groups, important from the chemical point of view, cannot be reliably detected using vibrational spectroscopy. Nevertheless, it is clear from the table that the range of diagnosable bonds and groups is rather wide and that it involves many important bonds and groups. Diagnostic operations are mostly performed on organic compound molecules, although there are also examples from inorganic chemistry; consequently, the groups in Table XLIV have been selected with greater emphasis on organic molecules.

For practical diagnostic operations a table of characteristic vibration band wavenumbers is naturally insufficient. Serious perturbations may appear in the spectra of compounds sensitive to the effects of mutual molecular interactions and random overlaps of bands of vibrations of various bonds and groups may occur with molecules with more complex structures. However, these problems were discussed in detail in Chapter 3 and in the beginning of this Chapter, so that only the individual tasks will be described below.

Because of the extreme variability of the molecular structures of chemical compounds, there also are many diagnostic procedures. However, it seems that there exist roughly four application ranges, into which the known diagnostic procedures can be classified.

The first group involves tasks during which the presence of certain bonds or groups is diagnosed in molecules of unknown structures. Mostly very fragmentary data are available about the test compounds (e.g. information on the summary formula obtained from the elemental analysis) and vibrational spectroscopic diagnostics is applied at the very beginning of attempts at determining the structure. If the compound molecules contain heteroatoms O, N, S, Si, etc., it is first required to find out in what form the heteroatoms are built into the molecules. A certain number of possible

Table XLIV

Regions of the characteristic vibrations of important bonds and groups (cm^{-1})

Vibration	Region of Occurence[a]	Note
ν(O—H)	3700—2500	narrower regions for the individual types
ν(N—H)	3500—3100	
ν(N^+—H)	3300—2000	strongly affected by the anion nature
ν(C—H)	around 3300	acetylene derivatives
ν(C—H)	3150—3000	ethylene and benzene derivatives
ν(C—H)	2970—2800	saturated hydrocarbons
ν(S—H)	2650—2500	
ν(Si—H)	2300—2100	
ν(X=Y=Z)	2300—2000	—OCN, —SCN, —NCS
ν(C≡C)	2280—2100	
ν(C≡N)	2280—2000	
ν(C≡0)	2100—1800	metal carbonyls
ν(C=C=C)	around 1950	
ν(C=O)	1900—1500	narrower regions for the individual types
ν(C=C)	1700—1480	unsaturated compounds, more complicated with aromatics
ν(C=N)	1700—1450	
amide II	1640—1500	coupled mode of sec. amides
$\nu(COO^-)_a$	1630—1500	salts of carboxylic acids
ν(N=O)	1700—1600	esters of nitrous acid, etc.
$\nu(NO_2)_a$	1650—1500	nitrocompounds
$\nu(COO^-)_s$	1430—1300	salts of carboxylic acids
$\delta(CH_3)_s$	1420—1350	hydrocarbons, alkyl groups
ν(C—N)	1450—1000	
ν(C—O)	1450—1000	
ν(C—F)	1400—1000	
$\nu(SO_2)_a$	1380—1300	
$\nu(NO_2)_s$	1350—1250	nitrocompounds
ν(P=O)	1350—1150	
$\nu(SO_2)_s$	1180—1140	
$\nu(Si—O—Si)_a$	1100—1000	siloxanes
ν(S=O)	1100—950	sulphoxides, sulphurous acid derivatives
ν(C—S)	730—600	thiols, sulphides
ν(C—Cl)	800—550	
$\nu(Si—O—Si)_s$	600—550	siloxanes
ν(C—Br)	600—450	

[a] Regions of occurence include common cases and do not record anomalies; detailed information is summarized in the publications cited, especially in Chapter 5 and 5.3 in connection with the discussion of characteristic vibrations.

(most probable) structural variants is usually assembled and the one to which the vibrational spectra correspond is sought; e.g. with a compound containing a single oxygen atom in the molecule, the spectral bands of characteristic vibrations of O—H or C=O bonds, ether groups, etc. are sought. Then the spectral manifestations of the molecular skeleton are examined (this is the first step if the test compound is a hydrocarbon), the presence and character of double or triple bonds, occurence of alkyl groups (methyl, tert. butyl), rings (cyclopropane, benzene), etc. With very simple molecules (and with more complex but highly symmetrical molecules) this structural diagnostic procedure may even lead to complete elucidation of the molecular structure. For example, a compound with a summary formula of $C_2H_5NO_2$ and with a melting point of 235 °C yields an infrared spectrum (Fig. 20) containing the bands of the $-NH_3^+$ group (the group of intense bands between 3250 and 2500 cm^{-1}, the bands at 1611, 1522, 1132 an 1114 cm^{-1}) and of the $-COO^-$ group (at 1591 and 1414 cm^{-1}). Together with the melting point value, this information leads to unambiguous determination of the structure of this molecule, glycine, $H_3N^+CH_2COO^-$ (in the zwitterion form).

The lack of general information on the molecular structure in performing diagnostic tasks of this type necessitates the requirement of obtaining as much information as possible from the vibrational spectra. Therefore, both infrared and Raman spectra are mostly measured. Infrared spectra are exceptionally suitable for diagnostic purposes; groups with heteroatoms are usually strongly polar and their infrared bands are quite intense. There are, however, many cases when the information obtained from the Raman spectrum suitably complements the data obtained by interpretation of infrared spectra and sometimes they even yield unique information. The thiol group, S—H, can serve as an example; the band of its stretching vibration is difficult to detect in infrared spectra because of its low intensity and sometimes, e.g. in mercaptocarboxylic acids (whose spectra contain strong absorption by the carboxylic group), band assignment is completely impossible. However, the ν(S—H) vibration is pronounced in the Raman spectrum because of the large polarizability of the S—H bond. Raman spectra can also be utilized to advantage to diagnose bonds and groups in very symmetrical molecules; e.g. molecules of ethylene, *trans*-2-butene and 2,3-dimethyl-2-butene have a centre of symmetry and hence no band of the C=C bond stretching vibration appears in their infrared spectra. However, it appears strongly in the Raman spectrum because of the high bond polarizability.

Not all structural-diagnostic operations are as straightforward as would follow from the text above; sometimes band parameters other than wavenumbers must be employed to detect the presence of bonds or groups in molecules and/or supplementary experiments must be carried out as described in Chapter 4.

The second group of diagnostic tasks are those in which infrared or Raman spectroscopic study does not open, but closes the determination of the structure. Here the molecular structures are basically known except for some details and spectra are employed to determine information which is lacking.

Of the many examples available the determination of the structure of a pure unsaturated hydrocarbon, isolated from the reaction mixture after the isomerization of 1-methylcyclohexene (I) will be given. This compound has the same skeleton as the cycloolephin (I) and thus differs only in the position of the double bond. Isomers (II) – (IV) can be considered:

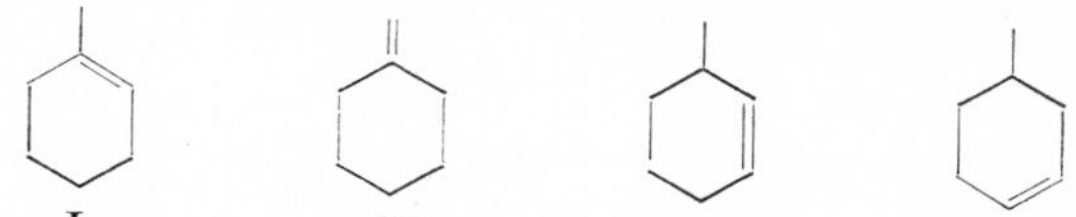

I II III IV

Using infrared spectra the type of substitution at the double bond can be determined and hydrocarbons I and II and III and IV be differentiated (both hydrocarbons III a IV contain a double bond with the same type of substitution).

The third group of problems include all the diagnostic procedures which determine the actual state of molecules capable of tautomeric changes etc. Thus, vibrational spectra can diagnose the presence of individual forms of β-ketoesters which display keto-enol tautomerism. Two bands appear in the infrared spectrum of the keto-form (V) of the esters of acetoacetic acid at 1740 and 1720 cm^{-1}, corresponding to the carbonyls in the carbalcoxylic and ketogroups. In the spectrum of the enol-forms (VI), both bands, of course, disappear but the hydroxyl group broad band appears around

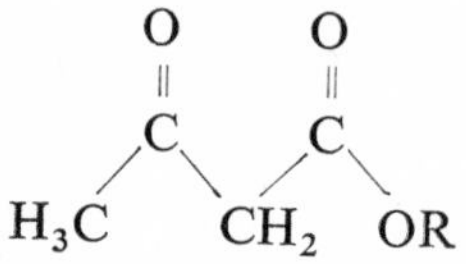

$H_3C-C(=O)-CH_2-C(=O)-OR$

V

$H_3C-C(-O-H\cdots O=)=CH-C-OR$

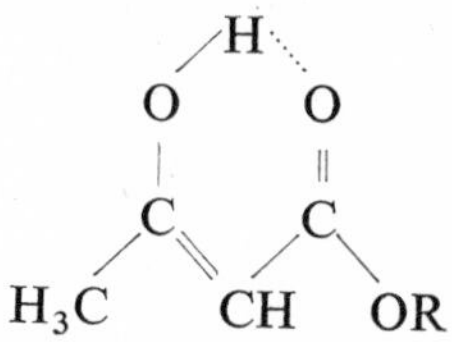

VI

3000 cm^{-1} (the hydroxyl group is hydrogen bonded to the carbonyl group oxygen) and the stretching vibration bands for the C=O and C=C bonds in the unsaturated conjugated ester appear at 1660 and 1640 cm^{-1}, respectively.

Applications concerned with the detection of the formation or the existence of complexes using chiefly infrared spectra, also belongs in this category. Principally, the formation of most complex types, i.e. coordination compounds, hydrogen-bonded complexes, charge-transfer complexes, weak (e.g. collision) complexes, etc., can be detected in the spectra.

The presence of complexes in certain systems is assumed from the presence of bands in the spectra which do not appear in the spectra of any component making up the system. Complex formation is best verified by experiments in which complex-forming compounds are mixed in various ratios. The spectra of the systems with the highest complex content must then exhibit the highest relative intensity of the bands, corresponding to the complex.

The bands according to which the presence of complexes in systems is diagnosed mostly belong to characteristic vibrations of bonds or groups that directly participate in the complex formation. The parameters of these bonds (e.g. bond strength or electron density distribution) differ in the parent compound and in the complex and thus their spectral manifestations also differ.

For example, the free and the bound forms of the methanol molecule, CH_3OH, can be distinguished in this way. The $\nu(O-H)$ fundamental frequency band of the free molecule appears around 3640 cm^{-1}, whereas the band for the molecule hydrogen-bonded to some basic molecule (an ether, amine etc.) appears at a substantially lower wavenumber. As can be seen from Fig. 49, with increasing concentration of the complex the intensity of the band of the given vibration of the complex increases, while the intensity of the band of the vibration of the original, free compound decreases.

Hence it follows that the bonds participating to the highest degree in the complex formation can be relatively reliably determined from the spectra. The most probable site of the interaction of sulfoxides R^1R^2SO or nitrosoderivatives RNO with complexing agents—the O atom—can be determined from the shifts to lower wavenumber values; shifts of $\nu(S=O)$ and $\nu(N=O)$ bands to larger wavenumbers indicate reagent attack on the S or N atoms.

In the spectra of complexes and coordination compounds originating from molecules of nitriles $RC \equiv N$, the $\nu(C \equiv N)$ band is shifted toward larger wavenumbers. With increasing interaction energy of the components, the $\nu(C{\equiv}N)$ band shifts to progressively larger values. The direction of the shift is given by the fact that nitrile molecules participate in the formation of complexes and coordination compounds through the lone electron pair on the nitrogen atom. In the free nitrile molecules, this lone electron pair is evidently somewhat delocalized into the $C{\equiv}N$ bond; on complex formation the original delocalization decreases, the order of the $C{\equiv}N$ bond increases and the stretching frequency also increases. This statement is verified by the very high $\nu(C{\equiv}N^+)$ wavenumber found in the spectrum of dimethylnitrilium cation $H_3C-C{\equiv}N^+-CH_3$, around 2430 cm^{-1}. In this cation, both the C atom and the N atom are bound by covalent bonds to the methylgroups and the high $\nu(C{\equiv}N^+)$ wavenumber reflects $sp - sp$ type hybridization. In free nitriles the hybrid character of the $C{\equiv}N$ bond will thus rather approach the $sp - sp^2$ state.

Vibrational spectroscopy was the basic method used for evaluation of the molecular structure of e.g. metal carbonyls, sandwich (e.g. ferrocene) and semisandwich molecules, metal chelates, etc.

Ion-pair formation in salts can also be disclosed in the infrared and Raman spectra. For example, ion-pairing between the anion and the cation may lead to a decrease in the local ion symmetry, compared with the free ions. In the spectra of polycrystalline sulphate samples the original high local symmetry of the SO_4^{2-} ion, $\mathscr{T}_d$, can be decreased by interaction with a cation; the infrared spectrum then exhibits the band of the originally forbiden totally symmetrical ν_1 vibration and the bands of the originally triply degenerate ν_3 and ν_4 vibrations are doubled or trebled. This explanation is not rigorous, as it employs the local symmetries of the ions, while the crystal unit cell symmetry is disregarded. However, even this interpretation often gives qualitative proof of ion-pairing with a sufficient degree of reliability.

It is quite interesting, though experimentally exacting, to follow the formation of surface complexes, i.e. molecules adsorbed and chemisorbed on the surfaces of catalysts, metals, zeolites, etc. From the many variants of these experiments, the process of dehydration of alcohols on acidic oxides has been selected as an example, as diagnostic applications of infrared spectroscopy are widely employed in it.

Acidic oxides, e.g. γ-alumina, contain hydroxyl groups bound to metal atoms. During thermal activation these oxides are gradually dehydrated, with pairs of neighbouring hydroxyl groups yielding one water molecule each; the remaining oxygen atom connects the two metal atoms by a bridge.

A small number of hydroxyl groups still remains on the surface of the catalyst prepared by thermal treatment of the oxide, as hydroxyl groups which are not close to any other hydroxyl groups are not dehydrated. If γ-alumina is activated at a temperature around 1100 °C, about 8 % of the hydroxyl groups remain on the surface.

If the pretreated catalyst is exposed to excess alcohol vapours, then after evacuation only the molecules adsorbed and chemisorbed on the catalyst surface remain. If the operation is performed in a reactor adapted for the measurement of infrared spectra, its individual phases can be monitored. First of all, the bands of the surface hydroxyls appear in the spectra; after adsorption of alcohols, the bands of their OH groups (mutually hydrogen-bonded) appear. On the increase in the temperature the molecules of the adsorbed alcohols are dehydrated, the water liberated re-hydrates the catalyst surface and olephins are liberated into the gaseous phase. The chemisorbed alcohols are bound to the surface as alcoxides metal-OR; the bands of the alcoxygroups appear in the spectra. Methanol is oxidized under the reaction conditions and surface formates are thus formed (the HCOO group is bound to two metal atoms through one of its oxygen atoms) and can be safely diagnosed spectroscopically. Some other reaction products, e.g. ethers, can also be readily distinguished.

The final, fourth group are applications in which the spectral data are utilized for evaluation of the nature of bonds and groups. Several examples will be given of the basic problem in this field, where the changes in the parameters of the characteristic vibrational bands form a basis for conclusions concerning the bond strength and/or polarity.

These applications can be very lucidly explained on the molecules of N-acetylderivatives of heteroaromatic azoles (pyrrole, imidazole etc.). In these molecules, the lone electron pair on the nitrogen atom that binds the acetyl group is delocalized both in a heteroaromatic ring and (e.g. in amides) in the acetyl group carbonyl. In the series of N-acetylazoles, in which the number of N atoms in the ring increases (pyrrole, imidazole, triazole, tetrazole), the mesomeric energy of the heteroaromatic ring increases and the lone pair on the nitrogen (bearing the acetyl group) is constantly strongly delocalized in the ring. The interaction with the acetyl group is mutually weekened and finally the N-acetyltriazole a N-acetyltetrazole molecules loose their amide character. The order of the C=O bond increases in the series of N-acetylazoles with increasing number of N atoms in the ring and consequently the ν(C=O) wavenumber of the N-acetylgroup also increases; as can be seen in Table XLV, the ν(C=O) wavenumber for

N-acetyltetrazole almost reaches the value for acid chlorides. The similarity in the wavenumber values is not quite formal, as both acetylchloride and N-acetyltetrazole are strong acetylating agents.

Table XLV

The wavenumbers of the $\nu(C{=}O)$ vibration for N-acetylderivatives of azoles (cm^{-1})

Compound	$\nu(C{=}O)$	Compound	$\nu(C{=}O)$
N-acetylpyrrole	1732	N-acetyltetrazole	1779
N-acetylimidazole	1747	Acetylchloride	1810
N-acetyltriazole	1765		

Bonding anomalies in the molecules of silyl- and germylketones, R_3SiCOR and R_3GeCOR have been discovered from the spectra, showing strong delocalization of the electrons in the $Si-C{=}O$ and $Ge-C{=}O$ groups. The $\nu(C{=}O)$ band for these compounds lies around 1620 cm^{-1}, i.e. cca. 100 cm^{-1} lower than for similar alkylketones.

The triple bond $C{\equiv}O$ was diagnosed in the acetylium ion CH_3CO^+ found in acetylating mixtures, on the basis of the high wavenumber value

Table XLVI

The values of force constants K for NO bonds in various compounds

Molecule, ion	$K\,(N\,.\,m^{-1})$[a]	Molecule, ion	$K\,(N\,.\,m^{-1})$[a]
Nitronium cation NO_2^+	1732	Nitrogen dioxide ONO	850
Nitrogen oxide NO	1540	Nitrite anion NO_2^-	580
Nitrous oxide NNO	1140		

[a] K, force constant, $1\ N\,.\,m^{-1} = 10^{-2}$ mdyne . $Å^{-1}$.

for the CO bond vibration (around 2300 cm^{-1}). The structure of the nitronium ion NO_2^+ could also be determined from the spectra; its antisymmetrical stretching vibration has a very high frequency (2360 cm^{-1}), similar to that of carbon dioxide, which indicates its linearity and non-formal similarity to CO_2.

The use of vibrational spectra for evaluation of the nature of some bonds is invaluable, e.g. for compounds containing an oxygen atom bonded to a nitrogen atom. The character of the NO bond in these compounds is very variable changing from a triple bond to an almost single bond cf. Table XLVI.

5.3.1.1.2 Empirical Correlation of Band Parameters

The dependence between the bond strengths and the appropriate wavenumbers (or other band parameters) described above can sometimes yield an empirical correlation, in which the spectral band parameters are correlated with quantities by means of which the effect of the remaining part of the molecule (or of a substituent) on the pertinent bond is described. Correlations are usually carried out with characteristically vibrating bonds/or groups.

Empirical correlations are usually very simple and can be classified into several categories. The most primitive correlations involve empirical parameters. An example is the correlation of the wavenumbers of the $\nu(P{=}O)$ vibrational bands for compounds of the $R^1R^2R^3P{=}O$ type, with empirical parameters π_i of substituent R^i, employing the relationship,

$$\tilde{\nu}(P{=}O) = 930 + \sum_i \pi_i \qquad (cm^{-1}). \tag{5.54}$$

Parameters π_i are determined from wavenumbers of the $\nu(P{=}O)$ band of compounds of the $R^1R^2R^3P{=}O$ type, $R^1 = R^2 = R^3 = Cl, CH_3, OCH_3$, etc., as $\tilde{\nu}(P{=}O)/3$.

Another, substantially larger group of correlations contains relations between the characteristic vibration band wavenumbers and the substituted electronegativity or the Taft constants for the substituent induction effect, σ_i or σ^*, which are generally linear,

$$\tilde{\nu}(XY) = K_1 + K_2 \sum_i \varkappa_i; \tag{5.55}$$

where K_1 and K_2 are empirical constants and $\varkappa_i$ is either the electronegativity or the inductive effect constant. Relationships of this type have been applied to the wavenumbers of the $Si{-}H$, $O{-}H$, $C{=}O$, NO_2, etc., bond vibration bands.

If the electronegativity is employed for evaluation of the group properties, frequently only the immediately neighbouring atoms are considered (e.g. the value for the carbon atom is substituted for the CH_3 group electronegativity). However, for more rigorous correlations the effective electronegativities of groups can be derived; e.g. the effective electronegativities, $\chi(R)$, for the CH_3 group and its halogenoderivatives are determined from the relationship,

$$\chi(R) = \chi(C)/2 + [\chi(A') + \chi(A'') + \chi(A''')]/6, \qquad (5.56)$$

where $\chi(C)$ is the carbon atom electronegativity and $\chi(A)'$ etc. are the electronegativities of the atoms bound to the carbon atom (in the Gordy scale). The electronegativities for common atoms are given in Table XLVII. Table XLVIII gives the values of Taft constants σ_R and σ_I for common substituents and Table XLIX the σ^* values.

Table XLVII

Electronegativities χ of atoms on the Gordy scale

H 2.13			
B 1.9	Al 1.5		
C 2.55	Si 1.8	Ge 1.7	Sn 1.7
N 2.98	P 2.1	As 2.0	Sb 1.8
O 3.45	S 2.53	Se 2.4	Te 2.1
F 3.95	Cl 2.97	Br 2.75	I 2.45

Correlations based on Eq. (5.55) are suitable for cases when the interaction of the vibrating group with the rest of the molecule can be satisfactorily described in terms of the induction effect. However, a non-linear correlation is better suited to systems in which conjugation effects also play a role. For example, wavenumbers $\tilde{\nu}(C{=}O)$ for the C=O bond stretching vibration in carboxylic acid derivatives RC=O(X) can be described by a Hammett-type relationship,

$$\tilde{\nu}(C{=}O) = 1514\,k\,.\,x \qquad (cm^{-1}), \qquad (5.57)$$

where k is an empirical constant describing the properties of residue R and x is an empirical constant for substituent X. The x a k values were calculated assuming that k equals unity for benzoic acid ($k = 1.000$).

The wavenumbers of the $\nu(O{-}H)$ band of phenol derivatives and the $\nu(C{=}O)$ band in acetophenone derivatives, etc., are usually well correlated with the Hammett constants σ of the substituents on the aromatic ring by relationships of the following type,

$$\tilde{\nu}(XY) = K^1 + K^2\sigma \qquad (cm^{-1}). \qquad (5.58)$$

The σ_p Hammett constants for para-derivatives and σ_m for meta-derivatives are given in Table XLVIII. The *p*-constants can also be employed for *o*-derivatives, provided that the *o*-substituents and the test group do not sterically affect one another too much or do not form a hydrogen bond. If the molecule contains more substituents in positions where they do not interfere with the

Table XLVIII

The values of Taft constants σ_I and σ_R for the description of the inductive and resonance properties of substituents and Hammett constants σ_m and σ_p

Substituent	Taft		Hammett	
	σ_I	σ_R	σ_m	σ_p
NH_3^+	0.86	0.00	0.88	0.82
NO_2	0.63	0.15	0.71	0.78
SO_2CH_3	0.59	0.14	0.60	0.72
CN	0.58	0.08	0.56	0.66
F	0.52	−0.46	0.34	0.06
$SOCH_3$	0.52	0.05	0.52	0.57
Cl	0.47	−0.24	0.37	0.23
Br	0.45	−0.22	0.39	0.23
I	0.39	−0.11	0.35	0.28
CF_3	0.42	0.13	0.43	0.54
OC_6H_5	0.38	−0.41	0.25	−0.03
$COOC_2H_5$	0.30	0.15	0.37	0.45
$COCH_3$	0.28	0.22	0.38	0.50
OCH_3	0.25	−0.51	0.05	−0.27
OH	0.25	−0.62	0.12	−0.37
NH_2	0.10	−0.76	−0.16	−0.66
$N(CH_3)_2$	0.10	−0.93	−0.21	−0.83
C_6H_5	0.10	−0.11	0.06	−0.01
H	0.00	0.00	0.00	0.00
CH_3	−0.05	−0.12	−0.07	−0.17
$t\text{-}C_4H_9$	−0.07	−0.17	−0.1	−0.20
$Si(CH_3)_3$	−0.42	0.16	−0.04	−0.07

test group (e.g. in positions 3, 4 and 5 on the benzene ring), Eq. (5.58) can be rewritten to give a form analogous to Eq. (5.55), substituting the Hammett constants for the substituents in the given positions for $\varkappa_i$. However, if the substituents interact at positions 3, 4 and 5, this relationship cannot be used. For example, the Hammett *p*-constant cannot describe the effect of a nitrogroup in position 4 of the benzene ring, if this group deviates from coplanarity with the benzene ring by bulky substituents in positions 3 a 5.

The wavenumbers of the vibrational bands have sometimes been correlated with less common parameters of groups and bonds. For example, the wavenumbers of the alcohol ν(O–H) bands have been correlated with

Table XLIX

The values of Taft constants σ^* for the description of inductive properties of substituents and groups

Group	σ^*	Group	σ^*	Group	σ^*
H	0.49	cycloC_6	–0.15	CH_2Cl	1.05
CH_3	0.00	cycloC_5	–0.20	$CHCl_2$	1.94
C_2H_5	–0.10	$CH_2C(CH_3)_3$	–0.14	CCl_3	2.65
n-C_3H_7	–0.115	$C(C_2H_5)_3$	–0.34	CH_2OCH_3	0.52
i-C_3H_7	–0.19	$CH_2CH_2C_6H_5$	0.08	$CH_2OC_6H_5$	0.85
n-C_4H_9	–0.13	$CH_2C_6H_5$	0.225	CH_2CN	1.30
i-C_4H_9	–0.125	C_6H_5	0.60	$COCH_3$	1.65
t-C_4H_9	–0.32	CH_2CH_2Cl	0.385	SO_2CH_3	1.32

the basicity of compounds forming hydrogen-bonded complexes with the alcohols, the wavenumbers of the ν(C=O) bands of esters $RCOOR^1$ with the pK_a values of alcohols R^1OH, etc. Characteristic vibration band wavenumbers have also been succesfully correlated with quantum-chemical parameters (atomic localization energies, bond orders, etc.).

Empirical correlations for the band intensities with substituent parameters are usually more complicated than wavenumber correlations. Their importance is smaller and thus they will not be treated in greater detail here.

Qualitative empirical correlations of the infrared band intensities for certain vibrations with the substituent parameters are of greater importance for evaluation of vibrating bond polarity. For example, the intensity of the ν(O–H) band of the hydroxyl group in phenol derivatives decreases with increasing donor strength of the substituent; this is caused by natural polarization of the O–H bond with excess negative charge on the more electronegative oxygen atom, which is weakened by the effect of a substituent with donor properties. Identical changes in the intensities of the stretching vibrations of the N–H bonds in aniline derivatives, the ≡C–H bonds in phenylacetylene derivatives and the N≡C bonds in phenylisocyanide derivatives on substitution indicate that all these bonds are of the same polarity type, with excess electron density on the bonded atom that is closer to the phenyl. The opposite intensity trend was found with the C–H, C≡N, C=O (in amide), P=O, etc., bonds.

In most cases, a decrease in the wavenumbers in a series of variously substituted compounds is accompanied by an increase in the intensity, e.g. with benzaldehyde derivatives, where the wavenumber of the ν(C=O) band decreases on substitution from 1708 cm^{-1} in *p*-nitroderivative to 1663 cm^{-1} in *p*-N,N-dimethylaminoderivative and the integrated band intensity increases from 210 to 290 l . mol^{-1} . cm^{-2}.

As the wavenumbers of the bands of some vibrations strongly depend on the state of aggregation of the test sample, the experimental conditions of the sample preparation must be exactly defined for each correlation. The magnitude of the difference is demonstrated by comparison of the equations employed for the correlation of the ν(P=O) band of phosphoryl derivatives $R^1R^2R^3P=O$ with the electronegativity of substituent R^i in the gaseous (5.59) and liquid (5.60) states

$$\tilde{\nu}(P=O) = 1039 + 31.3 \sum_i \pi_i \qquad (cm^{-1}), \qquad (5.59)$$

$$\tilde{\nu}(P=O) = 975 + 35.0 \sum_i \pi_i \qquad (cm^{-1}). \qquad (5.60)$$

Empirical correlations are reasonably valid when the effect of the rest of the molecule on the vibrating band or group can be considered as only a small perturbation; however, if there are stronger interactions between the two parts of the molecule, deviations from simple relationships appear.

As an example of unjustified correlation, substitution derivatives of aniline $RC_6H_4NH_2$ can be mentioned. Some substituents R can conjugate strongly with the $C_6H_4NH_2$ residue, affecting the hybrid character of the aminogroup N atom orbitals. With increasing conjugation with substituents which can attract electrons (e.g. groups NO_2, CN, etc.), the geometry of the CNH_2 group changes from the original pyramidal form (characteristic for the hybrid character of the sp^3-orbitals of the N atom) to an almost planar form (corresponding to sp^2 hybrid character of the N atom). Hence the substitution leads to a considerable change in the electron density of the vibrating group and even to a change in its geometry, which can no longer be considered as a small perturbation. The simple relationships, suitable e.g. for correlations of the wavenumbers of the ν(O–H) band of phenol derivative vibrations could thus not be employed for the description of the frequencies of the NH_2 group vibrations in aniline derivatives. More complex expressions have thus been proposed for the calculation of these frequencies, involving the force constant for the N–H bond, K, the magnitude of the H–N–H angle, ϑ, and the masses of atoms H (m_H) and N (m_N):

$$4\pi^2\tilde{\nu}_s^2 = K[1/m_H + (1 + \cos\vartheta)/m_N] \tag{5.61}$$

and

$$4\pi^2\tilde{\nu}_a^2 = K[1/m_H + (1 - \cos\vartheta)/m_N], \tag{5.62}$$

where $\tilde{\nu}_s$ and $\tilde{\nu}_a$ are the wavenumbers of the symmetrical and antisymmetrical stretching vibrations of the NH_2 group, respectively.

Parameters of the bands of some vibrations are closely related to the molecular geometric parameters. For example, the Bernstein linear relationship has been found to hold between the C–H bond vibration

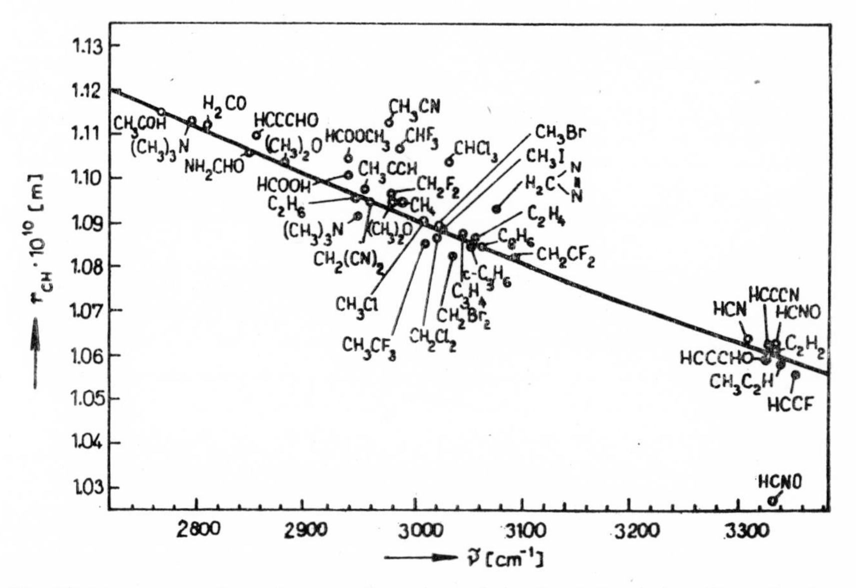

Fig. 68. Linear dependence between the values of the C—H bond stretching vibration band wavenumber $\tilde{\nu}$ and the bond length, r_{CH}.

wavenumber and the internuclear distance, r_0, of the atoms in this bond in various compounds. The two quantities agree very well, as can be seen from Fig. 68; internuclear distances can be estimated from the $\tilde{\nu}$(C–H) values with a precision of about $\pm 2.10^{-4}$ nm. The relationship is so rigorous that it helped in verifying the non-linearity of the HCNO molecule; the deviation following from Fig. 68 is caused by the fact that, in calculating the internuclear distance for the C—H bond from rotational spectra, the value corresponding actually to the projection of atom H onto the axis passing through the heavy atoms in the molecule was determined assuming linearity of the molecule and is lower than the real C—H bond length in the quasi-linear molecule. The HCN valence angle in the HCNO molecule estimated from the r_{CH} value assessed from the plot is 165°; this value has been verified by other procedures.

Correlations between the wavenumbers of the vibration bands of exocyclic double bonds C=C and C=O (i.e. of bonds projecting from the rings whose C atom is simultaneously one atom of the double-bond), are also well known. It generally holds that the wavenumber of the band of a stretching vibration of an exocyclic double-bond decreases with increasing ring size. This, of course, is only valid for three- to seven-membered rings, whereas more complex relationships hold for compounds with larger rings (the geometry of rings larger than seven-membered is rather complex).

For seven-membered and smaller cyclic ketones, a relationship has been proposed correlating the wavenumber $\tilde{\nu}(C{=}O)$ of the C=O bond stretching vibration with the CCC angle (with O double-bonded to the central C),

$$\tilde{\nu}(C{=}O) = 1278 + 0.68K(C{=}O) - 2.2\varphi \qquad (cm^{-1}; N.m^{-1}, °), \quad (5.63)$$

Table L

The dependence between the value of wavenumber $\tilde{\nu}$ for vibrations of the C=C and C=O bonds in olephins and ketones, whose double-bonds are directed out of the ring, and the ring size

Number of atoms in the ring	ν(C=C) (cm^{-1}) methylenecyclo-alkanes	ν(C=O) (cm^{-1})	
		ketones	lactams
3	1730	1820	1835
4	1675	1780	1750
5	1658	1751, 1730	1692
6	1652	1718	1672
7	1642	1707	1668

where $K(C{=}O)$ is the C=O bond force constant in $N \, . \, m^{-1}$ ($1 \, N \, . \, m^{-1}$ = $= 10^{-2}$ mdyne . $Å^{-1}$), taken from acetone and φ is the $\mathrm{C\overset{\overset{O}{\|}}{C}C}$ angle in degrees. The dependences of the wavenumbers for some exocyclic double-bonds on the pertinent ring magnitudes are given in Table L.

5.3.2 Determination of Molecular Structures

There are no qualitative differences among the diagnostic applications of vibrational spectroscopic methods and the determination of molecular structures; in the determination of structures, diagnostic procedures are employed as a part of a complex of physical (chiefly spectroscopic), separation and chemical methods.

It is outside the scope of this book to explain details of the procedures employed in the determination of molecular structures; here only the basic rules for application of infrared and Raman spectroscopy will be summarized. For determining molecular structures information on the substance itself is employed (molecular weight, elemental analysis and general formula, mass, vibrational, electronic, NMR spectra, etc.) and also information on compounds prepared from it by degradation (especially by degradation of the carbon skeleton), synthesis (building up of the carbon skeleton), or by procedures in which the carbon skeleton remains unchanged. The milestones are known (and identified) compounds which have been prepared from a compound with an unknown structure by a defined process.

The concept of structure is generally understood to mean the constitution, configuration and conformation of molecules. Vibrational spectroscopy is used in rather different ways when determining constitution, configuration and conformation and consequently applications will be discussed in further parts of Chapter 5 according this classification.

5.3.2.1 Determination of Constitution

In determining the constitution of a molecule, the atoms of the molecule which are bonded together and the character of the bonds are of interest. The difficulties connected with this operation logically increase with an increasing number of mutually bonded atoms in the molecule. Using modern instumentation, the constitution of simple compounds can be obtained almost immediately, but with more complex compounds the determination of the constitution may take a long time, even when the most modern and costly methods are employed. Very complicated molecules cannot be attacked at all without using modern (chiefly spectroscopic) instrumentation. The strychnine molecule mentioned above can serve as example; its constitution was not successfully solved until modern physical methods were applied.

Very varied paths are followed in determining constitutions, including application of various methods to various degrees. The procedures are naturally also determined by specific problems of molecules with unknown structures, the instrumentation available and the tradition of the laboratory.

Vibrational spectra are primarily used for diagnosing bonds and groups (cf. Section 5.3.1). It should be emphasized that structure-diagnostic data concerning carbon skeletons and parts thereof are especially important for determining the constitution of organic compounds.

The length of an unbranched carbon skeleton can sometimes be estimated from vibrational spectra. The rocking vibration of CH_2 groups is suitable for the determination of side-chain lengths from ethyl to n-pentyl. The

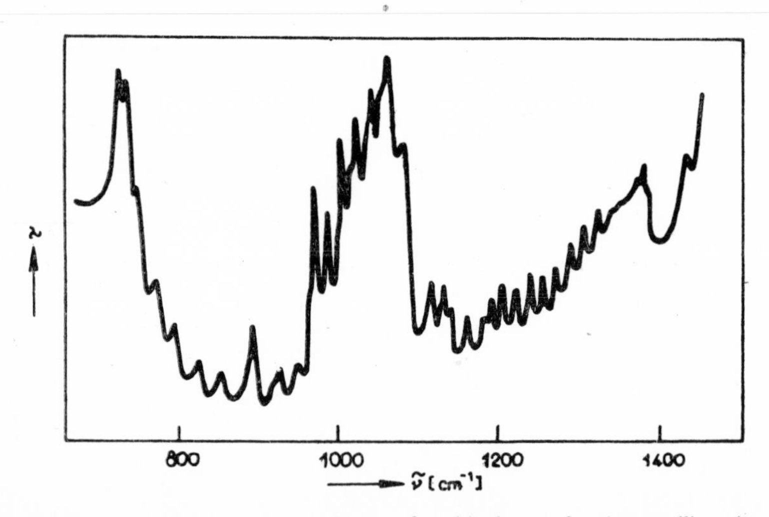

Fig. 69. The infrared absorption spectrum of a thin layer of polycrystalline eicosanol n-$C_{20}H_{41}OH$, exhibiting the typical progressions of the bands of coupled vibrations of the CH_2 groups and C—C bonds.

rocking vibration bands for n-pentyl (and other unbranched alkyl groups) were found in the spectra of liquid hydrocarbons at 723 cm^{-1} (n-pentyl), 730 cm^{-1} (n-butyl), 740 cm^{-1} (propyl) and 770 cm^{-1} (ethyl). The length of the carbon chain of n-alkanes and n-alkyl groups can also be estimated from the progression of the bands of coupled vibrations of CH_2 groups in crystalline samples. Some vibrations of CH_2 groups (e.g. rocking, wagging, twisting) and of the carbon skeleton are strongly coupled and this coupling is manifested by groups of absorption bands (cf. e.g. Fig. 69) called progressions. The number of bands in a progression and distances among the bands are determinated by the number of coupled CH_2 groups and this fact can be used for estimation of the chain length in n-alkanes and n-alkyls.

Vibrational spectroscopy yields very valuable information in determining the character of branching in hydrocarbons and alkyls. Branching is reflected e.g. in splitting of the symmetrical deformation vibration band of the CH_3 group around 1370 cm^{-1}. The isopropyl, gem. dimethyl and tert. butyl groups can be reliably distinguished from the nature of the splitting and from the distance between the splitted bands. The splitting is caused by nonbonding interaction of the methyl groups bound to a single carbon atom. The branch separation is cca 10 cm^{-1} for isopropyl and gem. dimethyl groups and as much as 30 cm^{-1} for the tert. butyl group.

Vibrational spectra permit indication of the presence of some rings in molecules, especially the three-membered, cyclopropane ring, whose skeletal carbon vibrations are significantly coupled and give rise to group characteristic vibrations. The same holds for the three-membered oxirane ring, etc. The vibrations of the ring atoms in aromatic and heteroaromatic compounds are very strongly coupled and consequently these rings give rise to large sets of group-characteristic vibrations in vibrational spectra. In benzene derivatives the vibrations of the hydrogen atoms on the benzene ring are also coupled; the types of benzene nucleus substitution can even be distinguished using their vibrations and the vibrations of the ring atoms.

Coupling of atoms in the aromatic rings is connected with the aromatic character of these compounds resulting from significant electrons delocalization. All other atom groupings exhibiting significant electron delocalization can give rise to coupled vibrations and group-characteristic vibrations. These are e.g. α,β-unsaturated ketones, in which the C=C and C=O bond vibrations are significantly coupled, especially when they are *s-cis* oriented (cf. Section 5.3.2.3).

Certain facts about the molecular constitution can be derived from the experimental observations on some vibrational modes. For example, vibrations can be expected to be alternately forbidden in the X=Y=X molecule; its totally symmetrical vibration does not produce absorption in the infrared, but gives a very strong (polarized) Raman line. On the other hand, a compound with the same general formula but with a different atoms arrangement, i.e. X=X=Y, will exhibit different absorption and scattering patterns. Hence the selection rules derived from the experimental spectra can yield some data on the symmetry and the constitution of molecules.

Linear and planar molecules can also be recognized on the basis of the selection rules and similarities in their spectra. However, statements on linearity and planarity must be made carefully, as some molecules can simulate the behaviour of linear and planar molecules. For example, the spectra of some derivatives of disiloxane, $X_3Si-O-SiX_3$, in which the Si–O–Si angle is around 160°, are very similar to the spectra of linear molecules; compounds with a slightly non-planar character behave analogously. Molecules of this type are termed pseudolinear and pseudoplanar. This pseudo-behaviour is caused by the fact that the molecules pass through linear or planar configurations during certain deformation vibrations.

Mutual similarities in the spectra can also be utilized for assessment of similarities in the corresponding molecules. For example, the vibrational spectra of borazole derivatives are, in principle, similar to those of benzene derivatives. The similarity in the spectra is not only formal, but actually reflects the molecular structure; the borazole molecule is planar and the alternating boron and nitrogen atoms form six equivalent BN bonds. Care must also be exercised during the determination of the molecular structure from information obtained from vibrational spectra through alternately forbidden bands. In larger molecules, especially if their parts are separated by relatively heavier atoms, e.g. S, Si, Ge, etc., the more distant parts are vibrationally independent to a certain extent (the isolation effect of the heavier atoms). The vibrational interactions cannot pass over heavier atoms (and longer distances) and hence these parts of the molecules can exhibit a certain vibrational independence including symmetry properties, which can be reflected to a certain extent in the selection rules.

This section can be concluded by an example of a procedure leading to the determination of the molecular constitution of the unsaturated ketone β-elemenone, which was separated from an essential oil. From the molecular mass of the compound and the results of its elemental analysis, the general formula of β-elemenone was derived, $C_{15}H_{22}O$. The compound evidently belongs among sesquiterpenes, i.e. it could be expected that its skeleton would consist of isoprene units. The compound absorbs in the ultraviolet spectral region at 255 nm (log ε = 3.82) and it is readily split hydrolytically (by formic acid) to the ketone $C_{12}H_{18}O$ and acetone.

The infrared spectrum of liquid β-elemenone (formula VII) exhibits characteristic bands of vibrations of the C=C and C=O bonds at 1690, 1648, 1620, 1007, 920 and 900 cm^{-1}. According to the ultraviolet spectrum and the bands in the infrared spectrum, this is a conjugated ketone. β-Elemenone was converted by a series of mild reactions to a saturated hydrocarbon (XI) with general formula $C_{15}H_{30}$ through intermediates, a saturated ketone (VIII), a saturated alcohol (IX) and an unsaturated hydrocarbon (X). The resultant monocyclic hydrocarbon (XI) was unambiguously identified as the sesquiterpenic hydrocarbon eleman, whose structure is known. The determination of the β-elemenone structure was then based on localization of three C=C double-bonds and the keto-group in the eleman skeleton, as the reactions used permitted the assumption that the original test compound skeleton had been preserved. According to the infrared spectra, two of the three C=C double-bonds are isolated in side chains (the bands at 1648, 1007 and 920 cm^{-1} correspond to the vinyl group vibrations; the isopropenyl group bands lie at 1648 and 900 cm^{-1}) and according to the infrared and ultraviolet spectra the third C=C bond and C=O bond are conjugated. The parameters of the infrared and ultraviolet bands and the behavior during hydrolysis indicate the presence of a six-membered ring ketone conjugated with a isopropylidene group; structure (XII) that alternatively satisfies all the experimental data can be excluded by indirect evidence. In the molecule of the saturated ketone (VIII) a CH_2 group is located beside the keto-group, whereas in the structure derived from ketone (XII) it is absent; the presence of this CH_2 group was detected on the basis of the scissoring deformation vibration at 1427 cm^{-1} occuring in the infrared spectrum of (VIII).

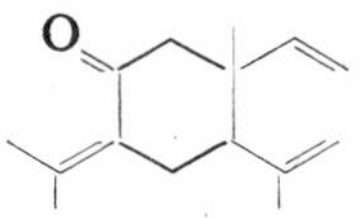

O VII

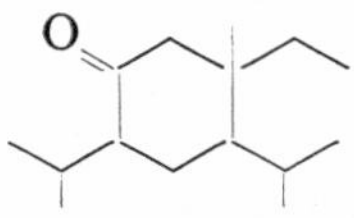

O VIII

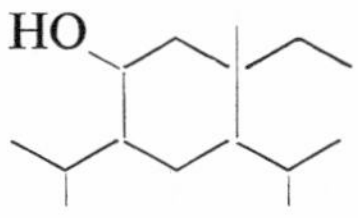

HO IX

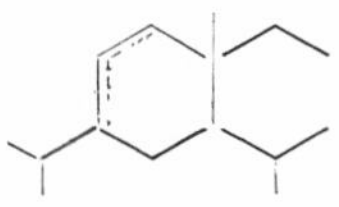

X

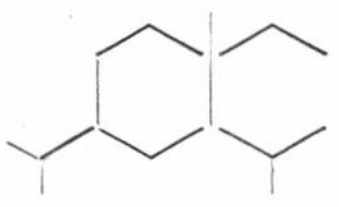

XI

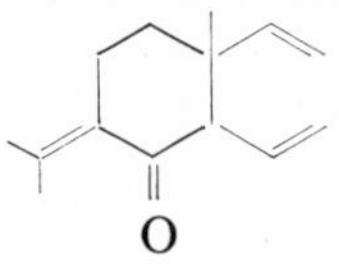

O XII

The actual procedure of the determination of the constitution of the β-elemenone molecule was naturally more complex and less straight forward; the verification of the structure thus determined also took some time and effort. However, the scheme given represents the logical sequence used in the operation in correct dimensions and demonstrates the close cooperation of chemical and physical methods.

5.3.2.2 Determination of Configuration

In some cases vibrational spectra can be utilized for determination of the configuration of molecules with known constitution. These are mostly molecules containing bonds or groups exhibiting characteristic vibrations, located in the immediate vicinity of the geometric isomerism centre. Hence only certain kinds of geometrical isomerism can be studied using vibrational spectra.

The situation is much more easier when the geometric isomerism of pairs of molecules is to be clarified using vibrational spectra; it is more difficult to determine the geometric isomerism when only one isomer of a pair is available. The use of knowledge of intramolecular hydrogen bonding in determining molecular configurations is discussed separately in Section 5.3.2.4.

The molecules of *cis*- and *trans*-disubstituted alkenes, $R^1CH{=}CHR^2$, are a classical example of the particular suitability of vibrational spectroscopy for differentiation among geometric isomers. If substituents R are simple, e.g. atoms D, halogens or CH_3 or $C(CH_3)_3$ groups, then in compounds with $R^1 = R^2$ the *trans*-isomers have a centre of symmetry and the infrared and Raman activity is alternatively forbidden in their spectra. The *cis*-isomer naturally does not have a centre of symmetry. The limitations for $R^1 \neq R^2$ = alkyls are not as rigorous (*trans*-isomers have no centre of symmetry), but still the isomers can be differentiated on almost the same principles; only very weak absorption appears rather than the band being completely forbidden. Some deformation vibrations of hydrogen nuclei are suitable for differentiation among isomers (an infrared band at 960 cm^{-1} is typical for *trans*-isomers and a band at 700 cm^{-1} for *cis*-isomers). The possibilities of differentiation among geometric isomers are connected with specific manifestation of the molecules of 1,2-disubstituted ethylene derivatives; the geometric isomerism of the molecules of 1,1,2-tri- and 1,1,2,2-tetra-substituted ethylene derivatives is no longer as easy to distinguish using the vibration spectra.

Another type of geometric isomers which can be distinguished using infrared spectra involves cyclohexane derivatives. These are chiefly compounds bearing polar substituents on vicinal carbon atoms, between which variously strong interaction can accur, according to the molecular configuration. Classical examples of this type of molecule are the α-bromoketones (e.g. steroid) that exhibit the Corey effect. In 2-α-bromo-3-cholestanone (XIII) the bromine atom is located equatorially, dipoles C–Br and C=O are almost paralell and significantly affect one another. Due to strong nonbonding interaction of the dipoles oriented in paralell (the field-effect), the strengths of the two bonds change and consequently their vibrational frequencies also change. The ν(C=O) vibration band shifts by 20 cm^{-1} to higher wavenumbers with respect to the ketone not containing bromine. In the β-isomer the bromine atom lies on the axial bond (XIV); the C–Br and C=O dipoles are almost perpendicular and hence interact only weakly. The ν(C=O) band for this geometric isomer is shifted by a mere 5 cm^{-1} to higher wavenumbers.

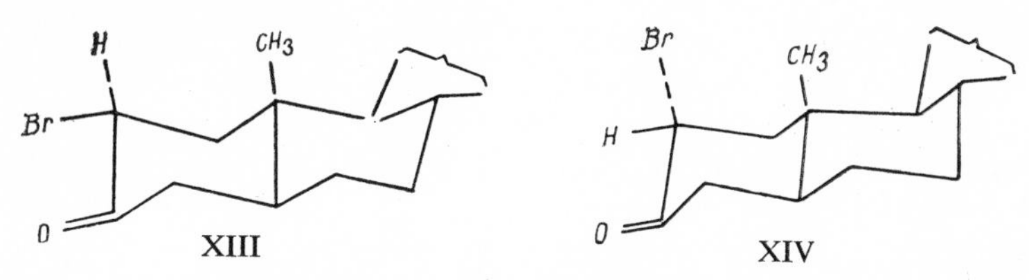

5.3.2.3 Internal Rotation in Molecules, Conformation

The manifestations of molecular internal rotation have already been frequently mentioned; here the basis of the phenomenon will be discussed, as well as the possibilities of using vibrational spectroscopy for studying rotational isomerism in molecules.

Internal rotation occurs around single bonds in molecules. One of the simplest examples of this phenomenon is disulphane H^1S-SH^2. The H^1SS and SSH^2 equilibrium angles are 91°32′; internal rotation is described as a change in dihedral (torsion) angle τ between the H^1SS and SSH^2 planes (the molecule in the Newman projection, in which the two sulphur atoms – whose connecting bond forms the rotation axis – are represented in one place, is given in Fig. 70).

On rotation of certain parts of molecules around single bonds the potential energy of the molecule changes. The changes are usually expressed in terms of a periodical function of dihedral (torsion) angle τ from 0° to 360°. With some molecules the potential energy changes very little during internal rotation, in units or tens of joules per mole (units or tens calories per mole); these are molecules with free or almost free internal rotation, such as nitromethane, toluene, dimethylmercury, dimethylacetylene, tetrafluorodiborane, etc., and work with them is somewhat different from that with other molecules. For example, their symmetry cannot be described in terms of classical point groups of symmetry, but by permutation-inversion groups, the width of some bands in their spectra is extraordinarily temperature-dependent, etc.

In most molecules capable of internal rotation the potential energy changes connected with internal rotation are quite pronounced. The dependence of the potential energy on the dihedral (torsion) angle τ for disulphane is depicted in Fig. 71. The plot exhibits two equally deep minima at about

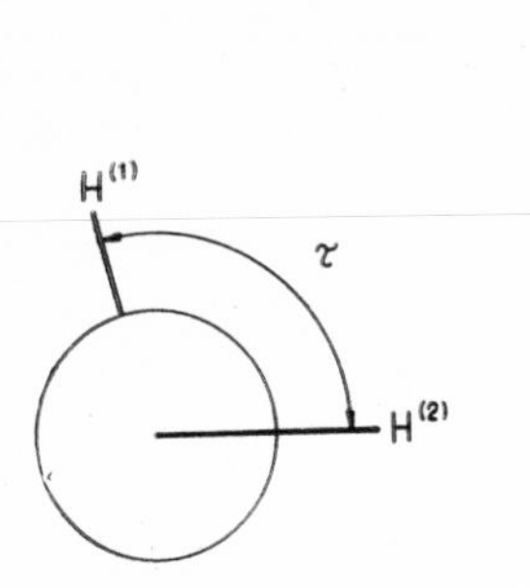

Fig. 70. The disulphane molecule HSSH in the Newman projection (the two S atoms overlap). τ is the torsion angle between the $H^{(1)}SS$ and $SSH^{(2)}$ planes.

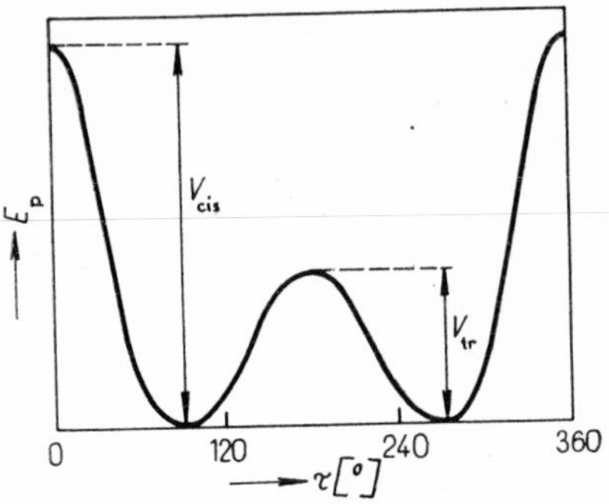

Fig. 71. Dependence of potential energy E_p of the disulphane molecule on torsion angle τ. V_{tr} and V_{cis} — barriers preventing internal rotation.

90° and 270° and two unequal maxima; the lower one at 180° represents the "*trans*"-barrier (the barrier is the potential energy difference between the maximum and the adjacent minimum on the curve), 8 kJ . mole^{-1}, i.e. 1.9 kcal . mole^{-1}, and the higher the "*cis*"-barrier, 31 kJ . mole^{-1}, i.e. 7.4 kcal . mole^{-1}. The contents of the individual forms of disulphane rotational isomers (or conformations, rotamers) can be derived from the plot in Fig. 71. The forms with the lowest potential energy will be populated most; the population of forms decreases exponentially with increasing potential energy. These considerations are valid for the gaseous state and laboratory conditions; certain shifts in the equilibrium may occur in the liquid state and especially in crystals (and sometimes a form can be completely frozen-out e.g. in crystals at very low temperatures).

The potential energy change vs. molecular dihedral angle plot for disulphane exhibits two minima, corresponding to forms which are virtually indistinguishable on the basis of their infrared spectra. However, there also are molecules exhibiting minima on the potential energy curve for species which differ in their infrared spectra. The content of the molecules in a given form is again determined by the statistical distribution laws; the decisive factors are the potential energy difference for the corresponding molecules and the temperature (this holds strictly only for the gaseous state). With decreasing temperature, the thermodynamically stable forms will start predominate and with increasing temperature less stable forms will begin to become important.

A molecule of this type is 1,2-dichloroethane. If the internal rotation around the C–C bond in its molecule is expressed by a dihedral (torsion) angle between the Cl^1CC and $CCCl^2$ planes in the $Cl^1CH_2CH_2Cl^2$ molecule (Fig. 72), the plot depicted in Fig. 73 is obtained. Three minima appear

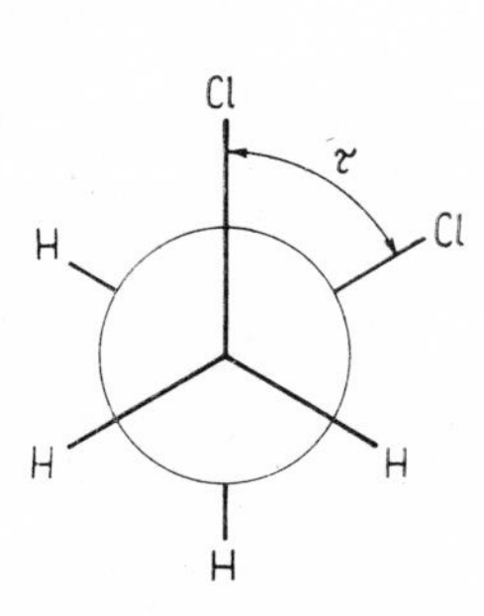

Fig. 72. The 1,2-dichloroethane molecule, $ClCH_2CH_2Cl$, in the Newman projection (the two C-atoms overlap). τ is the torsion angle between the $Cl^{(1)}CC$ and $CCl^{(2)}$ planes.

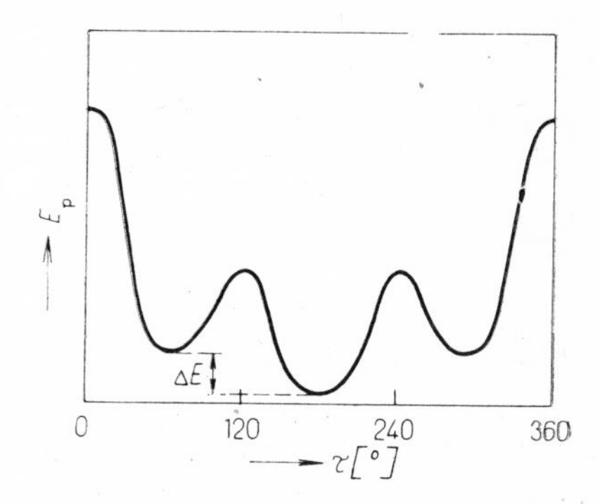

Fig. 73. Dependence of the potential energy of the 1,2-dichloroethane molecule, E_p, on torsion angle τ. ΔE— difference in the potential energies of the antiperiplanar ($\tau = 180°$) and synclinal ($\tau = 60°, 300°$) forms.

in the region from $\tau = 0°$ to $\tau = 360°$, two of them equal, different from the third, deepest one. The molecules corresponding to these dihedral angles do not differ very much in their potential energy, so that these forms will occur in the presence of one another in 1,2-dichloroethane samples.

Cyclohexane derivatives also exhibit rotational isomerism. This is a complex effect, when in conversion from one isomer into another internal rotation must occur around all C–C bonds of the cyclohexane ring. For example, in chlorocyclohexane this process leads to a substituent (the chlorine atom) inversion from the equatorial position (the more stable form, XV) to an axial position (XVI). All the hydrogen atoms, of course, also move,

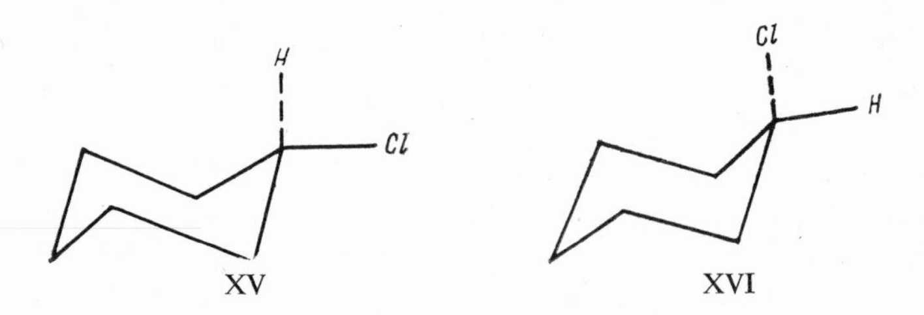

but they cannot be distinguished.*) If the cyclohexane is part of a molecule with a greater number of rings, as e.g. in steroids, synchronous rotation around all C–C bonds of the cyclohexane ring can no longer occur; molecules of this type are rigid**) and the substituents located on equatorial and axial bonds cannot be converted into one another other than by bond breakage. Thus rotational isomerism changes into geometric in these molecules.

Rotational isomerism was found in many types of aliphatic compounds, in hydrocarbons, alkylsilanes, dialkyl ethers, sulphides, etc.; it mostly concerns the movements of heavy atoms in the skeleton. Skeletal rotational isomerism can be exhibited only by compounds, that contain at least four heavy atoms in series (i.e. n-butane, methylethyl ether, etc.). These molecules have limited numbers of rotational isomers differing rather strongly in their spectra. The number of isomers sharply increases with an increasing number of linearly chained skeletal atoms; in the infrared spectra of compounds with a larger number of atoms in linear chains measured in the liquid state (or in vapours), only the sum curve appears, in which the individual contributions of the rotational isomers cannot be distinguished.

The spectra of alcohols also exhibit internal rotation. Skeletal internal rotation appears with n-alkanols with three or more carbon atoms in a linear chain. However, rotational isomerism originating from the rotation of the hydroxyl H-atom around the C–O single bond can also be discovered

*) The considerations of the cyclohexane ring assume its existence in the chair form. However, cyclohexane can also occur in the less stable boat form, for which, however, rather analogous rules are valid.

**) The cyclohexane ring in these compounds can only be converted from the chair into the boat configuration.

in the vibrational spectra. In the spectra of dilute solutions of methanol and tert. butanol in tetrachloromethane (under conditions when selfassociation no longer appears), very narrow bands of the stretching vibration $\nu(O-H)$ of the O–H bonds occur at 3644 cm^{-1} (methanol) or 3617 cm^{-1} (tert. butanol). However, in the spectra of ethanol and isopropanol measured under the identical conditions, doublets appear with branches at 3638 and 3627 cm^{-1} (ethanol, values obtained after separation of the overlapping bands) and 3627 and 3617 cm^{-1} (isopropanol). The simple structure of the $\nu(O-H)$ band found in the methanol and tert. butanol spectra results from the fact that all rotational isomers formed in hindered internal rotation around the C–O bond are indistinguishable. In the methanol molecule, the hydrogen of the hydroxyl group always occurs between two hydrogen atoms of the methyl group; in the tert. butanol it always occurs between two methyls of the tert. butyl group. However, the rotational isomers of ethanol and isopropanol can be distinguished, in dependence on whether the hydrogen of the hydroxyl group occurs between the two hydrogens of the CH_2 group or between one of these hydrogens and a methyl group (in ethanol), or between the hydrogen of the CH group and a methyl group or between two methyl groups (in isopropanol). Rotational isomerism can similarly occur with the N–H group, etc.

Rotational isomerism has also been discovered in phenol derivatives. The hydrogen of the hydroxyl group in phenols always lies in the same plane as all the atoms of the benzene ring and can thus be *cis* or *trans* with respect to substituents in positions 2 or 3. Rotational isomerism of this type occurs only occasionally, expecially if position 2 is substituted with a bulky alkyl group. In the spectrum of a dilute tetrachloromethane solution of 2-tert. butyl-4-methylphenol (under conditions excluding selfassociation), a weak band of the less abundant *cis*-form appears at 3650 cm^{-1} and a strong band of the *trans*-form at 3610 cm^{-1}. The spectrum of 2,6-ditert. butyl-4-methylphenol contains only a single $\nu(O-H)$ band at 3650 cm^{-1} and that of phenol also has a single band at 3612 cm^{-1} (measured under similar experimental conditions).

The spectra enable distinguishing of substituents on equatorial and axial bonds on a cyclohexane ring of non-rigid molecules. The stretching vibration band of the C–Cl bond of chlorocyclohexane was found at 728 cm^{-1} with the Cl atom in the equatorial position and at 683 cm^{-1} with the Cl atom in the axial position. Equatorial substitution is usually connected with a larger wavelength of the stretching vibration band than that for axial substitution; e.g., with bromocyclohexane the $\nu(C-Br)$ band for the equatorial form lies at 685 cm^{-1} and that for the axial form at 658 cm^{-1}. The two forms occur simultaneously in liquid samples of chloro- and bromocyclohexane; hence both the bands appear in the infrared (or Raman) spectrum of liquid samples (in the intensities representing their relative contents).

Rotational isomerism is also exhibited by compounds with two (or more) conjugated double bonds in the linear chain; internal rotation occurs around the single bond connecting the two double-bonds. Dienes, unsaturated, conjugated aldehydes (ketones) and α-dialdehydes (α-ketoaldehydes, α-diketones) can therefore occur in two arrangements (owing to the conjugation of the double-bonds both are planar) with *cis*- or *trans*-oriented bonds (denoted as *s-cis*- and *s-trans*-forms) XVII and XVIII. The forms differ in the strength of the vibration interaction (this is larger with the *s-cis*-forms) and consequently their spectra also differ.

X=C–C=X (s-cis) XVII

X=C–C=X (s-trans) XVIII

In the molecules of the *s-cis*-forms of unsaturated, conjugated ketones the two double bonds are vibrationally coupled more strongly than in the *s-trans*-form; the difference in the $\nu(C{=}O) - \tilde{\nu}(C{=}C)$ wavenumber values is always larger with the *s-cis*-form than with the *s-trans*-form and the ratio of the band integrated intensities, $A(C{=}O)/A(C{=}C)$ is always smaller with *s-cis*-forms than with *s-trans*-forms.

Rotational isomerism also occurs in the molecules of secondary amides and lactams, on the peptide bond, HN–CO. Owing to strong delocalization of the electrons from the nitrogen atom lone electron pair toward the carbonyl, the N–C bond has partial double-bond character, which leads to stabilization of rotational isomers with planar arangement of the HNCO atoms in the peptidic bond. The *s-trans*-arrangement (XIX) strongly predominantes in the molecules of secondary amides R^1NHCOR^2, provided that they are not sterically protected by bulky substituents R; the band called "amide II" is most typical of this arrangement. This is a relatively intense band in the infrared spectrum at 1515 cm^{-1}, belonging to a characteristic vibration of the peptidic bond with a large contribution from the in-plane deformation vibration of the hydrogen. This band is so sensitive that it can be used for testing the presence of *s-trans*-forms of lactams. The band does not exist in this region in the spectra of lactams with rings smaller than eight-membered in which the *s-cis*-arrangement (XX) is forced on the peptide bond. Spectra of lactams with larger rings, in which the peptide bond can exist in the *s-trans*-arrangement, contain a band corresponding to amide II.

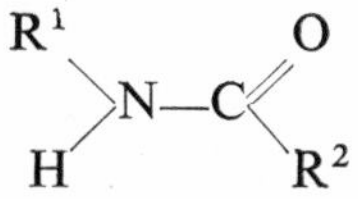

$R^1(H)N{-}C(=O)R^2$ (XIX) $R^1(H)N{-}C(R^2)=O$ (XX)

XIX 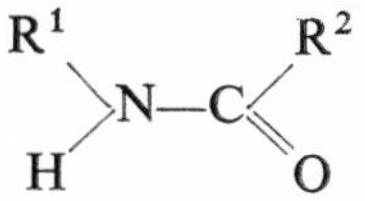 XX

With some secondary amides the *s-cis*- and *s-trans*-forms can also occur simultaneously; the specific manifestations of the two forms in the N–H bond stretching region yield reliable information. However, the ability of molecules with peptidic bond in the *s-cis*-form to produce very stable dimers (XXI) should be remembered. The arrangement and the stability of these dimers are similar to those of carboxylic acid dimers.

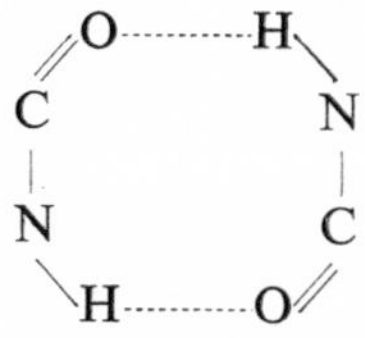

O·········H
C N
N C
H·········O

XXI

Rotational isomerism can, of course, occur with groups containing no carbon atoms. A typical example is nitrous acid and its derivatives, RO–NO, in which rotation occurs around the O–N single bond. With nitrous acid esters (R = alkyl), manifestations of *trans*-isomers were found between 1680–1670 cm^{-1} (the band of the N=O bond stretching vibration) and of *cis*-isomers at 1615 cm^{-1}.

5.3.2.4 Intramolecular Hydrogen Bond

In compounds with molecules containing atoms or groups with lone electron pairs (halogen atoms, OR, NR_2, C=O groups) close to a hydroxyl group (or NH group), an intramolecular hydrogen bond may be formed under certain circumstances. The rings formed are mostly five- or six-membered, but are not necessarily so; larger rings connected by intramolecular hydrogen bonds are formed in molecules where interacting ends can approach one another sufficiently as a result of internal rotation.

The hydrogen bond strength depends on the acidity of the donor group, the basicity of the acceptor group and the hydrogen bond geometry. Some of these properties can be derived from the manifestations of the intramolecular hydrogen bond in the vibrational spectra.

An intramolecular hydrogen bond appears in the infrared spectra as typical bands of stretching vibrations of the X–H bonds of the proton donor groups. The bands corresponding to vibrations of molecules with an intramolecular hydrogen bond occur in other regions than for molecules that do not contain a hydrogen bond. The stretching vibrations are shifted to lower wavenumbers, whereas deformation vibrations are shifted to higher wavenumbers. The bands of molecules with intramolecular hydrogen bonds are mostly very broad, especially the O–H and N–H bond stretching vibrations.

Information on intramolecular hydrogen bonding can be obtained only from the spectra of very dilute sample solutions; molecules capable of intramolecular interaction can also form intermolecular complexes in a liquid, crystal or concentrated solution and the spectra of these samples cannot always be interpreted correctly.

The general properties of molecules with an intramolecular hydrogen bond can be illustrated on the example of 2-chloroethanol, $ClCH_2CH_2OH$. Hindered internal rotation around the C–C and C–O bonds occurs in the molecule of this compound, so that various forms exist in dilute solutions of 2-chloroethanol, some of them being capable of intramolecular hydrogen bonding (XX), and some of them not (XXI, XXII). The infrared spectrum

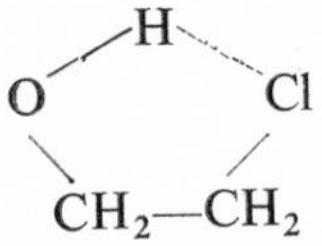

O–H···Cl (ring with CH_2—CH_2)

XX

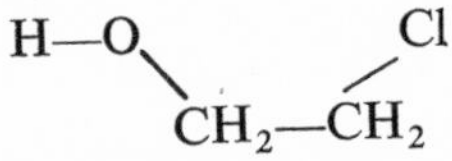

H—O, CH_2—CH_2, Cl

XXI

H—O, CH_2–CH_2, Cl

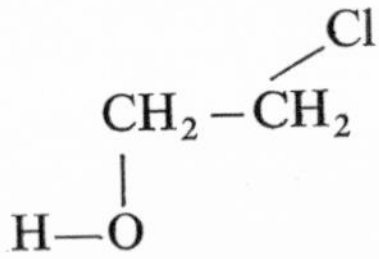

XXII

is the sum of the spectra of all existing forms; forms with unbound hydroxyl exhibit an ν(O–H) band at 3623 cm^{-1}, those with hydroxyl bound to the chlorine atom at 3597 cm^{-1} (measured in very dilute tetrachloromethane solution). The intensity of the two bands are proportional to the relative contents of the individual forms in the mixture of rotational isomers.

The possibility of formation of intramolecular hydrogen bonds depends on the distance between the interacting groups. In the 2-chloroethanol molecule the two groups forming an intramolecular hydrogen bond are, located on neighbouring carbons. If the chain between the two groups, $Cl-(CH_2)_n-OH$, lengthens, i.e. *n* equals three, four, etc., the number of forms capable of intramolecular hydrogen bonding decreases substantially. With increasing chain length, the number of internal rotation centres increases and the number of rotational isomers which do not form hydrogen bonds also increases. For $n = 4$, practically no evidence of intramolecular hydrogen bonding can be found in the spectra; only if a strongly basic NR_2 group is present in place of the chlorine atom can hydrogen bonding be observed up to $n = 6$.

In vicinal derivatives, such as 2-chloroethanol, a five-membered hydrogen bond is formed, while six-, seven- and more-membered rings are formed if the substituents are more distant. The forms in which the hydrogen-bonded atoms lie along a straight line are most stable; this situation cannot occur in a five-membered intramolecular hydrogen bond, which is consequently weaker than e.g. a six-membered bond. This can be demonstrated e.g. on aminoalcohols; the bound hydroxyl band in 3-amino-2-butanol with a five-membered hydrogen bond is shifted by 120 cm^{-1} compared with that of an unbound hydroxyl; this shifts amounts to as much as 250 cm^{-1} in 4-amino-2-pentanol containing a six-membered ring.

Another large group of compounds containing intramolecular hydrogen bonds are the *o*-derivatives of phenol. Forms with a free hydroxyl band in their spectra have the hydroxyl proton *trans*-arranged toward the substituent; species with a *cis*-hydroxyl form hydrogen bonds and exhibit the bound hydroxyl band.

For example, with 2-chlorophenol, the form with a hydrogen bent away from the chlorine atom yields a band at 3609 cm^{-1} and the form with a closed intramolecular hydrogen bond yields an $O-H$ bond stretching vibration band at 3547 cm^{-1} (in very dilute tetrachloromethane solution in order to exclude intermolecular association); in 2-chlorophenol the form with the intramolecular hydrogen bond is present in larger amounts and the band at 3547 cm^{-1} is consequently more intense.

On formation of intramolecular hydrogen bonds between hydroxyl groups and neighbouring substituents in phenol derivatives, not only does the wavenumber value of the $\nu(O-H)$ band change; the vibration band of the bound hydroxyl is also substantially broader and its integrated intensity is greater. The wavenumber of the $\nu(O-H)$ band decreases with increasing strength of the intramolecular hydrogen bond, while the integrated intensity and band half-width increase. However, it can happen that the intensity of the band maximum of the bound form is less than that for the free form; the increase in the integrated intensity is then a result of the extraordinary broadening of the band. In exceptional cases of hydroxycompounds with very strong intramolecular hydrogen bonds (e.g. some derivatives of hydroxyanthraquinone) the $\nu(O-H)$ band is shifted to the region below $3000\ cm^{-1}$ and is so broad that it is practically indistinguishable.

With compounds with strong intramolecular hydrogen bonds only one form (bound) can occur in measurable concentrations: the spectra of dilute solutions of these compounds then exhibit only the band of the bound form (usually broad). The methylester of salicylic acid, which forms a very stable intramolecular hydrogen bond, exhibits a broad $\nu(O-H)$ band at $3207\ cm^{-1}$. The stronger the intramolecular hydrogen bond, the smaller is the probability that the bond will be broken at higher concentrations or by basic solvents. For example, the hydrogen bond in 2-hydroxyacetophenone is so strong that it is not even broken in pyridine solutions.

Infrared spectra yield information about formation of intramolecular hydrogen bonds in quite specific cases. For example, intramolecular hydrogen bonding was detected in the molecules of phenoxyacetic acid (XXV), 2-phenylethanol (XXVI) and even in α-aminoacids (XXVII) in very dilute solutions in nonpolar solvents.

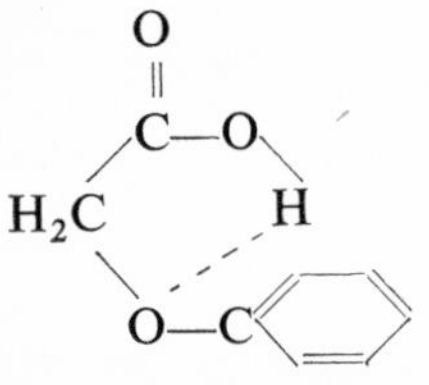

XXV 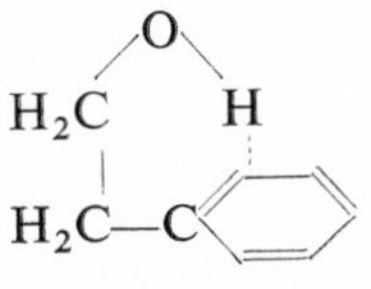 XXVI 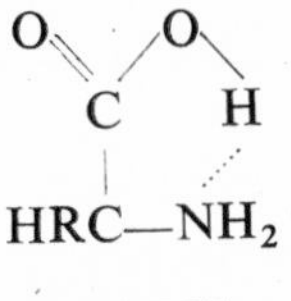 XXVII

Intramolecular hydrogen bonding in very dilute sulutions of the last compound type was assumed on the basis of the anomalous wavenumber for the carbonyl band, which is located in the same region as the carbonyl band in five-membered lactones (near $1770\ cm^{-1}$). In β-aminoacids, the corresponding form with a six-membered ring gives a band at lower wavenumber values; the content of the forms with an intramolecular hydrogen bond is naturally low.

Very elegant and practically useful procedures enable examination of the detailed spatial arrangement of molecules with the help of the data derived from the infrared spectra of species containing intramolecular hydrogen bonds. Studies of this kind are usually carried out with larger groups of compounds and the results are mostly only relatively valid.

The geometric arrangement of the molecule is indicated by certain interacting groups (which are identical for the whole series of compounds), so that the strength of the hydrogen bond reflects primarily the distance between the interacting groups and/or the geometry of the bond formed.

The magnitude of the interaction, i.e. the strength of the hydrogen bond, is most frequently described using quantity $\Delta\tilde{\nu}$, which is the wavenumber difference between the bands of chosen vibrations of the free and bound form (the bound form is that with an intramolecular hydrogen bond). Increasing $\Delta\tilde{\nu}$ values indicate greater stabilization energies and closer interacting groups or a more linear arrangement of the atoms in the hydrogen bond.

The practical importance of the described procedure will be illustrated on the results found in the measurement of the infrared spectra of very dilute solutions of 1,2-cycloalkanediols and 2-amino-1-cycloalkanols in tetrachloroethylene with different numbers of C-atoms in the rings. It can be seen from Table LI that the hydrogen bond is stronger in the *cis*-isomers of compounds with smaller rings, so that it can be assumed that the interacting groups are closer together in the *cis*-isomers than in the corresponding

Table LI

The differences in the wavenumbers, $\Delta\tilde{\nu}$, of the ν(O—H) vibration of free and bound hydroxyls in cyclic 1,2-cycloalkanediols and 2-amino-1-cycloalkanols[a] (cm^{-1})

Number of carbons in the ring	Diols		Aminoalcohols	
	cis	*trans*	*cis*	*trans*
5	61	0	175	0
6	38	33	127	93
7	44	37	139	123
8	51	43	133	151
9	49	45	134	139
10	44	45	123	146
11	—	—	117	160
12	38	51	102	163

[a] Values taken from the infrared spectra measured in very dilute solutions in tetrachloroethylene as the solvent.

trans-isomers (in agreement with assumptions on the geometry of these rings). However, the $\Delta\tilde{\nu}$ values indicate that the interaction is, on the contrary, stronger in the *trans*-isomers if the rings are larger.

5.4 Vibrational Spectra and Hypermolecular Structures, Macromolecules

The appearance of infrared and Raman spectra depends not only on the molecular structures but—as has already been mentioned—on their arrangement as a whole. The interaction of molecules in larger arrangements results in a change in the distribution of the electron density within the molecules; where orientation effects are important, these changes can result in a loss of symmetry of individual molecules.

Information on forces acting between molecules in the condensed phase is also contained specifically in vibrational spectra. In some cases spectral information can be employed relatively easily for the description of interactions (e.g. in the formation of complexes, linear or cyclic dimers, trimers etc.); in other cases, the manifestation can be used only to indicate the degree of interaction without employing spectral information to elucidate the character of the interaction. Information on molecular interaction is especially contained in the band shapes and in band widths; however, in many cases correct interpretation of experimental data is not yet possible.

In conclusion, it is necessary to mention the application of vibrational spectra in macromolecular chemistry. It should be noted that both infrared and Raman spectra have been used extensively in this branch of chemistry. Spectroscopy can yield extremely valuable information both in research and in production.

When dealing with macromolecular compounds, a certain structural inhomogenity must be expected (with both natural and synthetic polymers or copolymers). Compound samples may contain macromolecules of varying chain length, with different spatial arrangements of the heavy skeleton and even with different compositions. With commercial products, additives are most frequently present, such as fillers, stabilizers, antioxidants, etc. However, spectra can yieldbasic information in these areas and can frequently be employed to estimate the amounts of additives, etc.

The application of infrared and Raman spectra in macromolecular chemistry has been gradually developed in great detail, as can be seen from literature.[1-5]

Many problems in macromolecular chemistry are studied using molecular models taken from low molecular weight compounds. Information on the relationships between the spectra and the structures of model compounds, transferred to macromolecular systems, are identical with the data discussed in Chapter 5 in connection with classical chemical compounds.

5.4.1 Crystals and Reflection Spectroscopy

Neither infrared absorption spectra nor Raman spectra yield sufficient information for the description of the properties of solid substances and crystals (monocrystals). Their specific properties are, however, very well suited to reflectance spectra.

Here reflectance spectra will be understood to refer to the wavenumber dependence of the ratio of the spectral radiant power reflected during perpendicular reflection from the surface of the tested material to the incident radiant power. This quantity is termed the reflectance (reflection factor) and is designated*) $\varrho(\tilde{\nu})$.

The reflectance properties of materials are described by the complex quantity $\hat{r}(\tilde{\nu})$, called the reflectivity, which expresses the dependence of the vector of the electric field $\boldsymbol{E}_r$ of the reflected radiation on vector $\boldsymbol{E}_i$ of the incident radiation:

$$\boldsymbol{E}_r(\tilde{\nu}) = \hat{r}(\tilde{\nu}) \cdot \boldsymbol{E}_i(\tilde{\nu}). \tag{5.64}$$

The spectral power $\Phi_{\tilde{\nu}}$ is proportional to the square of the magnitude of the electric field vector; consequently, the reflectance, which can be measured experimentally, depends on the reflectivity according to the relationship

$$\varrho(\tilde{\nu}) = \Phi_{\tilde{\nu},r}(\tilde{\nu})/\Phi_{\tilde{\nu},i}(\tilde{\nu}) = |\boldsymbol{E}_r(\tilde{\nu})|^2/|\boldsymbol{E}_i(\tilde{\nu})|^2 = |\hat{r}(\tilde{\nu})|^2, \tag{5.65}$$

and is thus equal to the square of the absolute value of the reflectivity.

---

[1] Zbinden R.: Infrared Spectroscopy of High Polymers. Academic Press, New York 1964.

[2] Hummel D. O.: Infrared Spectra of Polymers in the Medium and Long Wavelength Regions. Wiley, New York 1966.

[3] Henniker C. J.: Infrared Spectrometry of Industrial Polymers. Academic Press, New York 1967.

[4] Dechant J.: Ultrarotspektroskopische Untersuchungen an Polymeren. Akademie-Verlag, Berlin 1972.

[5] Van Krevelen D. W., Hoftyzer P. J.: Properties of Polymers. Correlations with Chemical Structure. Elsevier, Amsterdam 1972.

*) In some papers the quantity $R(\tilde{\nu})$ is encountered; in this denotation $\varrho(\tilde{\nu})$ is reserved for the amplitude $\varrho(\tilde{\nu}) = \sqrt{R(\tilde{\nu})}$.

The complex reflectivity $\hat{r}(\tilde{\nu})$ is defined as

$$\hat{r}(\tilde{\nu}) = \sqrt{\varrho(\tilde{\nu})} \exp\left[i\Theta(\tilde{\nu})\right]; \quad (5.66)$$

quantity $\Theta(\tilde{\nu})$, giving the changes in the phase of the electric field of the electromagnetic wave during reflection, called the reflection phase angle, depends on optical constants $n(\tilde{\nu})$ and $k(\tilde{\nu})$ according to the relationship

$$\tan \Theta(\tilde{\nu}) = 2k(\tilde{\nu})/[n(\tilde{\nu})^2 + k(\tilde{\nu})^2 - 1]. \quad (5.67)$$

The optical constants represent the real and imaginary parts of the complex refractive index $\hat{N}(\tilde{\nu})$

$$\hat{N}(\tilde{\nu}) = n(\tilde{\nu}) + ik(\tilde{\nu}). \quad (5.68)$$

The complex reflectivity $\hat{r}(\tilde{\nu})$ depends on the complex refraction index $\hat{N}(\tilde{\nu})$ according to the relationship

$$\hat{r}(\tilde{\nu}) = [\hat{N}(\tilde{\nu}) - 1]/[\hat{N}(\tilde{\nu}) + 1]; \quad (5.69)$$

from which the reationship*)

$$\varrho(\tilde{\nu}) = \{[n(\tilde{\nu}) - 1]^2 + k(\tilde{\nu})^2\}/\{[n(\tilde{\nu}) + 1]^2 + k(\tilde{\nu})^2\}. \quad (5.70)$$

The optical constants $n(\tilde{\nu})$ and $k(\tilde{\nu})$ are obtained from the expresssions

$$n(\tilde{\nu}) = [1 - \varrho(\tilde{\nu})]/[1 + \varrho(\tilde{\nu}) - 2\sqrt{\varrho(\tilde{\nu})}\cos\Theta(\tilde{\nu})] \quad (5.71)$$

and

$$k(\tilde{\nu}) = [-2\sqrt{\varrho(\tilde{\nu})}\sin\Theta(\tilde{\nu})]/[1 + \varrho(\tilde{\nu}) - 2\sqrt{\varrho(\tilde{\nu})}\cos\Theta(\tilde{\nu})], \quad (5.72)$$

assuming that the reflectance $\varrho(\tilde{\nu})$ and the reflection phase angle $\Theta(\tilde{\nu})$ are known. The reflectance can be found experimentally; the reflection phase angle is calculated from the Kramers – Krönig relationship

$$\Theta(\tilde{\nu}) = -(\tilde{\nu}/\pi) \int_0^\infty \{[\ln \varrho(\xi)]/[\xi^2 - \tilde{\nu}^2]\} d\xi, \quad (5.73)$$

where ξ represents the integration variable. The phase angle can be found with greater precision, the greater the interval in which the dependence of ϱ is known.

If the values of the optical constants of solid substances or crystals can be found using equations (5.71) – (5.73), then further quantities can be

*) A simpler relationship (3.1) was employed in Chapter 3; there the reflectance $\varrho(\tilde{\nu})$ was expressed only as a function of quantity $n(\tilde{\nu})$. The simpler relationship can be used for the application described in Chapter 3 because the studied material does not absorb in the region of the validity of the relationship.

determined. First, the absorption factor $\alpha(\tilde{\nu})$ can be determined from the relationship

$$\alpha(\tilde{\nu}) = 4\pi k(\tilde{\nu}) \, . \, \tilde{\nu} \tag{5.74}$$

and thus the course of the absorption spectrum can be found. Primarily, however, the real $\varepsilon_1(\tilde{\nu})$ and imaginary $\varepsilon_2(\tilde{\nu})$ parts of the complex relative dielectric permittivity can be found

$$\hat{\varepsilon}(\tilde{\nu}) = \varepsilon_1(\tilde{\nu}) + i\varepsilon_2(\tilde{\nu}) \tag{5.75}$$

on the basis of its relationship to the complex refractive index

$$\hat{\varepsilon}(\tilde{\nu}) = [\hat{N}(\tilde{\nu})]^2. \tag{5.76}$$

It follows that

$$\varepsilon_1(\tilde{\nu}) = n(\tilde{\nu})^2 - k(\tilde{\nu})^2 \tag{5.77}$$

and

$$\varepsilon_2(\tilde{\nu}) = 2n(\tilde{\nu}) \, . \, k(\tilde{\nu}). \tag{5.78}$$

The advantages of the study of these phenomena for the determination of the properties of solid substances and crystals follow from this short description of basic information and relationships between quantities appearing in perpendicular reflection from the surface. However, reflection spectra exhibit a number of experimental pecularities. Primarily in the wave-number region where the studied material does not absorb, reflection can and does occur not only on the front surface of the sample, but also on the back surface (e.g. with thin plate) and the value of the reflectance measured is thus distorted. Strong interference also frequently occurs under the described conditions. Details on the theory and experimental technique in this field can be found in the literature.[1-4]

There are further procedures in which the reflection from the surface of the material is measured at angles at which complete reflection occurs. These are primarily the ATR (Attenuated Total Reflection) and FMIR (Frustrated Multiple Internal Reflection) methods. In these experiments the absorption of radiation by a sample placed beyond the reflecting surface is measured; the sample is applied to the inner surface of the reflecting plate (in the FMIR arrangement) and the radiation passes through the crystal.

---

1 Harrick N. J.: Internal Reflection Spectroscopy. Interscience, New York 1967.

2 Wendlandt W. W., Hecht H. G.: Reflectance Spectroscopy. Wiley, New York 1966. New edition: Wendlandt W. W.: Modern Aspects of Reflectance Spectroscopy. Plenum Press, New York 1968.

3 Kortüm G.: Reflexions-Spektroskopie. Springer, Berlin 1969.

4 Abelès F. (editor): Optical Properties of Solids. North-Holland, Amsterdam 1972.

If the totally reflected radiation is registered, it does not contain the wavenumbers of the absorbed radiation. The results of these experiments are thus spectra analogous to absorption spectra.

The reflection plates are made of material with a high refraction index so that the total reflection angle need not be too great (e.g. germanium, KRS-5).

These methods have, of course, nothing in common with methods measuring real reflection spectra, described in the first part of section 5.4.1. Nonetheless, they are widely used, primarily because they can be employed to determine absorption spectra of materials which are difficult to measure by direct passage of radiation. They are also used for substances with very high absorption (which would be difficult to prepare in a thin layer) and for substances with a specific consistency (pastes, putty, woven materials, etc.).

5.5 Documentation of Vibrational Spectroscopy

Original works and monographs or textbooks on infrared and Raman spectroscopy and on some special questions in this field have been cited frequently in this book. In this section, sources of information for systematic literature survey and for review purposes will be summarized.

For the sake of completeness if should be pointed out that the basic sources of information about spectra and spectroscopy are the abstracting journals, Chemical Abstracts, Referativnyi Zhurnal and Chemisches Zentralblatt. Recently, computer excerpts from Chemical Abstracts have been utilized especially frequently. Some information can also be obtained from the Spectroscopic Abstracts.

Many journals also publish regular or frequent reviews of original papers in certain spectroscopic fields, e.g. Analytical Chemistry (yearly or biyearly very detailed reviews) or periodical publications such as Advances in Spectroscopy, Advances in Physical Chemistry, etc.

The titles of spectroscopic works are periodically recorded in the series, Schnellinformation über die IR- und Raman Spektroskopie.[1] The titles of works on infrared and Raman spectroscopy can be found classified according to fields in the punched-card documentation "Junior Index"

---

[1] Published by Deutsche Akademie der Wissenschaften zu Berlin, Institut für Optik und Spektroskopie, Informationsbüro Spektroskopie.

with references to the Literature List (Documentation of Molecular Spectroscopy, Butterworths, London).

Data on published spectra are collected in indices, An Index of Published Infrared Spectra, Ministry of Aviation, Technical Information and Library Service H.M.S.O., London, Vol. I and II 1960, Vol. III 1964 and Hershenson H. M.: Infrared Absorption Spectra Index for 1945 – 1957, Academic Press, New York 1959. The spectra are surveyed on the basis of spectra collections and of the original literature in Molecular Formula List of Compounds, Names and References to Published Infrared Spectra, ASTM Special Technical Publication No. 331, Philadelphia 1962.

Critically selected wavenumber values taken from the analysis of the vibrational spectra of many molecules can be found in the publications periodically edited by T. Shimanouchi, Tables of Molecular Vibrational Frequencies, Vol. I – III, 1968 – 1969 and other works.

References to works dealing with the spectroscopy of the compounds of the IVB main subgroup of elements are gathered in the publications of Licht K., Reich P.: Literature Data for IR, Raman and NMR Spectroscopy of Si, Ge, Sn and Pb Organic Compounds, Deutsche Verlag der Wissenschaften, Berlin 1971 and Bažant V. et al., Organosilicon Chemistry, Dekker, New York 1975.

Practically all large manufacturers of infrared and Raman spectrometers publish a lot of interesting information in their journals, often dealing with more general problems than those involving their products.

Students and teachers can find a number of collections of problems dealing with the interpretation of spectra and the determination of molecular structure.[1–9]

---

1 Cairns T.: Spectroscopic Problems in Organic Chemistry. Heyden, London 1965.

2 Baker A. J., Eglinton G., Cairns T., Preston F. J.: More Spectroscopic Problems in Organic Chemistry. Heyden, London 1967.

3 Trost B. M.: Problems in Spectroscopy: Organic Structure Determination by NMR, IR, UV and Mass-Spectra. Addition—Wesley, New York 1967.

4 Phillips L.: Interpreted Infrared Spectra. Cit. from Chem. Ind. 1969, 21, 689—690.

5 Van der Maas J. H.: Basic Infrared Spectroscopy. Heyden, London 1969.

6 Colthup N. B.: Interpretation of Infrared Spectra (Includes Five Cassette Tapes). American Chem. Soc., New York 1971.

7 Cook B. W., Jones K.: A Programmed Introduction to Infrared Spectroscopy Heyden, London 1971.

8 Van der Maas J. H., Willis H. A.: Interpretation of Infrared Spectra—an Audiovisual Course. Heyden, London 1974.

References

Collections of Spectra

1. American Society for Testing Materials: 92,000 IR Spectra Cards. ASTM, Philadelphia.
2. The Infrared Spectral Data, American Petroleum Institute Research Project 44. Carnegie Institute of Technology, Pittsburgh.
3. Stadtler Standard Spectra. The Stadtler Research Laboratories, Philadelphia.
4. Documentation on Molecular Spectroscopy. Butterworths, London.
5. Dobriner K., Katzenellenbogen E. R., Jones R. N.: Infrared Absorption Spectra of Steroids. An Atlas. Interscience, New York 1953.
6. Roberts G., Gallagher B. S., Jones R. N.: Infrared Absorption Spectra of Steroids. An Atlas. Interscience, New York 1958.
7. Plíva J., Horák M., Herout V., Šorm F.: Die Terpene, Sammlung der Spektren und physikalischen Konstanten. Teil I.: Sesquiterpene. Akademie Verlag, Berlin 1960.
8. Horák M., Motl O., Plíva J., Šorm F.: Die Terpene, Sammlung der Spektren und physikalischen Konstanten. Teil II.: Monoterpene. Akademie Verlag, Berlin 1963.
9. Moenke H.: Mineralspektren. Teil I und II. Akademie Verlag, Berlin 1962.
10. Sunshine I., Gerber S. R.: Spectrophotometric Analysis of Drugs Including Atlas of Spectra. Thomas, Springfield (Illinois, USA) 1963.
11. Neudert W., Röpke H.: Steroid-Spektrenatlas. Springer, Berlin 1965.
12. Holubek J. et al.: Spectral Data and Physical Constants of Alkaloids. Vol. I (1965) — VIII (1970). ČSAV Publ. House, Prague.
13. Mecke R., Langenbucher F.: Infrared Spectra of Selected Chemical Compounds. Heyden, London 1967.
14. Bentley F. F., Smithson L. D., Rozek A. L.: Infrared Spectra and Characteristic Frequencies 700 — 300 cm^{-1}. Interscience, New York 1968.
15. Hummel D. O.: Infrared Spectra and Analysis of Polymers, Resins and Additives. Fibres and Resins, an Atlas, Vol. I. Wiley, New York 1969.
16. Pristera F., Fredericks W.: Compilation of Infrared Spectra of Ingredients of Propellants and Explosives. Report No. AD-859-846. Picatinny Arsenal, Dover, N. J. 1969.
17. Pouchert C. J.: The Aldrich Library of Infrared Spectra (8000 IR-Spectra in One Convenient Volume). Emanuel, Middlessex 1970.
18. Nyquist R. P., Kagel R. O.: Infrared Spectra of Inorganic Compounds (3000 — 45 cm^{-1}). Academic Press, New York 1971.
19. Anonym: Atlas der IR-Spektren von Zellstoff Papier und Hilfstoffen. Institut für Makromolekulare Chemie der Techn. Hochschule, Darmstadt 1972.
20. Graselli J. G. (ed.): Atlas of Spectral Data and Physical Constants for Organic Compounds. The Chemical Rubber Co., Cleveland 1973.
21. Schrader B., Meier W. (eds.): Raman/IR-Atlas organischer Verbindungen. Verlag Chemie, Weinheim 1974.

Quantitative Analysis

1. Kössler I.: Methoden der Infrarot Spektroskopie in der Chemischen Analyse. Quantitative Analyse. Akademie Verlag, Leipzig 1966.

Interpretation of Spectra, Molecular Structure

1. Randall H. M., Fowler R. G., Fuson N., Dangl J. R.: Infrared Determination of Organic Structures. Van Nostrand, New York 1949.
2. Mizushima S. I.: Structure of Molecules and Internal Rotation. Academic Press, New York 1954.
3. Cross A. D.: An Introduction to Practical Infrared Spectroscopy. Butterworths, London 1960.
4. Nakanishi K.: Infrared Absorption Spectroscopy — Practical. Holden-Day, San Francisco 1962.
5. Bellamy L. J.: The Infrared Spectra of Complex Molecules. Methuen, London 1962.
6. Jones R. N., Sandorfy C.: The Application of Infrared and Raman Spectroscopy to the Elucidation of Molecular Structure. In the book, Technique of Organic Chemistry (A. Weissberger, ed.), Vol. IX. Interscience, New York 1963 (reedition 1969).
7. Flett M. S. C.: Characteristic Frequencies of Chemical Groups in the Infrared, Elsevier, Amsterdam 1963.
8. Rao C. N. R.: Chemical Applications of Infrared Spectroscopy. Academic Press. New York 1964.
9. Colthup N. B., Daly L. H., Wiberley S. E.: Introduction to Infrared and Raman Spectroscopy. Academic Press, New York 1964.
10. Szymanski H. A.: Infrared Band Handbook. Vol. I (1964) — IV (1967). Plenum Press, New York.
11. Freeman S. K.: Interpretative Spectroscopy. Reinhold, New York 1965.
12. Mathieson D. W. (ed.): Interpretation of Organic Spectra. Academic Press, New York 1965.
13. Zhbankov R. G., Stepanov B. I. (eds.): Infrared Spectra of Cellulose and its Derivatives. Plenum Press, New York 1966.
14. Little L. H.: Infrared Spectra of Adsorbed Species. Academic Press, New York 1966.
15. Brand J. C., Eglinton G.: Applications of Spectroscopy to Organic Chemistry. Davey, New York 1967.
16. Bolshakov G. F., Glebovskaya J. A., Kaplan E. G.: Infrakrasnyie Spektry i Rentgenogrammy Heteroorganicheshikh Soyedinenii. Khimiya, Leningrad 1967.
17. Wheatley P. J.: The Determination of Molecular Structure. Clarendon Press, London 1968.
18. Bellamy L. J.: Advances in Infrared Group Frequencies. Methuen, London 1968.
19. Venkstern T. V., Baev A. A.: Spectra of Nucleic Acid Compounds. Plenum Press, New York 1968.
20. Szymanski H. A.: Correlation of Infrared and Raman Spectra of Organic Compounds. Hettilon Press, Cambridge Springs (Pennsylvania, USA) 1969.
21. Welti D.: Infrared Vapour Spectra. Group Frequency Correlations. Sample Handling and the Examination of Gas Chromatographic Fractions. Heyden, London 1969.
22. Belanato J., Hidalgo A.: Infrared Analysis of Essential Oils. Heyden, London 1969.
23. Varsányi G.: Vibrational Spectra of Benzene Derivatives. Akad. Kiadó, Budapest 1969.

24. Alpert N. L., Kaiser W. E., Szymanski H. A.: Theory and Practice of Infrared Spectroscopy. Plenum Press, New York 1970.
25. Parker F. S.: Applications of Infrared Spectroscopy in Biochemistry, Biology and Medicine. Plenum Press, New York 1971.
26. Yamaguchi K.: Spectral Data of Natural Products. Elsevier, New York 1970.
27. Avram M., Mateescu G. D.: La spectroscopie infrarouge et ces applications en chimie organique. Dunod, Paris 1971.
28. Winstead M. B.: Organic Chemistry Structure Problems. Heyden, London 1971.
29. Steele D.: The Interpretation of Vibrational Spectra. Chapman and Hall, London 1971.
30. Simon W., Clere T.: Structural Analysis of Organic Compounds by Spectroscopic Methods. MacDonald, London 1971.
31. Chumaevskii N. A.: Kolebatelnyie Spektry Elementorganicheskikh Soiedinenii Grup IVB i VB. Nauka, Moscow 1971.
32. Volkmann H.: Infrarot-Spektroskopie. Verlag Chemie, Weinheim 1972.
33. O'Connor R. T. O. (ed.): Instrumental Analysis of Cotton Cellulose and Modified Cotton Cellulose. Dekker, New York 1972.
34. Siggia S. (ed.): Instrumental Methods of Organic Functional Group Analysis. Wiley, New York 1972.
35. Weitkamp H., Barth R.: Infrarot-Strukturanalyse. Ein dualistisches Interpretationsschema. Thieme, Stuttgart 1972.
36. Durig J. R. (ed.): Vibrational Spectra and Structure. Dekker, New York 1972.
37. Cyvín S. J. (ed.): Molecular Structure and Vibrations. Elsevier, Amsterdam 1972.
38. Nachod F. C., Zuckermann J. J. (eds.): Determination of Organic Structures by Physical Methods. Academic Press, New York 1973.
39. Thomas L. C.: Interpretation of the Infrared Spectra of Organophosphorus Compounds. Heyden, London 1974.
40. Horák M., Papoušek D. and coworkers: Infrared Spectra and Molecular Structure. ČSAV Publ. House, Prague 1976 (in Czech).

Crystals, Ionic and Inorganic Compounds

1. Lawson K.: Infrared Absorption of Inorganic Substances. Reinhold, New York 1961.
2. Nakamoto K.: Infrared Spectra of Inorganic and Coordination Compounds. Wiley, New York 1963.
3. Siebert H.: Anwendung der Schwingungsspektroskopie in der anorganischen Chemie. Springer, Berlin 1966.
4. Adams D. M.: Metal-ligand and Related Vibrations: A Critical Survey of the Infrared and Raman Spectra of Metallic and Organometallic Compounds. St. Martin's Press, New York 1968.
5. Hepdra P. J., Hallam H. J.: Inorganic Spectroscopic Problems. Heyden, London 1969.
6. Rao C. N. R., Ferraro J. R.: Spectroscopy in Inorganic Chemistry. Vol. I (1970) and II (1971). Academic Press, New York.
7. Ferraro J. R.: Low-Frequency Vibrations of Inorganic and Coordination Compounds. Plenum Press, New York 1971.

8. Greenwood N. N.: Spectroscopic Properties of Inorganic and Organometallic Compounds. The Chemical Society, London.
9. Jones L. H.: Inorganic Vibrational Spectroscopy. Dekker, New York 1971.
10. Abeles F. (ed.): Optical Properties of Solids. North-Holland, Amsterdam 1972.
11. Greenwood N. N., Ross E. J. F., Straughan B. P.: Index of Vibrational Spectra of Inorganic and Organometallic Compounds. Vol. I (1945—1960). Butterworth, London 1972.
12. Sherwood P. M. A.: Vibrational Spectroscopy of Solids. Cambridge University Press, London 1972.
13. Vlasova A. G., Florinskoi V. A.: Infrakrasnyie Spektry Inorganicheskikh Kristalov. Khimiya, Leningrad 1972.
14. Newman R. C.: Infrared Studies of Crystal Defects. Taylor and Francis, London 1973.

APPENDIX

Description of Computer Programs and Subroutines

In this appendix, some programs and subroutines are described which can be utilized in spectroscopic practice for handling the data obtained. The programs are constructed in the building-block form, so that individual users can simply modify them for their specific requirements.

The programs are written in the FORTRAN IV language, in an extent roughly corresponding to the level of compiler G in the IBM System/360 or System/370.[1,2] The source listings for the individual modules are given in Part e of this Appendix. In the left-hand column, four digit numbers of the individual statements, which, however, are not part of the program itself, are given; beyond a vertical line follows the program test itself and in the right-hand column the card identification serial number (this can be given on the card in columns 73 to 80; however, it is not necessary).

a. General Hints for Users

As has already been mentioned, the programs given here should be applicable without any modifications with IBM 360 and IBM 370 computers, provided with compiler FORTRAN IV (G). The compilers of other computers (e.g. IBM 7040, the computers from the CDC Company, etc.) have certain peculiarities which may necessitate some minor adjustments to the programs. If the user is experienced in programming, he can carry out these modifications himself; otherwise he will have to consult the system programmers of the given computing centre. In any case, the following points should be kept in mind:

1. The programs assume that the data are read through channel*)

---

1 "IBM System/360. FORTRAN IV Language". IBM Systems Reference Library C28-6515-7. New York 1968.

2 "IBM System/360 Operating System. FORTRAN IV (G and H) Programmers' Guide". IBM System Reference Library C28-6817-0. New York 1968.

*) In the IBM nomenclature: file identification number.

No. 5 and printed through channel No. 6. In the programs for separation, variables ICRQ and ILPR are given in READ and WRITE statements instead of constants 5 and 6 and values 5 and 6, respectively, are assigned to them in the introduction of the program (statements 0008 and 0009). If the given compiler requires input or output through another channel, or if it does not accept the use of an integer variable for denoting the channel number, appropriate systematic changes must be made.

2. In the programs, both literal constants (the texts between apostrophes) and the Hollerith constants (H-formats) are employed in parallel in FORMAT statements. However, some compilers do not accept the use of literal constants.

3. In the SMOO subroutine (statement No. 0003), the definition of the variable initial values is used in specification statement REAL. If the compiler does not permit such a definition, the above specification must be replaced by the following sequence of statements:

```
REAL CØ, C(3), NF
CØ = 7.
C(1) = 6.
C(2) = 3.
C(3) = -2.
NF = 21.
```

4. The programs commonly employ logical IF statement.

5. No procedure or program, however, uses less common extensions of the FORTRAN language, such as mixed arithmetical expressions, cycles with a real variable or loops with limits given in the form of arithmetic expressions.

In addition to these questions immediately concerned with the program structure itself, the user inexperienced in computing techniques must further consult several further points with system programmers in the computing centre.

6. It must be determined whether the centre prepares the programs for translation (punching and checking, if necessary) directly from the program reproductions, whether the reproductions must be rewritten in preprinted forms or whether the user has to arrange the punching elsewhere.

7. If the programs are prepared outside the centre, it must be known in what physical form the program can be provided (on punched cards or punched tape, in what code, in what form special characters are punched, such as plus and minus signs, equation signs, decimal points, commas, parentheses, apostrophes, division signs, asterisks, etc.).

8. In rewriting the programs and entering the initial data into forms the letter "o" and digit zero must be distinguished. Practice in various centres is very different. Most often symbol 0 is used for the letter and symbol Ø for the numeral, but the practice in the given centre must be determined (or at the place where the programs or data will be prepared).

9. The way of controlling the job treatment should be known. With small computers this is mostly done directly by the computer operator; with medium and large computers it is usually the duty of the user to attach job control statements, whose concrete form changes in dependence on the computer type and the system in which it works, to the program itself. This information is quite useful even if these statements are added by the centre technicians.

b. Preparation of Programs and Data and their Treatment on IBM 360 and IBM 370 Computers

The most common way of presenting programs and data to the computer employs eighty-column cards of the Hollerith system, in the given case in EBCDIC (Extended Binary Coded Decimal Interchange Code).

The cards are punched so that the individual program statements start at the earliest in the 7th column and end at the latest in the 72nd column of the card. If the statement does not fit in a single line (has more than 65 characters), it can be continued on a following card, punching unity (or another nonblank character) in the 6th column and continuing the text of the statement from the 7th column. An example is statement 0072 in the separation program. In any place in the statement an arbitrary number of blanks (i.e. empty columns) can be inserted to improve the program graphical form, except for texts between apostrophes and in the H-formats.

Statement number (not to be confused with the serial number of statements to the left of the vertical line or with the serial numbers of the cards in the ends of the lines) are punched anywhere in columns 1 to 5.

If the first sign on the card is C (comment), the text on the card is considered as a commentary that does not affect the program flow (see e.g. the card following statement 0042 in the separation program).

Arbitrary symbols can be punched in columns 73 to 80 (e.g. numbering of the cards or their other identification).

In data punching, their location on the card must be maintained exactly. This is prescribed in programs by FORMAT statements, whose individual formats are given in the description of the input data of individual programs. The format transcription generally consists of a number (which may sometimes be absent), a letter and a number, sometimes followed by a decimal point and another number, e.g. 10F12.3. The first number is optional and specifies how many times the given format is successively repeated. If this number is not given, the format occurs only once (e.g. 1I5 would be identical with I5). The letter specifies what kind of input data is involved: A denotes alphanumerical data (i.e. texts), I numerical data with whole numbers without decimal points and F or E non-integral numerical data. The first number following the letter specifies how many columns on the card are available for the given value (called the "field width"). The point and further numeral occur only with formats F and E and their significance will be described in more detail below.

Format A serves mostly for reading the job identification. For example, 20A4 denotes that twenty groups with four arbitrary alphameric characters each (letters, numbers, interpunction signs, etc.) will be read from the card.

Format I describes how to read integer numbers (they must not contain a decimal point); units are always placed in the rightmost column in the assigned space. For example, format I5 would mean that the value punched as (*b* will denote a blank)

*bb*5*bb*

would be interpreted in the program as 500. The only correct transcription containing the value 5 is

*bbbb*5

Formats introduced by the letter F denote transcription of a rational (non-integer) value. In such a case the decimal point can be given explicitly, in any place within the ascribed field on the card. Format Fm . n denotes that m characters are reserved for the transcription. For example, the value 14.35 against format F12.3 can be written as

14.35*bbbbbbb*

or

*bbb*14.35*bbbb*

or

*bbbbbbb*14.35

However, if the decimal point is not given, the number n in format Fm . n gains importance and specifies how many places are to be expected beyond the decimal point, Therefore, for

*bbbbbbb*1435*b*,
↑

the computer will assume the decimal point to be at the place indicated by the arrow, for format F12.3, and the value will be interpreted correctly (i.e. 14.35). On the other hand, transcription

*bbbbbbbb*1435

would be considered as the value 1.435. It can therefore be recommended to always write the decimal point, to avoid unnecessary errors.

Formats starting with symbol E permit writing exponents (i.e. "times ten to a power"). E.g. for format E2Ø.6 can be punched

*bbbbbbbbbbbbbb*1.5E-7

which means 1.5×10^{-7}. If the letter E (and the following sign) is not given in the transcription, the number is interpreted in the same way as with format F. It is also recommended here to always write the decimal point.

With IBM 360 and IBM 370 computers with operating systems OS and OS-VS, the job control statements for compilation of the program punched on eighty-column cards in the EBCDIC code and for computation itself with the input data also stored on cards look as follows:

```
//jobname  JOB  account-number,programmer-name,further-data
//EXEC  FORTGCLG
//FORT.SYSIN  DD  *
  cards of the main program
  cards of the first subroutine
  cards of the second subroutine
              .
              .
              .
              .
              .
  cards of the last subroutine
  /*
  //GO.SYSIN  DD  *
  cards with the data
  /*
```

In the first statement, the identification of the job is written in the place "name", which must always start with a letter and must not contain anything else than letters and numerals (e.g. SEPAR). In place of account-number the order code given to the user in the centre is written (say 999–999). In place of "programmer name" the programmer's name is given (e.g. SMITH). In place of "further-data", any other data required are given according to instructions from the system programmer (e.g. CLASS=A). The operation defining word JOB must be separated by at least one blank from the job name and the account number; on the other hand, no spaces are permitted in other parts of the statement. The introductory statement (JOB-statement) will thus be e.g.

```
//SEPAR   JOB   999-999,SMITH,CLASS=A.
```

The exact form of the job statement must always be specified by the computing centre.

The two slashes with which the job control statements start must always be in the first two columns of the card. Blanks can occur only in the places indicated above. Details can be found in the literature[2] on p. 340.

c. Generally Useful Subroutines

Subroutine SMI

The subroutine for inversion of a symmetrical matrix is stored in the memory in the economical form as the upper triangle in the form of a one-dimensional field (AMAT). Matrix **A** elements are stored successively, $a_{1,1}, a_{1,2}, \ldots, a_{1,n}, a_{2,2}, a_{2,3}, \ldots, a_{2,n}, a_{3,3}, \ldots, a_{n,n}$. The inverted matrix is stored in the same field as the original matrix, which means that the original matrix is rewritten during the computation. If this matrix is to be stored for further computations, it must be rewritten into another field before employing subroutine SMI.

The auxiliary field dimensions in the subroutine enable inversion of matrices up to dimensions of 50×50. The subroutine controls whether the matrix is regular in the mathematic (not physical) sense. If the matrix is singular (i.e. hasn't an inversion) the computation is terminated and the subroutine returns the program control back to the calling program.

The subroutine is called up by

```
CALL SMI(AMAT, IRDER, ISING)
```

Parameters:

AMAT – one-dimensional field containing the matrix to be inverted at the entry into the subroutine and the inversion matrix at the exit from the subroutine;

IRDER – integer expression, representing the dimension of the matrix to be inverted;

ISING – integer variable, which, at the end of the subroutine, contains 0 if the matrix to be inverted is regular (the inversion was successfully completed) or 1 if the matrix to be inverted is singular (inversion cannot be carried out).

Subroutine MUTV

A subroutine for multiplying a vector by a symmetrical matrix stored in the economical form (see the description of subroutine SMI).

Calling up of the subroutine:

CALL MUTV(N,A,X,Y)

Parameters:

N – integer expression specifying the dimension of the vectors and of the matrix;
A – one dimensional field containing matrix $\mathbf{A}$;
X – one dimensional field containing vector $\boldsymbol{x}$ to be multiplied;
Y – one dimensional field containing the resultant vector, $\boldsymbol{y} = \mathbf{A} \, . \, \boldsymbol{x}$ at the exit from the subroutine.

Subroutine PRTCVM

A subroutine for computation of a covariance matrix and its printing. The subroutine is started with the inversion matrix to a normal equation matrix and with the value of the sum of the squares of the deviation, from which the elements of the covariance matrix are computed using Eq. (3.60) or (3.82).

Calling up of the subroutine:

CALL PRTCVM(N,B,SS,M).

Parameters:

N – integer expression, the covariance matrix dimension;
B – one dimensional field containing the inversion matrix to the normal equation matrix in an economical form at the entry into the subroutine (see the description of subroutine SMI) and the covariance matrix in the same form at the exit from the subroutine;
SS – real expression equal to the sum of the squares of the deviations;
M – integer expression, the number of experimental points.

Output:

On the printer (or in the output ILPR channel) the covariance matrix is printed row by row under the heading COVARIANCE MATRIX. In each row, the row index is given first and then all elements of this line from the matrix main diagonal to the end of the row. The individual rows are separated by an empty line.

Subroutine SMOO

A subroutine for smoothing statistical noise from the data read at equidistant points, using the method described in Chapter 3.2.2.3 (Eq.(3.86)). In the specification statement, REAL (statement No. 0003, see also p. 341), the values from Table XIV are given for a thirteen-point smoothing function (N = 7); if the user wishes to employ another function, he gives analogous values in this statement, taken from this table or from Table XV.

Calling up of the subroutine:

CALL SMOO (NP, R, S, N1, N2)

Parameters:

NP – The total number of points in the vector of the experimental values to be smoothed;

R – one dimensional field, vector of the values to be smoothed;

S – one dimensional field, vector of the smoothed values;

N1 – integer expression, the index of the first point to be smoothed;

N2 – integer expression, the index of the last point to be smoothed.

Subroutine SMOOTH

A subroutine for smoothing (or interpolation) by using cubic spline function (3.87). The necessary condition is that vector values $\boldsymbol{x}$ (the independent variable) should either decrease or increase continuously, i.e. that for each I it should hold that X(I) < X(I + 1) or X(I) > X(I + 1). Using the subroutine, the values of coefficients a, b, c, d stored in fields A, B, C, D are computed so that the curve constructed is as smooth as possible and the sum of the squares of the deviations equals a chosen non-negative value S. If the estimates of the standard deviations for the individual experimental points are stored in field DY, the expected value of the sum of the squares of the deviations equals the number of experimental points less one (N – 1).

Calling up of the subroutine:

CALL SMOOTH (N, X, Y, DY, S, A, B, C, D)

Parameters:

N – The number of experimental points;

X – one dimensional field, containing the independent variable values;

Y – one dimensional field, containing the dependent variable values;

DY – one dimensional field, containing values proportional to the independent variable standard deviation;

S – the required sum of the squares of the deviations;

A, B, C, D – one dimensional fields, containing the coefficients of the cubic parabolas representing the individual segments (see Eq. (3.87)).

Note: If the subroutine for smoothing is used, then the values of smoothed functional values are contained in field A.

d. Program Descriptions

Programs for common statistical treatment of data are contained in the libraries of standard programs in every computing centre. For example, in centres equipped with IBM 360 or IBM 370 machines standard libraries of scientific subroutines with detailed description can be found[1].

The simplest statistical manipulations with measured data, e.g. evaluation of the mean value, standard deviation etc., can be carried out with the data screening program[2], DASCR, multiple linear regression with the program[3], REGRE, polynomial regression with program[4] named POLRG. Analogous programs can also be found with other computers or in various collections of published computer programs (e.g.[5]).

A very useful collection of fifty programs for infrared spectroscopy was published recently by Jones[6].

Therefore, only programs which are not commonly accessible are described in this section.

Program for Band Separation in Spectra

The program consists of the main program and subroutines SMI, NEWPAR, SPAR, MUTV, NEQM, TISK and function SQ. The type of

---

1 "System/360 Scientific Subroutine Package (360A-CM-03X). Version III. Application Description." H20-0166-5. IBM Corp., White Plains 1968 (or later revised editions).

2 Op. cit., p. 400.

3 Op. cit., p. 404.

4 Op. cit., p. 408.

5 "Collected Algorithms from CACM". Association of Computing Machinery, New York, 1966 and subsequent supplements. Algorithms are regularly prepublished in the journal "Communications of the ACM".

6 Jones R. N.: "Computer Programs for Infrared Spectrophotometry" National Research Council Canada, Ottawa 1976.

function used for the description of the spectral band can be changed by a change in subroutine NEQM and function SQ. Two variants are given here, for the Cauchy function and for the fractional rational function.

The algorithm mathematical description is given in Chapter 3.3.4.3. During the iteration process the convergence is examined and the damping factor is automatically adjusted so that the sum of the squares of the deviation continuously decreases. The iteration process is terminated when either the preset sum of the squares of the deviations is attained (GQ), the maximum number of iterations is exceeded (NIT) or ten successive modifications of the damping factor do not lead to convergence of the process.

It is assumed that the input data have already been transformed to the absorbance scale and each point is accompanied by the appropriate weighing factor. The maximum number of experimental points is 500, the maximum number of refined and unrefined parameters is 50 altogether. During reading of the initial estimates of the spectral band and background parameters, it is simultaneously determined whether a certain parameter will be refined or not.

Input data:

Ident.	Columns	Format	Significance
	Input channel ICRQ (usually card reader)		
1st logical record			
NAME	1—60	15A4	Alphanumerical job identification
IDENT	71—75	I5	Numerical job identification
2nd logical record			
NB	1—5	I5	The number of bands
NG	6—10	I5	Polynomial order[a] in the fractional rational function, Eq. (2.30)
NPOL	11—15	I5	Polynomial order[b] of the approximating background
NIT	16—20	I5	The maximum numbers of iterations
GQ	21—40	E20.6	The final sum of the squares of the deviations[c]
DUMP	41—60	E20.6	Damping factor[d]
3rd and following logical records[e]			
P(J)	1—12	F12.3	Parameter estimate[f]
	16—27		
	31—40		
	46—57		
	61—72		

FIX(J)	14—15 29—30 44—45 59—60 74—75	I2	If a non-zero value is given in this field, the parameter given in the previous field is not refined during the computation
The last logical record			
IN	1—5	I5	The number of the input channel through which the experimental data are read
Channel IN			
1st logical record			
ISPEC	1—60	15A4	Alphanumerical spectrum identification
IDSPEC	71—75	I5	Numerical spectrum identification
2nd logical record			
NP	1—5	I5	The number of experimental points
CONC	6—15	F10.4	Concentration[g]
THICK	16—25	F10.4	Cell thickness[g]
3rd and following logical records[h]			
X(I)	1—10	F10.5	Experimental point wavenumber
Y(I)	11—20	F10.5	Experimental point absorbance
W(I)	21—30	F10.5	Statistical weight[i]

a. If not specified, this is assumed to equal unity.

b. If not specified, background parallel with the wavenumber axis is assumed (zero polynomial order).

c. If not specified, the maximum number of iterations (NIT) is carried out.

d. If not specified, a value of 2.0 is automatically substituted.

e. The number of records depends on the number of bands (NB); at most 5 parameters are given in one record. Parameters for each band always start at the beginning of a record, even if the previous record does not contain the full number (i.e. five groups) of data. If there are more than five parameters for one band, they continue in immediately following records.

f. The order of parameters in the individual records is $\tilde{\nu}_{max}$, D_{max}, b_1 to b_{NG} (if the Cauchy function is used, then $b_1 = b_c$); if NG > 3, then coefficients b_4 and following are given in the following record. The last record (if NPOL > 4, then last several records) contains the estimates of the parameters of the polynomial function approximating the background in the order, absolute, linear, quadratic, etc. terms.

g. If not specified, the value 1.0 is substituted.

h. The number of records equals the number of experimental points (NP).

i. If not specified, the value 1.0 is substituted.

Output:

For control, the input parameter estimates, the experimental spectrum and the difference between the experimental and the calculated spectra are printed out under the heading STARTING PARAMS. During the iteration process, the iteration number, the number of the attempt to attain convergence in the given iteration cycle, the sum of the squares of the deviations attained, damping (the product of the sum of the squares of the deviations in the previous cycle and the damping factor), the damping factor and information about the normal equation matrix (1 = singular, 0 = regular) are printed.

If convergence is not attained after ten attempts in a single iteration cycle, information ITERATION FAILED IN CYCLE n is printed (n = the iteration cycle number).

Under the heading ADJUSTED PARAMETERS, the adjusted parameter values are pointed out, followed by a table of the experimental absorbance values and the differences between the experimental and calculated absorbances; finally, the covariance matrix is printed under heading COVARIANCE MATRIX.

Special Subroutines Used

Subroutine NEWPAR

The subroutine calculates new parameter estimates for a selected damping value (DUMP). By labelled common area COMMON /CB/, the normal equation matrix (field B) is transferred into the subroutine after its adjustment according to Eq. (3.80) and inversion by subroutine SMI. If the adjusted normal equation matrix is regular, the initial values of parameters P (transferred into the subroutine in blank common area) are corrected by corrections DP, calculated as the product of the vector of the right-hand sides G (COMMON /CB/) and inverted matrix C. This product is obtained in subroutine MUTV. For the resultant corrected vector of parameters PQ, the differences from the experimental values (DQ) are calculated, as well as the sum of the squares of the deviations FI, using function SQ (this function subroutine depends on the type of functions employed for the description of the band profile and the background). The values of PQ, DQ and FI are transferred into the calling program by means of the parameters. Labelled common area COMMON /CF/ contains vector IFIX, which carries information whether a certain parameter is being refined (IFIX (I) = 0 or not (IFIX(I) = 1).

Subroutine NEWPAR can be used without modifications for any type of function describing the band shape.

Calling up of the subroutine:

CALL NEWPAR(DUMP, PQ, DQ, FI, IWR)

Parameters:

DUMP – real expression, the value of damping d in Eq. (3.80);
PQ – one dimensional field containing the corrected parameter values at the output;
DQ – one dimensional field containing the difference between the experimental and calculated absorbance for the values of parameters PQ at the output;
FI – real variable containing the sum of the weighed squares of the deviations for the values of parameters PQ;
IWR – integer variable, containing information at the output on whether the modified normal equation matrix was regular (IWR = 0) or singular (IWR = 1).

Subroutine SPAR

The subroutine stores the vector of the refined parameters (PQ), the differences between the experimental and computed absorbance and the sum of the squares of the deviations, in place of the original (i.e. unrefined) values. Thus these values, vector P, vector D and variable FILAST in blank common area, become the initial values for the further iteration cycle.

Parameters:

PQ – one dimensional field, containing the refined parameter;
DQ – one dimensional field, containing the differences between the experimental and computed absorbance for the values of parameters PQ;
FI – real expression, whose value is the sum of the squares of the deviations for, the values of parameters PQ.

Subroutine TISK

The subroutine prints the parameter values (P) and optionally a table of the wavenumber, experimental absorbance and the difference between the experimental and computed absorbance values (successively X, Y, D). All these values are transferred into the subroutine by blank common area.

Calling up of the subroutine:

CALL TISK(FF)

Parameters:

FF – logical expression; if its value is .TRUE., the table of absorbances is printed, if its value is .FALSE., it is not printed.

Function SQ

Function subroutine SQ computes the differences between the experimental absorbance (Y) and the computed absorbance (YCALC) for the given wavenumber values (X), the selected function describing the band shape and the given parameter values (PQ), and stores the resultant values into the difference vector (DQ). It simultaneously computes the weighed sum of the squares of the deviations (S), which is then transferred into the calling program as functional value SQ. The values of the number of bands (NB), the number of parameters (NPRS), the number of coefficients in the denominator of the rational function (NG), the number of coefficients in the polynomial describing the background (NPOL), the field of wavenumber values (X) and the field of experimental absorbances (Y) are transferred into the functional subroutine in blank common area. The field containing the experimental point statistical weights (W) is transferred into the function in labelled common area COMMON /WT/.

In this subroutine, two versions of the sum of the squares computation algorithm are given simultaneously, namely, that for the transmittance scale and that for the absorbance scale. The proper algorithm is chosen by deleting either statement No. 0040 (if the computation in absorbance scale is desired) or statement No. 0039 (computation in transmittance scale), respectively.

Calling up of the function:

FI = SQ(PQ, DQ)

Parameters:

PQ – one dimensional field, containing the values of the parameters of the functions describing the bands and the spectral background;

DQ – one dimensional field, containing the vector of the differences between the experimental and computed absorbances at the end of the subroutine.

Subroutine NEQM

The subroutine generates the normal equation matrix (B) in an economical form (see the description of subroutine SMI) for the selected type of function describing the bands in the spectrum, simultaneously with the vector of the right-hand sides (G). The subroutine has no parameters and all data between it and the calling program are transferred by means of common areas. In blank common area are transferred the number of experimental points (NP), the number of bands (NB), the order of the fractional rational polynomial function (NG), the total number of the parameters (NPRS), the number of parameters to be refined (PQ), the order of the polynomial describing the background (NPOL), the parameters of the functions describing the bands and the background (P), wavenumbers (X), absorbances (Y) and the differences between the experimental and calculated absorbance (D) for the individual points of the spectrum. In labelled common area COMMON /CB/, normal equation matrices (B) and the vector of the right-hand sides of the normal equations (G) are stored. In labelled common area COMMON /WT/, the vector of the statistical weights of the experimental points (W) is stored. The vector of information (IFIX) on whether the individual parameters will be refined (IFIX(I) = 0) or not (IFIX(I) = 1) is stored in block COMMON /CF/.

Calling up of the subroutine

```
CALL NEQM
```

The subroutine has no parameters.

Subroutines SMI, MUTV and PRTCVM are described independently (see. p. 345).

Program for Pseudodeconvolution of Spectra

The program consists of the main program and subroutine CONVOL.

The algorithm is mathematically described in Chapter 3.3.3.3.4. For the sake of economy it is assumed that the spectrum can be divided into several sections, while an instrument function which is constant over the section can be used in each section. The number of these sections in a spectrum handled is arbitrary.

The necessary condition is equidistant distribution of the points describing the spectrum, which is given in terms of the vector of the transmittance values. At the end of the program, the vector of the deconvoluted spectrum is obtained, supplemented by the vectors of the original spectrum and a control recording of the spectrum obtained by the method described in Chapter 3.2.2.3 (Eq. (3.86)).

Calling up of the subroutine:

CALL CONVOL(N, DS, CS, M, G, N1, N2)

Parameters:

N – integer expression, the number of points of the spectrum;

DS – one dimensional field, the transmittance values of the original spectrum;

CS – one dimensional field, the transmittance values after convolution of the original spectrum (DS);

M – integer expression, the number of points of the vector describing the instrument function (or the smoothing function coefficients)

G – one dimensional field, the vector describing the instrument function (or the smoothing function coefficients);

N1 – index of the first point of the spectrum being deconvoluted (or smoothed);

N2 – index of the last point of the spectrum being deconvoluted (or smoothed).

e. Listings of Programs, Subroutines and Test Data

In this part of Appendix are given listings of the individual modules of the programs discussed in Parts c and d.

```
0001          SUBROUTINE SMI (AMAT, IRDER, ISING)                          00025100
      C       THIS ROUTINE UPDATED AND/OR RESEQUENCED JANUARY 8, 1972      00025200
      C       FORTRAN 4 VERSION ALGORITHM 150 SYMIN2-UPPER TRI STORAGE     00025300
      C       AMAT DIMENSIONS MUST BE SPECIFIED IN THE CALLING PROGRAM.    00025400
0002          DIMENSION RBOO(50), AMAT(600), QAR(33), PAR(33)              00025500
0003          LOGICAL RBOO                                                 00025600
0004          ISING = 0                                                    00025700
0005          DO 1 IMAT=1, IRDER                                           00025800
0006        1 RBOO(IMAT) = .TRUE.                                          00025900
      C       GRAND LOOP                                                   00026000
0007          DO 11 IMAT=1, IRDER                                          00026100
      C       SEARCH FOR PIVOT                                             00026200
0008                BIG = 0.                                               00026300
0009                INCR = IRDER                                           00026400
0010                LRP = 1                                                00026500
0011                DO 3 JPP=1,IRDER                                       00026600
0012                      SIG = ABS(AMAT(LRP))                             00026700
0013                            IF(RBOO(JPP).AND.SIG.GT.BIG) GO TO 22      00026800
0014          GO TO 2                                                      00026900
0015       22 BIG = SIG                                                    00027000
0016                      KAT = JPP                                        00027100
0017                      KLRP = LRP                                       00027200
0018        2             LRP = LRP+INCR                                   00027300
0019        3       INCR = INCR-1                                          00027400
0020                IF (BIG) 5, 4, 5                                       00027500
0021        4       ISING = 1                                              00027600
0022                GO TO 12                                               00027700
      C       PREPARATION OF ELIMINATION STEP 1                            00027800
0023        5       RBOO(KAT) = .FALSE.                                    00027900
0024                BIG = 1./AMAT(KLRP)                                    00028000
```

```
0025          QAR(KAT) = BIG                                00281000
0026          PAR(KAT) = 1.                                 00028200
0027          AMAT(KLRP) = 0.                               00028300
0028          KKAT = KAT-1                                  00028400
0029          LRP = KAT                                     00028500
0030          IF (KKAT.LT.1) GO TO 7                        00028600
0031          INCR = IRDER                                  00028700
0032          DO 6 JPP=1, KKAT                              00028800
0033              SIG = AMAT(LRP)                           00028900
0034              PAR(JPP) = SIG                            00029000
0035              IF (RBOO(JPP)) SIG = —SIG                 00029100
0036              QAR(JPP) = SIG*BIG                        00029200
0037              AMAT(LRP) = 0.                            00029300
0038              INCR = INCR—1                             00029400
0039      6   LRP = LRP+INCR                                00029500
0040      7   KKKAT = KAT+1                                 00029600
0041          IF (KKKAT.GT.IRDER) GO TO 9                   00029700
0042          DO 8 JPP=KKKAT,IRDER                          00029800
0043              LRP = LRP+1                               00029900
0044              SIG = —AMAT(LRP)                          00030000
0045              QAR(JPP) = SIG*BIG                        00030100
0046              IF (RBOO(JPP)) SIG = —SIG                 00030200
0047              PAR(JPP) = SIG                            00030300
0048      8   AMAT(LRP) = 0.                                00030400
        C   ELIMINATION PROPER                              00030500
0049      9   LRP = 0                                       00030600
0050          DO 10 JPP=1,IRDER                             00030700
0051              DO 10 KAT=JPP,IRDER                       00030800
0052              LRP = LRP+1                               00030900
0053     10   AMAT(LRP) = AMAT(LRP)+PAR(JPP)*QAR(KAT)       00031000
```

```
0054         11 CONTINUE                                                        00031100
0055         12 RETURN                                                          00031200
0056            END                                                             00031300

0001            SUBROUTINE MUTV (N,A,X,Y)                                       00036100
          C  ************************************************************      00036200
          C  *   MULTIPLICATION OF THE VECTOR X BY THE MATRIX A;          *     00036300
          C  *                          Y=AX                               *     00036400
          C  *    A IS A REAL AND SYMMETRIC MATRIX, VECTOR STO-           *     00036500
          C  *       RED AS THE UPPER TRIANGLE IN THE ARRAY A             *     00036600
          C  ************************************************************      00036700
0002            DIMENSION A(1), X(N), Y(N)                                      00036800
0003            NI = N—1                                                        00036900
0004            IF (N1) 7,1,1                                                   00037000
0005          1 DO 2 I=1,N                                                      00037100
0006          2 Y(I) = 0.                                                       00037200
0007            L = 1                                                           00037300
0008            IF (N1) 6,6,3                                                   00037400
0009          3 DO 5 I=1,N1                                                     00037500
0010                  S = Y(I)+A(L)*X(I)                                        00037600
0011                  L = L+1                                                   00037700
0012                  I1 = I+1                                                  00037800
0013                  DO 4 J=I1,N                                               00037900
0014                        S = S+A(L)*X(J)                                     00038000
0015                        Y(J) = Y(J)+A(L)*X(I)                               00038100
0016          4       L = L+1                                                   00038200
0017                  Y(I) = S                                                  00038300
0018          5 CONTINUE                                                        00038400
```

```
0019 |     6 Y(N) = Y(N)+A(L)*X(N)                              00038500
0020 |     7 RETURN                                             00038600
0021 |       END                                                00038700

0001 |       SUBROUTINE PRTCVM(M,B,FI,NP)
0002 |       DIMENSION B(1)
0003 |       ILPR=6
0004 |       S=FI/FLOAT(NP—M)
0005 |       NTRI=M*(M+1)/2
0006 |       DO 1 I=1,NTRI
0007 |     1 B(I)=B(I)*S
0008 |   951 FORMAT(1H1//' COVARIANCE MATRIX'//)
0009 |       WRITE(ILPR,951)
0010 |       M1=M
0011 |       L2=0
0012 |       DO 2 I=1,M
0013 |       M1=M1—1
0014 |       L1=L2+1
0015 |       L2=L1+M1
0016 |   952 FORMAT(/1X,I5,2X,5E20.6,9(/8X,5E20.6))
0017 |     2 WRITE(ILPR,952) I,(B(J),J=L1,L2)
0018 |       RETURN
0019 |       END

0001 |       SUBROUTINE SMOO(X,S,N)
0002 |       DIMENSION X(N),S(N)
```

```
0003          REAL  C0/7./,C(3)/6.,3.,—2./,NF/21./
0004          M=3
      C     7 POINT SMOOTH
0005          DO 1 I=1,N
0006          Y=C0*X(I)
0007          DO 2 J=1,M
0008          I1=I—J
0009          IF(I1.LT.1)I1=1
0010          I2=I+J
0011          IF(I2.GT.N)I2=N
0012        2 Y=C(J)*(X(I1)+X(I2))+Y
0013        1 S(I)=Y/NF
0014          RETURN
0015          END
```

```
0001          SUBROUTINE SMOOTH(N,X,Y,DY,S,A,B,C,D)                           SMO   1
      C  ******************************************************               SMO   2
      C  **   FOR DESCRIPTION SEE:                                **           SMO   3
      C  **   CRISTIAN H. REINSCH: "SMOOTHING BY SPLINE           **           SMO   4
      C  **      FUNCTIONS", NUM. MATH. 10 (1967) 177—183         **           SMO   5
      C  **    MAX. NO. OF  POINTS = 200                          **           SMO   6
      C  ******************************************************               SMO   7
0002          DIMENSION X(N),Y(N),DY(N),A(N),B(N),C(N),D(N)                   SMO   8
0003          DIMENSION R(202),R1(202),R2(202),T(202),T1(202),U(202),V(202)   SMO   9
0004          COMMON /CSMOO/ IFLAG,R,R1,R2,T,T1,U,V                           SMO  10
0005          IFLAG=0                                                         SMO  11
      C    ** TEST FOR MONOTONICITY **                                        SMO  12
0006          SRNB=X(2)—X(1)                                                  SMO  13
```

```
0007          IF(SRNB) 75,74,75                          SMO 14
0008       75 M2=N—1                                     SMO 15
0009          DO 750 I=2,M2                              SMO 16
0010          IF(SRNB*(X(I+1)—X(I)))754,752,750          SMO 17
0011      750 CONTINUE                                   SMO 18
0012          SRNB=SIGN(1.,SRNB)                         SMO 19
0013          M2=N+1                                     SMO 20
0014          R(1)=0.                                    SMO 21
0015          R(2)=0.                                    SMO 22
0016          R1(M2)=0.                                  SMO 23
0017          R2(M2)=0.                                  SMO 24
0018          R2(M2+1)=0.                                SMO 25
0019          U(1)=0.                                    SMO 26
0020          U(2)=0.                                    SMO 27
0021          U(M2)=0.                                   SMO 28
0022          U(M2+1)=0.                                 SMO 29
0023          P=0.                                       SMO 30
0024          M2=N—1                                     SMO 31
0025          H=X(2)—X(1)                                SMO 32
0026          F=(Y(2)—Y(1))/H                            SMO 33
0027          DO 100 I=2,M2                              SMO 34
0028          I1=I+1                                     SMO 35
0029          G=H                                        SMO 36
0030          H=X(I1)—X(I)                               SMO 37
0031          E=F                                        SMH 38
0032          F=(Y(I1)—Y(I))/H                           SMO 39
0033          A (I) = F — E                              SMO 40
0034          T(I1)=.6666667*(G+H)                       SMO 41
0035          T1(I1)=.3333333*H                          SMO 42
0036          R2(I1)=DY(I—1)/G                           SMO 43
```

```
0037          R(I1)=DY(I1)/H                                          SMO  44
0038          R1(I1)=—DY(I)/G—DY(I)/H                                 SMO  45
0039      100 CONTINUE                                                SMO  46
0040          DO 110 I=2,M2                                           SMO  47
0041          I1=I+1                                                  SMO  48
0042          I2=I+2                                                  SMO  49
0043          B(I)=R(I1)**2+R1(I1)**2+R2(I1)**2                       SMO  50
0044          C(I)=R(I1)*R1(I2)+R1(I1)*R2(I2)                         SMO  51
0045          D(I)=R(I1)=R2(I+3)                                      SMO  52
0046      110 CONTINUE                                                SMO  53
0047          F2=—S                                                   SMO  54
0048      199 CONTINUE                                                SMO  55
0049          DO 200 I=2, M2                                          SMO  56
0050          11=I+1                                                  SMO  57
0051          12=I—1                                                  SMO  58
0052          R1(I)=F*R(I)                                            SMO  59
0053          R2(I2)=G*R(I2)                                          SMO  60
0054          R(I1)=1./(P*B(I)+T(I1)—F*R1(I)—G*R2(I2))                SMO  61
0055          U(I1)=A(I)—R1(I)*U(I)—R2(I2)*U(I2)                      SMO  62
0056          F=P*C(I)+T1(I1)—H*R1(I)                                 SMO  63
0057          G=H                                                     SMO  64
0058          H=D(I)*P                                                SMO  65
0059      200 CONTINUE                                                SMO  66
0060          I1=N                                                    SMO  67
0061          DO 210 I=2, M2                                          SMO  68
0062          U(I1)=R(I1)*U(I1)—R1(I1)*U(I1+1)—R2(I1)*U(I1+2)         SMO  69
0063      210 I1=I1—1                                                 SMO  70
0064          E=0.                                                    SMO  71
0065          H=0.                                                    SMO  72
0066          DO 220 I=1, M2                                          SMO  73
```

```
0067          I1=I+1                                   SMO  74
0068          I2=I+2                                   SMO  75
0069          G=H                                      SMO  76
0070          H=(U(I2)—U(I1))/(X(I1)—X(I))             SMO  77
0071          V(I1)=(H—G)*DY(I)**2                     SMO  78
0072          E=E+V(I1)*(H—G)                          SMO  79
0073      220 CONTINUE                                 SMO  80
0074          G=—H*DY(N)**2                            SMO  81
0075          V(N+1)=G                                 SMO  82
0076          E=E—G*H                                  SMO  83
0077          G=F2                                     SMO  84
0078          F2=E*P*P                                 SMO  85
0079          IF(F2.GE.S .OR. F2.LE.G) GO TO 400       SMO  86
0080          F=0.                                     SMO  87
0081          H=(V(3)—V(2))/(X(2)—X(1))                SMO  88
0082          DO 300 I=2, M2                           SMO  89
0083          I1=I+1                                   SMO  90
0084          12=I+2                                   SMO 100
0085          IM2=I—1                                  SMO 101
0086          G=H                                      SMO 102
0087          H=(V(I2)—V(I1))/(X(I1)—X(2))             SMO 103
0088          G=H—G—R1(I)**2—R2(IM2)**2                SMO 104
0089          F=F+G*G*R(I1)                            SMO 105
0090          R(I1)=G                                  SMO 106
0091      300 CONTINUE                                 SMO 107
0092          H=E—P*F                                  SMO 108
0093          IF(H) 400, 400, 301                      SMO 109
0094      301 P=P+(S—F2)/((SRNB*SQRT(S/E)+P)*H)        SMO 110
0095          GO TO 199                                SMO 111
0096      400 CONTINUE                                 SMO 112
```

```
0097  |        DO 410 I=1, N                                                      SMO 113
0098  |        I1=I+1                                                             SMO 114
0099  |        A(I)=Y(I)—P*V(I1)                                                  SMO 115
0100  |        C(I)=U(I1)                                                         SMO 116
0101  |    410 CONTINUE                                                           SMO 117
0102  |        DO 420 I=1, M2                                                     SMO 118
0103  |        I1=I+1                                                             SMO 119
0104  |        H=X(I1)—X(I)                                                       SMO 120
0105  |        D(I)=(C(I1)—C(I))/(3.*H)                                           SMO 121
0106  |        B(I)=(A(I1)—A(I))/H—(H*D(I)+C(I))*H                                SMO 122
0107  |    420 CONTINUE                                                           SMO 123
0108  |        RETURN                                                             SMO 124
0109  |    754 IF(SRNB) 755, 74, 751                                              SMO 125
0110  |     74 I=1                                                                SMO 126
0111  |    752 WRITE(6, 9752) I                                                   SMO 127
0112  |   9752 FORMAT('ZERO DIFFERENCE BETWEEN ABSCISSAE AFTER POINT', I4)        SMO 128
0113  |        GO TO 753                                                          SMO 129
0114  |    751 WRITE(6, 9751) I                                                   SMO 130
0115  |   9751 FORMAT('SEQUENCE OF ABSCISSAE NOT INCREASING AFTER POINT', I4)     SMO 131
0116  |        GO TO 753                                                          SMO 132
0117  |    755 WRITE(6, 9755) I                                                   SMO 133
0118  |   9755 FORMAT('SEQUENCE OF ABSCISSAE NOT DECREASING AFTER POINT', I4)     SMO 134
0119  |    753 IFLAG=1                                                            SMO 135
0120  |        RETURN                                                             SMO 136
0121  |        END                                                                SMO 137
```

```
      | C      BAND SEPARATION — MAIN PROGRAM
0001  |        INTEGER FIX(50)
```

```
0002          DIMENSION NAME(15), ISPEC(15)
0003          DIMENSION P0(50), D0(500)
0004          COMMON NP,NB,NG,NPRS,NPQ,NPOL,P(50),X(500),Y(500),D(500),FILAST
0005          COMMON /CF/ FIX
0006          COMMON /WT/ W(500)
0007          COMMON /CD/ CONC,THICK,IDSPEC
0008          ICRQ=5
0009          ILPR=6
0010      100 CONTINUE
0011          READ(ICRQ,901) NAME,IDENT,NB,NG,NPOL,NIT,GQ,DUMP
0012          IF(NB.EQ.0) STOP
0013          IF(DUMP.EQ.0.) DUMP=2.
0014          IF(NG.EQ.0) NG=1
0015          NPOL=NPOL+1
0016      901 FORMAT(15A4,10X,15/4I5,2E20.6)
0017          NPRS=NB*(2+NG)+NPOL
0018          NPQ=0
0019          L=1
0020          DO 1 I=1,NB
0021          K=L+NG+1
0022          READ(ICRQ,902) (P(J),FIX(J), J=L, K)
0023      902 FORMAT(5(F12.3,1X,I2))
0024     9299 FORMAT(9H0SPECTRUM/(1X,F10.2,3F10.4))
0025     9210 FORMAT(1H1/5H JOB ,15A4/10H SPECTRUM ,15A4//16H STARTING PARAMS/)
0026          DO 11 J=L,K
0027          IF(FIX(J).EQ.0) NPQ=NPQ+1
0028       11 CONTINUE
0029        1 L=K+1
0030          K=L+NPOL—1
0031          READ(ICRQ,902) (P(J),FIX(J),J=L,K)
```

```
0032          DO 12 J=L,K
0033          IF(FIX(J).EQ.0) NPQ=NPQ+1
0034       12 CONTINUE
0035          READ(ICRQ,903) IN
0036      903 FORMAT(16I5)
0037          READ(IN,904) ISPEC,IDSPEC,NP,CONC,THICK
0038      904 FORMAT(15A4,10X,15/15,2F10.4)
0039      905 FORMAT(3F10.5)
0040          DO 200 I=1,NP
0041          READ(IN,905) X(I),Y(I),W(I)
0042      200 CONTINUE
      C       INITIATE FILAST
0043          FILAST=SQ(P,D)
0044          WRITE(ILPR,9210) NAME,ISPEC
0045          CALL TISK(.TRUE.)
0046          WRITE(ILPR,9299)(X(I),Y(I),W(I),D(I),I=1,NP)
0047     9211 FORMAT(15H0STARTING SQ = ,E20.6/)
0048          WRITE(ILPR,9211) FILAST
0049     9212 FORMAT('0 ICYC ITRY        FI0        DUMP*FILAST',
             0     '          DUMP        IWR'/)
0050          WRITE(ILPR,9212)
0051          READ(ICRQ,903) NIT,IC3,IZFIX
0052          DO 300 ICYC=1,NIT
0053          ITRY=1
0054          CALL NEQM
0055      301 CONTINUE
0056          CALL NEWPAR(DUMP*FILAST,P0,D0,FI0,IWR)
0057     9399 FORMAT(1X,215,3E20.6,15)
0058          DFL1=DUMP*FILAST
0059          WRITE(ILPR,9399) ICYC,ITRY,F10,DFL1,DUMP,IWR
```

```
0060          IF(FI0.LT.FILAST .AND. IWR.EQ.0) GOTO 350
0061          DUMP=2.*DUMP
0062          ITRY=ITRY+1
0063          IF(ITRY.LT.10) GOTO 301
0064          WRITE(ILPR,9301) ICYC
0065          GOTO 400
0066     9301 FORMAT(26H0ITERATION FAILED IN CYCLE,I3///)
0067      350 CONTINUE
0068          CALL SPAR(P0,D0,FI0)
0069      300 CONTINUE
0070      400 CONTINUE
0071          WRITE(ILPR,9400) NAME,ISPEC,IDSPEC
0072     9400 FORMAT(1H1/7H JOB = ,15A4/12H SPECTRUM = ,15A4,13H IDENT. NO = ,
            1 I6//20H ADJUSTED PARAMETERS//)
0073          CALL TISK(.FALSE.)
0074          WRITE(ILPR,9299)(X(I),Y(I),W(I),D(I),I=1,NP)
0075          CALL SMI(B,NPQ,IWR)
0076          IF(IWR.NE.0) GOTO 501
0077          CALL PRTCVM(NPQ,B,FILAST,NP)
0078      501 CONTINUE
0079          GOTO 100
0080          END
```

```
0001          SUBROUTINE NEWPAR (DUMP,PQ,DQ,FI,IWR)                       00031400
        C     ************************************************************ 00031500
        C     *  COMPUTATION OF A NEW SET OF PARAMETERS (PQ) FOR         * 00031600
        C     *  THE CHOSEN VALUE OF THE DAMPING FACTOR (DUMP)           * 00031700
        C     ************************************************************ 00031800
```

```
0002        DIMENSION PQ(50), DQ(500), DP(50)                                   00031900
0003        COMMON NP,NB,NG,NPRS,NPQ,NPOL,P(50),X(500),Y(500)                   00032000
0004        COMMON /CB/ B(1275),G(50),C(1275)                                   00032100
0005        COMMON /CF/ IFIX(50)                                                00032200
0006        D1 = 1.+DUMP                                                        00032300
0007        IK = NPQ*(NPQ+1)/2                                                  00032400
0008        DO 1 I=1,IK                                                         00032500
0009      1 C(I) = B(I)                                                         00032600
0010        IK = NPQ                                                            00032700
0011        L = 1                                                               00032800
0012        DO 2 I=1,NPQ                                                        00032900
0013              C(L) = C(L)+DUMP                                              00033000
0014              L = L+IK                                                      00033100
0015      2 IK = IK—1                                                           00033200
0016        CALL SMI (C,NPQ,IWR)                                                00033300
0017        IF (IWR) 7,3,7                                                      00033400
0018      3 CALL MUTV (NPQ,C,G,DP)                                              00033500
0019        L = 1                                                               00033600
0020        DO 6 I=1,NPRS                                                       00033700
0021              IF (IFIX(I)) 4,5,4                                            00033800
0022      4       PQ(I) = P(I)                                                  00033900
0023              GO TO 6                                                       00034000
0024      5       PQ(I) = P(I)+DP(L)                                            00034100
0025              L = L+1                                                       00034200
0026      6 CONTINUE                                                            00034300
0027        FI = SQ(PQ,DQ)                                                      00034400
0028      7 RETURN                                                              00034500
0029        END                                                                 00034600
```

```
0001        SUBROUTINE SPAR (PQ,DQ,FI)                                             00034700
       C    ************************************************************           00034800
       C    *     STORE THE BEST ESTIMATE OF PARAMETERS AND THE        *           00034900
       C    *     CORRESPONDING VECTOR OF DIFFERENCES                  *           00035000
       C    ************************************************************           00035100
0002        COMMON NP,NB,NG,NPRS,NPQ,NPOL,P(50),X(500),Y(500),D(500),FILAST        00035200
0003        DIMENSION PQ(50), DQ(500)                                              00035300
0004        DO 1 I=1,NPRS                                                          00035400
0005      1 P(1) = PQ(I)                                                           00035500
0006        DO 2 I=1,NP                                                            00035600
0007      2 D(I) = DQ(I)                                                           00035700
0008        FILAST = FI                                                            00035800
0009        RETURN                                                                 00035900
0010        END                                                                    00036000
```

```
0001        SUBROUTINE TISK(FF)
0002        LOGICAL FF
0003        COMMON NP,NB,NG,NPRS,NPQ,NPOL,P(50),X(500),Y(500),D(500),FILAST
0004        WRITE(6,900) (P(I),I=1,NPRS)
0005    900 FORMAT(7X,3E18.6)
0006        IF(FF) GOTO 500
0007        RETURN
0008    500 CONTINUE
0009        WRITE(6,900) (X(I), Y(I),D(I),I=1,NP)
0010        RETURN
0011        END
```

```
0001          FUNCTION SQ (PQ,DQ)                                      00039600
        C     *****************************************************    00039700
        C     *  COMPUTATION OF THE SUM OF SQUARES FOR THE        *    00039800
        C     *  PARTICULAR SET OF PARAMETERS (PQ). THE VECTOR    *    00039900
        C     *  OF DIFFERENCES (DQ = Y — YCALC) IS ALSO STORED;  *    00040000
        C     *****************************************************    00040100
0002          COMMON NP,NB,NG,NPRS,NPQ,NPOL,P(50),X(500),Y(500)        00040200
0003          COMMON /WT/ W(500)/CB/XYZ(1325),B2(50)                   00040300
0004          DIMENSION PQ(50) ,DQ(500) ,PY(500)                       00040400
        C   *****   GENERALIZED CAUCHY (LORENTZ) FUNCTION *****        00040500
0005          DO 1 I=1,NPRS                                            00040600
0006        1 B2(I) = PQ(I)**2                                         00040700
0007          S1=0.                                                    00040800
0008          S = 0.                                                   00040900
0009          DO 7 I=1,NP                                              00041000
0010             YCALC = 0.                                            00041100
0011             L = 0                                                 00041200
0012             DO 3 J=1,NB                                           00041300
0013                L = L+2                                            00041400
0014                A = PQ(L)                                          00041500
0015                X0 = PQ(L—1)—X(I)                                  00041600
0016                X0 = X0*X0                                         00041700
0017                XP = X0                                            00041800
0018                DEN = 1.                                           00041900
0019                DO 2 K=1,NG                                        00042000
0020                   L = L+1                                         00042100
0021                   DEN = DEN+B2(L)*XP                              00042200
0022        2          XP = XP*X0                                      00042300
0023                YCALC = A/DEN+YCALC                                00042400
0024        3    CONTINUE                                              00042500
```

```
0025 |             IF (NPOL) 6,6,4                                   00042600
0026 |        4    XP = 1.                                           00042700
0027 |             X0 = X(I)                                         00042800
0028 |             DO 5 K=1,NPOL                                     00042900
0029 |                  L = L+1                                      00043000
0030 |                  YCALC = YCALC+PQ(L)*XP                       00043100
0031 |        5    XP = XP*X0                                        00043200
     |  C                                                            00043300
0032 |             DD = 1./10.**Y(I) —1./10.**YCALC                  00043400
0033 |        6    D = Y(I)—YCALC                                    00043500
     |  C                                                            00043600
0034 |             DQ(I) = D                                         00043700
0035 |        S=DD*DD+S                                              00043800
0036 |        S1=D*D +S1                                             00043900
0037 |  C                                                            00044000
0038 |       7 CONTINUE                                              00044100
0039 |        SQ=S1                                                  00044200
0040 |        SQ = S                                                 00044300
0041 |        RETURN                                                 00044400
0042 |        END                                                    00044500

0001 |        SUBROUTINE NEOM                                        00044600
     |  C     ***************************************************************  00044700
     |  C     *  COMPUTATION OF THE NORMAL EQUATIONS MATRIX B = J'WJ *        00044800
     |  C     *  AND OF THE RIGHT SIDE VECTOR G = J'WD               *        00044900
     |  C     *  FOR THE GENERALIZED CAUCHY (LORENTZ) FUNCTION       *        00045000
     |  C     ***************************************************************  00045100
0002 |        DIMENSION XJ(50), AUX(10), B2(50)                      00045200
```

```
0003        COMMON NP,NB,NG,NPRS,NPQ,NPOL,P(50),X(500),Y(500),D(500)          00045300
0004        COMMON /CB/ B(1275),G(50),B2,AUX,XJ                                00045400
0005        COMMON /WT/ W(500)                                                 00045500
0006        COMMON /CF/ IFIX(50)                                               00045600
0007        L = 0                                                              00045700
0008        DO 1 I=1,NPQ                                                       00045800
0009             G(I) = 0.                                                     00045900
0010             DO 1 J=I,NPQ                                                  00046000
0011                  L = L+1                                                  00046100
0012      1 B(L) = 0.                                                          00046200
0013        DO 2 I=1,NPRS                                                      00046300
0014      2 B2(I) = P(I)**2                                                    00046400
0015        NH = NG+2                                                          00046500
0016        DO 14 I=1,NP                                                       00046600
0017             XI = X(I)                                                     00046700
0018             WI = W(I)                                                     00046800
0019             DI = D(I)                                                     00046900
     C   *****   JACOBIAN FOR THE GENERALIZED CAUCHY FUNCTION *****            00047000
0020             L = 0                                                         00047100
0021             M = 0                                                         00047200
0022             M1 = 0                                                        00047300
0023             DD 7 J=1,NB                                                   00047400
0024                  L = L+2                                                  00047500
0025                  A = P(L)                                                 00047600
0026                  XP = P(L—1)—XI                                           00047700
0027                  X0 = XP*XP                                               00047800
0028                  XQ = X0                                                  00047900
0029                  DEM = 0.                                                 00048000
0030                  DEN = 1.                                                 00048100
0031                  DO 3 K=1,NG                                              00048200
```

```
0032               L = L+1                                  00048300
0033            AUX(K+2) = P(L)*XQ                          00048400
0034            DEN = DEN+B2(L)*XQ                          00048500
0035            DEM = DEM+FLOAT(K)*B2(L)*XP                 00048600
0036            XQ = XQ*X0                                  00048700
0037            XP = XP*X0                                  00048800
0038      3     CONTINUE                                    00048900
0039            DEN = 1./DEN                                00049000
0040            FPR = —2.*A*DEN*DEN                         00049100
0041            AUX(1) = DEM*FPR                            00049200
0042            AUX(2) = DEN                                00049300
0043            DO 4 K=3,NH                                 00049400
0044      4     AUX(K) = AUX(K)*FPR                         00049500
0045            DO 6 K=1,NH                                 00049600
0046                M = M+1                                 00049700
0047                IF (IFIX(M)) 6,5,6                      00049800
0048      5         M1 = M1+1                               00049900
0049                XJ(M1) = AUX(K)                         00050000
0050      6     CONTINUE                                    00050100
0051      7 CONTINUE                                        00050200
     C  *****  BACKGROUND  *****                            00050300
0052        IF (NPOL) 11,11,8                               00050400
0053      8 XP = 1.                                         00050500
0054        DO 10 K=1,NPOL                                  00050600
0055            M = M+1                                     00050700
0056            IF (IFIX(M)) 10,9,10                        00050800
0057      9     M1 = M1+1                                   00050900
0058            XJ(M1) = XP                                 00051000
0059     10 XP = XP*XI                                      00051100
0060     11 L = 0                                           00051200
```

```
0061           IF (M1.NE.NPQ.AND.I.EQ.1) WRITE (6,15) NPQ,M1                    00051300
        C  *****  CONTRIBUTIONS TO THE N. E. MATRIX AND R. H. VECTOR  *****    00051400
                                                                                00051500
0062           DO 13 J=1,NPQ                                                    00051600
0063              XJW = WI*XJ(J)                                                00051700
0064              G(J) = G(J)+DI*XJW                                            00051800
0065              DO 12 K=1,NPQ                                                 00051900
0066                 L = L+1                                                    00052000
0067                 B(L) = B(L)+XJW*XJ(K)                                      00052100
0068        12    CONTINUE                                                      00052200
0069        13    CONTINUE                                                      00052300
0070        14 CONTINUE                                                         00052400
0071           RETURN                                                           00052500
        C                                                                       00052600
0072        15 FORMAT (7H NPQ = ,I3,8H M1 = ,I3)                                00052700
0073           END
```

Example

As a test example, part of the methanesulphonyl chloride spectrum in a region from 1150 to 1199 cm^{-1} is given. It is assumed for separation purposes that all points have the same (unit) statistical weight, that only two bands are located in the given spectral region and that their profile can be represented by the simple Cauchy function (i.e. a fractional rational function with NG = 1). As the initial parameter estimates, the following values were accepted:

1st band – maximum wavenumber, $\tilde{\nu}_{max} = 1170\ cm^{-1}$
maximum absorbance, $D_{max} = 0.45$
width parameter, $b_1 = 0.17$ cm

2nd band – $\tilde{\nu}_{max} = 1177\ cm^{-1}$
$D_{max} = 0.4$
$b_1 = 0.15$

The background is assumed to be linear, parallel with the wavenumber axis (NPOL = 0, $a_0 = 0.002$).
The data are then as follows:

```
TEST RUN — CH3SO2CL
   2   1     0     5           9.999994E—07           2.000000E+00
   1170.000          0.450           0.170
   1177.000          0.400           0.150
      0.002
TEST RUN — CH3SO2CL
  50    0.0010       1.0000
1150.00000     0.06793     1.00000
1151.00000     0.07444     1.00000
1152.00000     0.08194     1.00000
1153.00000     0.09063     1.00000
1154.00000     0.10075     1.00000
1155.00000     0.11264     1.00000
1156.00000     0.12668     1.00000
1157.00000     0.14342     1.00000
1158.00000     0.16352     1.00000
1159.00000     0.18784     1.00000
1160.00000     0.21752     1.00000
1161.00000     0.25400     1.00000
1162.00000     0.29909     1.00000
1163.00000     0.35494     1.00000
1164.00000     0.42392     1.00000
```

1165.00000	0.50797	1.00000
1166.00000	0.60723	1.00000
1167.00000	0.71749	1.00000
1168.00000	0.82668	1.00000
1169.00000	0.91348	1.00000
1170.00000	0.95329	1.00000
1171.00000	0.93235	1.00000
1172.00000	0.85810	1.00000
1173.00000	0.75337	1.00000
1174.00000	0.64169	1.00000
1175.00000	0.53814	1.00000
1176.00000	0.44910	1.00000
1177.00000	0.37547	1.00000
1178.00000	0.31567	1.00000
1179.00000	0.26739	1.00000
1180.00000	0.22837	1.00000
1181.00000	0.19669	1.00000
1182.00000	0.17070	1.00000
1183.00000	0.14944	1.00000
1184.00000	0.13171	1.00000
1185.00000	0.11687	1.00000
1186.00000	0.10434	1.00000
1187.00000	0.09369	1.00000
1188.00000	0.08458	1.00000
1189.00000	0.07672	1.00000
1190.00000	0.06991	1.00000
1191.00000	0.06397	1.00000
1192.00000	0.05877	1.00000
1193.00000	0.05418	1.00000
1194.00000	0.05012	1.00000
1195.00000	0.04651	1.00000
1196.00000	0.04328	1.00000
1197.00000	0.04039	1.00000
1198.00000	0.03779	1.00000
1199.00000	0.03544	1.00000

The resultant parameters correspond to the values given in the left-half o Table V for bands Nos 9 and 8.

```
      C         PSEUDODECONVOLUTION — MAIN PROGRAM
0001            DIMENSION DS(500),CS(500),RS(500),SCS(500)
0002            DIMENSION NAME(15),G(50),H(25)
0003            ICRQ=5
0004            ILPR=6
0005        900 FORMAT(15A4,5X,I5)
0006        901 FORMAT(3I5,E20.6)
0007        902 FORMAT(8F10.4)
0008            READ(ICRQ,901) NS
0009            READ(ICRQ,902)(H(I),I=1,NS)
0010            READ(ICRQ,902) HNORM
0011            DO 1 I=1,NS
0012          1 H(I)=H(I)/HNORM
0013            READ(ICRQ,900) NAME,IDSP
0014            READ(ICRQ,901) NP
0015            READ(ICRQ,902)(DS(I),I=1,NP)
0016            DO 2 I=1,NP
0017          2 CS(I)=DS(I)
0018            N2=0
0019        100 N1=N2+1
0020            READ(ICRQ,901) M,N2,NIT,EPS
0021            READ(ICRQ,902)(G(I),I=1,M)
0022            ITN=0
0023        101 CALL CONVOL(NP,DS,SCS,NS,H,N1,N2)
0024            CALL CONVOL(NP,SCS,RS,M,G,N1,N2)
0025            S=0.
0026            DO 102 I=N1,N2
0027        102 S=(CS(I)-RS(I))**2+S
0028            ITN=ITN+1
0029            IF(ITN.GE.NIT .OR. S.LE.EPS) GOTO 200
0030            DO 103 I=N1,N2
0031        103 DS(I)=SCS(I)+CS(I)—RS(I)
0032            GOTO 101
```

```
0033    200 IF(N2.LT.NP) GOTO 100
0034        WRITE(ILPR,951) NAME,IDSP
0035    951 FORMAT(1H1//23H DECONVOLUTED SPECTRUM ,15A4,
            5X,15//
          1    ' DECONV     EXPER          CALC            DIFF'/)
0036    952 FORMAT(1X,3F10.3,2E20.6)
0037        S=0.
0038        DO 201 I=1,NP
0039        D1=CS(I)—RS(I)
0040        D2=D1*D1
0041        S=D2+S
0042        WRITE(ILPR,952) DS(I),CS(I),RS(I),D1,D2
0043    201 CONTINUE
0044    853 FORMAT(/' TOTAL SUM OF SQUARES ',E20.6)
0045        WRITE(ILPR,953) S
0046        END

0001        SUBROUTINE CONVOL(N,DS,CS,M,G,N1,N2)
0002        DIMENSION DS(N),CS(N),G(M)
0003        DO 2 I=N1,N2
0004        S=G(1)*DS(I)
0005        J1=I
0006        J2=I
0007        DO 1 J=2,M
0008        IF(J1.GT.1) J1=J1—1
0009        IF(J2.LT.N) J2=J2+1
0010      1 S=G(J)*(DS(J1)+DS(J2))
0011      2 CS(I)=S
0012        RETURN
0013        END
```

The List of Symbols, Quantities and Units in Spectroscopy

Symbol	Quantity	Unit SI	Dimension of SI unit	Auxiliary Unit[a]	Conversion factor[b]
a_X	activity of substance X	$mol . m^{-3}$	$m^{-3} . mol$	$mol . l^{-1}$	10^3
$\hat{a}_X$	relative activity of substance X	—	1		
A_{abs}	absolute integrated intensity	$m^2 . s^{-1}$	$m^2 . s^{-1}$	$cm^2 . s^{-1}$	10^{-4}
A_{pract}	(practical) integrated intensity	$m . mol^{-1}$	$m . mol^{-1}$	$l . mol^{-1} . cm^{-2}$	10
A_{sec}	secondary integrated intensity	m^2	m^2	cm^2	10^{-4}
b_c	width parameter of the Cauchy function	m	m	cm	10^{-2}
b_g	width parameter of the Gauss function	m	m	cm	10^{-2}
B_{abs}	absolute apparent integrated intensity	$m^2 . s^{-1}$	$m^2 . s^{-1}$	$cm^2 . s^{-1}$	10^{-4}
B_{pract}	(practical) apparent integrated intensity	$m . mol^{-1}$	$m . mol^{-1}$	$l . mol^{-1} . cm^{-2}$	10
B_{sec}	secondary apparent integrated intensity	m^2	m^2	cm^2	10^{-4}
c	light velocity in vacuo, $c = (2.997\,925 \pm 0.000\,003) \times 10^8\ m . s^{-1}$	$m . s^{-1}$	$m . s^{-1}$		
c_X	concentration of substance X	$mol . m^{-3}$	$m^{-3} . mol$	$mol . l^{-1}$	10^3
$\hat{c}_X$	relative concentration of substance X	—	1		
d	cell thickness, the beam path length in the absorbing medium	m	m	cm	10^{-2}
D	absorbance (internal transmission density)	—	1		
D_{max}	band maximum absorbance	—	1		
D'_{max}	apparent band maximum absorbance	—	1		
E	energy of a quantum of radiation, vibrational level energy	J	$m^2 . kg . s^{-2}$	eV (erg)	$1.602\,10 \times 10^{-19}$ 10^{-7}
f_X	activity coefficient of substance X	—	1		
F	Helmholtz function	J	$m^2 . kg . s^{-2}$	(kcal)	$4.186\,8 \times 10^3$
F_m	molar Helmholtz function	$J . mol^{-1}$	$m^2 . kg . s^{-2} . mol^{-1}$	$(kcal . mol^{-1})$	$4.186\,8 \times 10^3$

Symbol	Quantity	Unit SI	Dimension of SI unit	Auxiliary Unit[a]	Conversion factor[b]
g	degeneration of state	—	1		
$g(\tilde{\nu}_m, \tilde{\nu})$	instrument function	—	1		
G	Gibbs free energy (Gibbs function)	J	$m^2 . kg . s^{-2}$	(kcal)	$4.186\,8 \times 10^3$
G_m	molar free energy	$J . mol^{-1}$	$m^2 . kg . s^{-2} . mol^{-1}$	$(kcal . mol^{-1})$	$4.186\,8 \times 10^3$
h	Planck constant, $h = (6.625\,6 \pm 0.000\,5) \times 10^{-33}$ J . s	J . s	$m^2 . kg . s^{-1}$		
H	enthalpy	J	$m^2 . kg . s^{-2}$	(kcal)	$4.186\,8 \times 10^3$
H_m	molar enthalpy	$J . mol^{-1}$	$m^2 . kg . s^{-2} . mol^{-1}$	$(kcal . mol^{-1})$	$4.186\,8 \times 10^3$
I	Raman line intensity	$mol^{-1} . sr^{-1}$	$mol^{-1} . sr^{-1}$		
J	rotational quantum number	—	1		
k	Boltzmann constant, $k = (1.380\,54 \pm 0.000\,18) \times 10^{-23}$ $J. K^{-1}$	$J . K^{-1}$	$m^2 . kg . s^{-2} . K^{-1}$		
n_X	amount of substance X	mol	mol		
n_X	number of molecules of substance X in a unit volume	m^{-3}	m^{-3}	cm^{-3}	10^6
N_A	Avogadro's constant, $N_A = (6.025\,2 \pm 0.000\,28) \times 10^{23}$ mol^{-1}	mol^{-1}	mol^{-1}		
p	pressure	Pa	$m^{-1} . kg . s^{-2}$	$(kp . cm^{-2} = at)$	$9.806\,65 \times 10^4$
				(Torr)	$1.333\,22 \times 10^2$
				bar	10^5
p_o	normal pressure, $p_o = 1.013\,25 \times 10^5$ Pa	Pa	$m^{-1} . kg . s^{-2}$		
$P(\tilde{\nu})$	band profile function	—	1		
Q	energy of radiation flux	J	$m^2 . kg . s^{-2}$		
Q	partition function	—	1		
$Q(\tilde{\nu})$	apparent band profile function	—	1		

Symbol	Quantity	Unit SI	Dimension of SI unit	Auxiliary Unit[a]	Conversion factor[b]
R	molar (universal) gas constant, $R =$ $= (8.314\,3 \pm 0.001\,2)\ J . K^{-1} . mol^{-1} \doteq$ $= 1.985\,8 \times 10^{-3}\ kcal . mol^{-1} . K^{-1}$	$J . K^{-1} . mol^{-1}$	$m^2 . kg . s^{-2} .$ $. K^{-1} . mol^{-1}$	$(kcal . mol^{-1} .$ $. K^{-1})$	$4.186\,8 \times 10^3$
R	reflectivity coefficient	—	1		
R	resolution	—	1		
s	spectral slit width	m^{-1}	m^{-1}	cm^{-1}	10^2
S	entropy	$J . K^{-1}$	$m^2 . kg . s^{-2} .$ $. K^{-}.^{i}$	$(kcal . K^{-1})$	$4.186\,8 \times 10^3$
S_m	molar entropy	$J . K^{-1} . mol^{-1}$	$m^2 . kg . s^{-2} .$ $. K^{-1} . mol^{-1}$	$(kcal . K^{-1} .$ $. mol^{-1})$	$4.186\,8 \times 10^3$
T	absolute temperature	K	K		
T	period	s	s		
U	internal energy	J	$m^2 . kg . s^{-2}$	(kcal)	$4.186\,8 \times 10^3$
U_m	internal molar energy	$J . mol^{-1}$	$m^2 . kg . s^{-2} .$ $. mol^{-1}$	(kcal . mol)	$4.186\,8 \times 10^3$
v	vibrational quantum number	—	1		
V	volume	m^{-3}	m^{-3}	1	10^{-3}
w_{ef}	effective slit width	m	m	μm	10^{-6}
w_{tr}	true (geometric) slit width	m	m	μm	10^{-6}
x_X	mole fraction of substance X	—	1		
α	absorptance	—	1	%	10^{-2}
$\Delta\tilde{\nu}_{\frac{1}{2}}$	band half-width	m^{-1}	m^{-1}	cm^{-1}	10^2
$\Delta\tilde{\nu}'_{\frac{1}{2}}$	apparent band half-width	m^{-1}	m^{-1}	cm^{-1}	10^2
$\varepsilon(\tilde{\nu})$	molar linear absorption coefficient (for solutions)	$m^2 . mol^{-1}$	$m^2 . mol^{-1}$	$1 . mol^{-1} . cm^{-2}$	10^{-1}
$\varepsilon_g(\tilde{\nu})$	molar linear absorption coefficient (for gases)	$Pa^{-1} . m^{-1}$	$kg^{-1} . s^2$	$(cm . kp^{-1})$ $(Torr^{-1} . cm^{-1})$	$1.019\,72 \times 10^{-3}$ $7.500\,64 \times 10^{-1}$

Symbol	Quantity	Unit SI	Dimension of SI unit	Auxiliary Unit[a]	Conversion factor[b]
ε	permitivity	$F . m^{-1}$	$m^{-3} . kg^{-1} . s^4 . A^2$		
ε_0	permitivity of a vacuum, $\varepsilon_0 = (8.854\,118 \pm 0.000\,002) \times 10^{-12}$ F . m^{-1}	$F . m^{-1}$	$m^{-3} . kg^{-1} . s^4 . A^2$		
λ	wavelength	m	m	µm (µ) nm (Å)	10^{-6} 10^{-9} 10^{-10}
μ	dipole moment	C . m	m . s . A	D	$3.335\,64 \times 10^{-30}$
ν	frequency	Hz	s^{-1}	THz GHz	10^{12} 10^{9}
$\tilde{\nu}$	wavenumber	m^{-1}	m^{-1}	cm^{-1}	10^{2}
$\tilde{\nu}_{max}$	band maximum wavenumber	m^{-1}	m^{-1}	cm^{-1}	10^{2}
ϱ	reflectance	—	1		
ϱ	depolarization factor	—	1		
ϱ	density (specific weight)	$kg . m^{-3}$	$m^{-3} . kg$	$g . cm^{-3}$	10^{3}
τ	transmittance	—	1	%	10^{-2}
Φ	radiant power	W	$m^2 . kg . s^{-3}$		
Φ_λ	spectral radiant power (on the wavelength scale)	$W . m^{-1}$	$m . kg . s^{-3}$	$W . \mu m^{-1}$	10^{6}
Φ_ν	spectral radiant power (on the frequency scale)	$W . Hz^{-1}$	$m^2 . kg . s^{-2}$		
$\Phi_{\tilde{\nu}}$	spectral radiant power (on the wavenumber scale)	W . m	$m^3 . kg . s^{-3}$	W . cm	10^{-2}

a. Units given in parentheses are obsolete and their use is not recommended.

b. Coefficient by which the numerical value of the quantity expressed in an auxiliary unit must be multiplied in order to obtain the numerical value for expression in SI units.

Recalculation of wavenumbers $\tilde{\nu}$ (cm^{-1}) to wavelengths λ (μm)

$\tilde{\nu}$ (cm^{-1})	0	1	2	3	4	5	6	7	8	9
	λ (μm)									
400	25.000	24.938	24.876	24.814	24.752	24.691	24.631	24.570	24.510	24.450
410	24.390	24.331	24.272	24.213	24.155	24.096	24.038	23.981	23.923	23.866
420	23.809	23.753	23.697	23.641	23.585	23.529	23.474	23.419	23.364	23.310
430	23.256	23.202	23.148	23.095	23.041	22.988	22.936	22.883	22.831	22.779
440	22.727	22.676	22.624	22.573	22.523	22.472	22.422	22.371	22.321	22.272
450	22.222	22.173	22.124	22.075	22.026	21.978	21.930	21.882	21.834	21.786
460	21.739	21.692	21.645	21.598	21.552	21.505	21.459	21.413	21.367	21.322
470	21.277	21.231	21.186	21.142	21.097	21.053	21.008	20.964	20.920	20.877
480	20.833	20.790	20.747	20.704	20.661	20.619	20.576	20.534	20.492	20.450
490	20.408	20.367	20.325	20.284	20.243	20.202	20.161	20.121	20.080	20.040
500	20.000	19.960	19.920	19.881	19.841	19.802	19.763	19.724	19.685	19.646
510	19.608	19.569	19.531	19.493	19.455	19.417	19.380	19.342	19.305	19.268
520	19.231	19.194	19.157	19.120	19.084	19.048	19.011	18.975	18.939	18.904
530	18.868	18.832	18.797	18.762	18.727	18.692	18.657	18.622	18.587	18.553
540	18.518	18.484	18.450	18.416	18.382	18.349	18.315	18.282	18.248	18.215
550	18.182	18.149	18.116	18.083	18.051	18.018	17.986	17.953	17.921	17.889
560	17.857	17.825	17.794	17.762	17.730	17.699	17.668	17.637	17.606	17.575
570	17.544	17.513	17.483	17.452	17.422	17.391	17.361	17.331	17.301	17.271
580	17.241	17.212	17.182	17.153	17.123	17.094	17.065	17.036	17.007	16.978
590	16.949	16.920	16.892	16.863	16.835	16.807	16.779	16.750	16.722	16.694
600	16.667	16.639	16.611	16.584	16.556	16.529	16.502	16.474	16.447	16.420
610	16.393	16.367	16.340	16.313	16.287	16.260	16.234	16.207	16.181	16.155
620	16.129	16.103	16.077	16.051	16.026	16.000	15.974	15.949	15.924	15.898
630	15.873	15.848	15.823	15.798	15.773	15.748	15.723	15.699	15.674	15.649
640	15.625	15.601	15.576	15.552	15.528	15.504	15.480	15.456	15.432	15.408
650	15.385	15.361	15.337	15.314	15.291	15.267	15.244	15.221	15.198	15.175
660	15.152	15.129	15.106	15.083	15.060	15.038	15.015	14.993	14.970	14.948
670	14.925	14.903	14.881	14.859	14.837	14.815	14.793	14.771	14.749	14.728
680	14.706	14.684	14.663	14.641	14.620	14.599	14.577	14.556	14.535	14.514
690	14.493	14.472	14.451	14.430	14.409	14.388	14.368	14.347	14.327	14.306
700	14.286	14.265	14.245	14.225	14.205	14.184	14.164	14.144	14.124	14.104
710	14.085	14.065	14.045	14.025	14.006	13.986	13.966	13.947	13.928	13.908
720	13.889	13.870	13.850	13.831	13.812	13.793	13.774	13.755	13.736	13.717
730	13.699	13.680	13.661	13.643	13.624	13.605	13.587	13.569	13.550	13.532
740	13.514	13.495	13.477	13.459	13.441	13.423	13.405	13.387	13.369	13.351
750	13.333	13.316	13.298	13.280	13.263	13.245	13.228	13.210	13.193	13.175
760	13.158	13.141	13.123	13.106	13.089	13.072	13.055	13.038	13.021	13.004
770	12.987	12.970	12.953	12.937	12.920	12.903	12.887	12.870	12.853	12.837
780	12.821	12.804	12.788	12.771	12.755	12.739	12.723	12.706	12.690	12.674
790	12.658	12.642	12.626	12.610	12.594	12.579	12.563	12.547	12.531	12.516
800	12.500	12.484	12.469	12.453	12.438	12.422	12.407	12.392	12.376	12.361

	0	1	2	3	4	5	6	7	8	9
$\tilde{\nu}$ (cm^{-1})					λ (μm)					
810	12.346	12.330	12.315	12.300	12.285	12.270	12.255	12.240	12.225	12.210
820	12.195	12.180	12.165	12.151	12.136	12.121	12.107	12.092	12.077	12.063
830	12.048	12.034	12.019	12.005	11.990	11.976	11.962	11.947	11.933	11.919
840	11.905	11.891	11.876	11.862	11.848	11.834	11.820	11.806	11.792	11.779
850	11.765	11.751	11.737	11.723	11.710	11.696	11.682	11.669	11.655	11.641
860	11.628	11.614	11.601	11.587	11.574	11.561	11.547	11.534	11.521	11.507
870	11.494	11.481	11.468	11.455	11.442	11.429	11.416	11.403	11.390	11.377
880	11.364	11.351	11.338	11.325	11.312	11.299	11.287	11.274	11.261	11.249
890	11.236	11.223	11.211	11.198	11.186	11.173	11.161	11.148	11.136	11.123
900	11.111	11.099	11.086	11.074	11.062	11.050	11.038	11.025	11.013	11.001
910	10.989	10.977	10.965	10.953	10.941	10.929	10.917	10.905	10.893	10.881
920	10.870	10.858	10.846	10.834	10.823	10.811	10.799	10.787	10.776	10.764
930	10.753	10.741	10.730	10.718	10.707	10.695	10.684	10.672	10.661	10.650
940	10.638	10.627	10.616	10.604	10.593	10.582	10.571	10.560	10.549	10.537
950	10.526	10.515	10.504	10.493	10.482	10.471	10.460	10.449	10.438	10.428
960	10.417	10.406	10.395	10.384	10.373	10.363	10.352	10.341	10.331	10.320
970	10.309	10.299	10.288	10.277	10.267	10.256	10.246	10.235	10.225	10.215
980	10.204	10.194	10.183	10.173	10.163	10.152	10.142	10.132	10.121	10.111
990	10.101	10.091	10.081	10.070	10.060	10.050	10.040	10.030	10.020	10.010
1000	10.000	9.990	9.980	9.970	9.960	9.950	9.940	9.930	9.921	9.911
1010	9.901	9.891	9.881	9.872	9.862	9.852	9.843	9.833	9.823	9.814
1020	9.804	9.794	9.785	9.775	9.766	9.756	9.747	9.737	9.728	9.718
1030	9.709	9.699	9.690	9.681	9.671	9.662	9.653	9.643	9.634	9.625
1040	9.615	9.606	9.597	9.588	9.579	9.569	9.560	9.551	9.542	9.533
1050	9.524	9.515	9.506	9.497	9.488	9.479	9.470	9.461	9.452	9.443
1060	9.434	9.425	9.416	9.407	9.398	9.390	9.381	9.372	9.363	9.355
1070	9.346	9.337	9.328	9.320	9.311	9.302	9.294	9.285	9.276	9.268
1080	9.259	9.251	9.242	9.234	9.225	9.217	9.208	9.200	9.191	9.183
1090	9.174	9.166	9.158	9.149	9.141	9.132	9.124	9.116	9.107	9.099
1100	9.091	9.083	9.074	9.066	9.058	9.050	9.042	9.033	9.025	9.017
1110	9.009	9.001	8.993	8.985	8.977	8.969	8.961	8.953	8.945	8.937
1120	8.929	8.921	8.913	8.905	8.897	8.889	8.881	8.873	8.865	8.857
1130	8.850	8.842	8.834	8.826	8.818	8.811	8.803	8.795	8.787	8.780
1140	8.772	8.764	8.757	8.749	8.741	8.734	8.726	8.718	8.711	8.703
1150	8.696	8.688	8.681	8.673	8.666	8.658	8.651	8.643	8.636	8.628
1160	8.621	8.613	8.606	8.598	8.591	8.584	8.576	8.569	8.562	8.554
1170	8.547	8.540	8.532	8.525	8.518	8.511	8.503	8.496	8.489	8.482
1180	8.475	8.467	8.460	8.453	8.446	8.439	8.432	8.425	8.418	8.410
1190	8.403	8.396	8.389	8.382	8.375	8.368	8.361	8.354	8.347	8.340
1200	8.333	8.326	8.319	8.313	8.306	8.299	8.292	9.285	8.278	8.271
1210	8.264	8.258	8.251	8.244	8.237	8.230	8.224	8.217	8.210	8.203
1220	8.197	8.190	8.183	8.177	8.170	8.163	8.157	8.150	8.143	8.137
1230	8.130	8.123	8.117	8.110	8.104	8.097	8.091	8.084	8.078	8.071

	0	1	2	3	4	5	6	7	8	9
$\tilde{\nu}$ (cm^{-1})					λ (μm)					
1240	8.065	8.058	8.052	8.045	8.039	8.032	8.026	8.019	8.013	8.006
1250	8.000	7.994	7.987	7.981	7.974	7.968	7.962	7.955	7.949	7.943
1260	7.937	7.930	7.924	7.918	7.911	7.905	7.899	7.893	7.886	7.880
1270	7.874	7.868	7.862	7.855	7.849	7.843	7.837	7.831	7.825	7.819
1280	7.812	7.806	7.800	7.794	7.788	7.782	7.776	7.770	7.764	7.758
1290	7.752	7.746	7.740	7.734	7.728	7.722	7.716	7.710	7.704	7.698
1300	7.692	7.686	7.680	7.675	7.669	7.663	7.657	7.651	7.645	7.639
1310	7.634	7.628	7.622	7.616	7.610	7.605	7.599	7.593	7.587	7.582
1320	7.576	7.570	7.564	7.559	7.553	7.547	7.541	7.536	7.530	7.524
1330	7.519	7.513	7.508	7.502	7.496	7.491	7.485	7.479	7.474	7.468
1340	7.463	7.457	7.452	7.446	7.440	7.435	7.429	7.424	7.418	7.413
1350	7.407	7.402	7.396	7.391	7.386	7.380	7.375	7.369	7.364	7.358
1360	7.353	7.348	7.342	7.337	7.331	7.326	7.321	7.315	7.310	7.305
1370	7.299	7.294	7.289	7.283	7.278	7.273	7.267	7.262	7.257	7.252
1380	7.246	7.241	7.236	7.231	7.225	7.220	7.215	7.210	7.205	7.199
1390	7.194	7.189	7.184	7.179	7.174	7.168	7.163	7.158	7.153	7.148
1400	7.143	7.138	7.133	7.128	7.123	7.117	7.112	7.107	7.102	7.097
1410	7.092	7.087	7.082	7.077	7.072	7.067	7.062	7.057	7.052	7.047
1420	7.042	7.037	7.032	7.027	7.022	7.018	7.013	7.008	7.003	6.998
1430	6.993	6.988	6.983	6.978	6.973	6.969	6.964	6.959	6.954	6.949
1440	6.944	6.940	6.935	6.930	6.925	6.920	6.916	6.911	6.906	6.901
1450	6.897	6.892	6.887	6.882	6.878	6.873	6.868	6.863	6.859	6.854
1460	6.849	6.845	6.840	6.835	6.831	6.826	6.821	6.817	6.812	6.807
1470	6.803	6.798	6.793	6.789	6.784	6.780	6.775	6.770	6.766	6.761
1480	6.757	6.752	6.748	6.743	6.739	6.734	6.729	6.725	6.720	6.716
1490	6.711	6.707	6.702	6.698	6.693	6.689	6.684	6.680	6.676	6.671
1500	6.667	6.662	6.658	6.653	6.649	6.645	6.640	6.636	6.631	6.627
1510	6.623	6.618	6.614	6.609	6.605	6.601	6.596	6.592	6.588	6.583
1520	6.579	6.575	6.570	6.566	6.562	6.557	6.553	6.549	6.545	6.540
1530	6.536	6.532	6.527	6.523	6.519	6.515	6.510	6.506	6.502	6.498
1540	6.494	6.489	6.485	6.481	6.477	6.472	6.468	6.464	6.460	6.456
1550	6.452	6.447	6.443	6.439	6.435	6.431	6.427	6.423	6.418	6.414
1560	6.410	6.406	6.402	6.398	6.394	6.390	6.386	6.382	6.378	6.373
1570	6.369	6.365	6.361	6.357	6.353	6.349	6.345	6.341	6.337	6.333
1580	6.329	6.325	6.321	6.317	6.313	6.309	6.305	6.301	6.297	6.293
1590	6.289	6.285	6.281	6.277	6.274	6.270	6.266	6.262	6.258	6.257
1600	6.250	6.246	6.242	6.238	6.234	6.231	6.227	6.223	6.219	6.215
1610	6.211	6.207	6.203	6.200	6.196	6.192	6.188	6.184	6.180	6.177
1620	6.173	6.169	6.165	6.161	6.158	6.154	6.150	6.146	6.143	6.139
1630	6.135	6.131	6.127	6.124	6.120	6.116	6.112	6.109	6.105	6.101
1640	6.098	6.094	6.090	6.086	6.083	6.079	6.075	6.072	6.068	6.064
1650	6.061	6.057	6.053	6.050	6.046	6.042	6.039	6.035	6.031	6.028
1660	6.024	6.020	6.017	6.013	6.010	6.006	6.002	5.999	5.995	5.992

	0	1	2	3	4	5	6	7	8	9
$\tilde{\nu}$ (cm^{-1})					λ (μm)					
1670	5.988	5.984	5.981	5.977	5.974	5.970	5.967	5.963	5.959	5.956
1680	5.952	5.949	5.945	5.942	5.938	5.935	5.931	5.928	5.924	5.921
1690	5.917	5.914	5.910	5.907	5.903	5.900	5.896	5.893	5.889	5.886
1700	5.882	5.879	5.875	5.872	5.869	5.865	5.862	5.858	5.855	5.851
1710	5.848	5.845	5.841	5.838	5.834	5.831	5.828	5.824	5.821	5.817
1720	5.814	5.811	5.807	5.804	5.800	5.797	5.794	5.790	5.787	5.784
1730	5.780	5.777	5.774	5.770	5.767	5.764	5.760	5.757	5.754	5.750
1740	5.747	5.744	5.741	5.737	5.734	5.731	5.727	5.724	5.721	5.718
1750	5.714	5.711	5.708	5.705	5.701	5.698	5.695	5.692	5.688	5.685
1760	5.682	5.679	5.675	5.672	5.669	5.666	5.663	5.659	5.656	5.653
1770	5.650	5.647	5.643	5.640	5.637	5.634	5.631	5.627	5.624	5.621
1780	5.618	5.615	5.612	5.609	5.605	5.602	5.599	5.596	5.593	5.590
1790	5.587	5.583	5.580	5.577	5.574	5.571	5.568	5.565	5.562	5.559
1800	5.556	5.552	5.549	5.546	5.543	5.540	5.537	5.534	5.531	5.528
1810	5.525	5.522	5.519	5.516	5.513	5.510	5.507	5.504	5.501	5.498
1820	5.495	5.491	5.488	5.485	5.482	5.479	5.476	5.473	5.470	5.467
1830	5.464	5.461	5.459	5.456	5.453	5.450	5.447	5.444	5.441	5.438
1840	5.435	5.432	5.429	5.426	5.423	5.420	5.417	5.414	5.411	5.408
1850	5.405	5.402	5.400	5.397	5.394	5.391	5.388	5.385	5.382	5.379
1860	5.376	5.373	5.371	5.368	5.365	5.362	5.359	5.356	5.353	5.350
1870	5.348	5.345	5.342	5.339	5.336	5.333	5.330	5.328	5.325	5.322
1880	5.319	5.316	5.313	5.311	5.308	5.305	5.302	5.299	5.297	5.294
1890	5.291	5.288	5.285	5.283	5.280	5.277	5.274	5.271	5.269	5.266
1900	5.263	5.260	5.258	5.255	5.252	5.249	5.247	5.244	5.241	5.238
1910	5.236	5.233	5.230	5.227	5.225	5.222	5.219	5.216	5.214	5.211
1920	5.208	5.206	5.203	5.200	5.198	5.195	5.192	5.189	5.187	5.184
1930	5.181	5.179	5.176	5.173	5.171	5.168	5.165	5.163	5.160	5.157
1940	5.155	5.152	5.149	5.147	5.144	5.141	5.139	5.136	5.133	5.131
1950	5.128	5.126	5.123	5.120	5.118	5.115	5.112	5.110	5.107	5.105
1960	5.102	5.099	5.097	5.094	5.092	5.089	5.086	5.084	5.081	5.079
1970	5.076	5.074	5.071	5.068	5.066	5.063	5.061	5.058	5.056	5.053
1980	5.051	5.048	5.045	5.043	5.040	5.038	5.035	5.033	5.030	5.028
1990	5.025	5.023	5.020	5.018	5.015	5.013	5.010	5.008	5.005	5.003
2000	5.000	4.998	4.995	4.993	4.990	4.988	4.985	4.983	4.980	4.978
2010	4.975	4.973	4.970	4.968	4.965	4.963	4.960	4.958	4.955	4.953
2020	4.950	4.948	4.946	4.943	4.941	4.938	4.936	4.933	4.931	4.929
2030	4.926	4.924	4.921	4.919	4.916	4.914	4.912	4.909	4.907	4.904
2040	4.902	4.900	4.897	4.895	4.892	4.890	4.888	4.885	4.883	4.880
2050	4.878	4.876	4.873	4.871	4.869	4.866	4.864	4.861	4.859	4.857
2060	4.854	4.852	4.850	4.847	4.845	4.843	4.840	4.838	4.836	4.833
2070	4.831	4.829	4.826	4.824	4.822	4.819	4.817	4.815	4.812	4.810
2080	4.808	4.805	4.803	4.801	4.798	4.796	4.794	4.792	4.789	4.787
2090	4.785	4.782	4.780	4.778	4.776	4.773	4.771	4.769	4.766	4.764

	0	1	2	3	4	5	6	7	8	9
$\tilde{\nu}$ (cm^{-1})					λ (μm)					
2100	4.762	4.760	4.757	4.755	4.753	4.751	4.748	4.746	4.744	4.742
2110	4.739	4.737	4.735	4.733	4.730	4.728	4.726	4.724	4.721	4.719
2120	4.717	4.715	4.713	4.710	4.708	4.706	4.704	4.701	4.699	4.697
2130	4.695	4.693	4.690	4.688	4.686	4.684	4.682	4.679	4.677	4.675
2140	4.673	4.671	4.669	4.666	4.664	4.662	4.660	4.658	4.655	4.653
2150	4.651	4.649	4.647	4.645	4.643	4.640	4.638	4.636	4.634	4.632
2160	4.630	4.627	4.625	4.623	4.621	4.619	4.617	4.615	4.613	4.610
2170	4.608	4.606	4.604	4.602	4.600	4.598	4.596	4.593	4.591	4.589
2180	4.587	4.585	4.583	4.581	4.579	4.577	4.575	4.572	4.570	4.568
2190	4.566	4.564	4.562	4.560	4.558	4.556	4.554	4.552	4.550	4.548
2200	4.545	4.543	4.541	4.539	4.537	4.535	4.533	4.531	4.529	4.527
2210	4.525	4.523	4.521	4.519	4.517	4.515	4.513	4.511	4.509	4.507
2220	4.505	4.502	4.500	4.498	4.496	4.494	4.492	4.490	4.488	4.486
2230	4.484	4.482	4.480	4.478	4.476	4.474	4.472	4.470	4.468	4.466
2240	4.464	4.462	4.460	4.458	4.456	4.454	4.452	4.450	4.448	4.446
2250	4.444	4.442	4.440	4.439	4.437	4.435	4.433	4.431	4.429	4.427
2260	4.425	4.423	4.421	4.419	4.417	4.415	4.413	4.411	4.409	4.407
2270	4.405	4.403	4.401	4.399	4.398	4.396	4.394	4.392	4.390	4.388
2280	4.386	4.384	4.382	4.380	4.378	4.376	4.374	4.373	4.371	4.369
2290	4.367	4.365	4.363	4.361	4.359	4.357	4.355	4.354	4.352	4.350
2300	4.348	4.346	4.344	4.342	4.340	4.338	4.337	4.335	4.333	4.331
2310	4.329	4.327	4.325	4.323	4.322	4.320	4.318	4.316	4.314	4.312
2320	4.310	4.308	4.307	4.305	4.303	4.301	4.299	4.297	4.296	4.294
2330	4.292	4.290	4.288	4.286	4.284	4.283	4.281	4.279	4.277	4.275
2340	4.274	4.272	4.270	4.268	4.266	4.264	4.263	4.261	4.259	4.257
2350	4.255	4.254	4.252	4.250	4.248	4.246	4.244	4.243	4.241	4.239
2360	4.237	4.235	4.234	4.232	4.230	4.228	4.227	4.225	4.223	4.221
2370	4.219	4.218	4.216	4.214	4.212	4.211	4.209	4.207	4.205	4.203
2380	4.202	4.200	4.198	4.196	4.195	4.193	4.191	4.189	4.188	4.186
2390	4.184	4.182	4.181	4.179	4.177	4.175	4.174	4.172	4.170	4.168
2400	4.167	4.165	4.163	4.161	4.160	4.158	4.156	4.155	4.153	4.151
2410	4.149	4.148	4.146	4.144	4.143	4.141	4.139	4.137	4.136	4.134
2420	4.132	4.131	4.129	4.127	4.125	4.124	4.122	4.120	4.119	4.117
2430	4.115	4.114	4.112	4.110	4.108	4.107	4.105	4.103	4.102	4.100
2440	4.098	4.097	4.095	4.093	4.092	4.090	4.088	4.087	4.085	4.083
2450	4.082	4.080	4.078	4.077	4.075	4.073	4.072	4.070	4.068	4.067
2460	4.065	4.063	4.062	4.060	4.058	4.057	4.055	4.054	4.052	4.050
2470	4.049	4.047	4.045	4.044	4.042	4.040	4.039	4.037	4.036	4.034
2480	4.032	4.031	4.029	4.027	4.026	4.024	4.023	4.021	4.019	4.018
2490	4.016	4.014	4.013	4.011	4.010	4.008	4.006	4.005	4.003	4.002
2500	4.000	3.998	3.997	3.995	3.994	3.992	3.990	3.989	3.987	3.986
2510	3.984	3.982	3.981	3.979	3.978	3.976	3.975	3.973	3.971	3.970
2520	3.968	3.967	3.965	3.964	3.962	3.960	3.959	3.957	3.956	3.954

	0	1	2	3	4	5	6	7	8	9
$\tilde{\nu}$ (cm^{-1})					λ (μm)					
2530	3.953	3.951	3.949	3.948	3.946	3.945	3.943	3.942	3.940	3.939
2540	3.937	3.935	3.934	3.932	3.931	3.929	3.928	3.926	3.925	3.923
2550	3.922	3.920	3.918	3.917	3.915	3.914	3.912	3.911	3.909	3.908
2560	3.906	3.905	3.903	3.902	3.900	3.899	3.897	3.896	3.894	3.893
2570	3.891	3.890	3.888	3.887	3.885	3.883	3.882	3.880	3.879	3.877
2580	3.876	3.874	3.873	3.871	3.870	3.868	3.867	3.865	3.864	3.862
2590	3.861	3.860	3.858	3.857	3.855	3.854	3.852	3.851	3.849	3.848
2600	3.846	3.845	3.843	3.842	3.840	3.839	3.837	3.836	3.834	3.833
2610	3.831	3.830	3.828	3.827	3.826	3.824	3.823	3.821	3.820	3.818
2620	3.817	3.815	3.814	3.812	3.811	3.810	3.808	3.807	3.805	3.804
2630	3.802	3.801	3.799	3.798	3.797	3.795	3.794	3.792	3.791	3.789
2640	3.788	3.786	3.785	3.784	3.782	3.781	3.779	3.778	3.776	3.775
2650	3.774	3.772	3.771	3.769	3.768	3.766	3.765	3.764	3.762	3.761
2660	3.759	3.758	3.757	3.755	3.754	3.752	3.751	3.750	3.748	3.747
2670	3.745	3.744	3.743	3.741	3.740	3.738	3.737	3.736	3.734	3.733
2680	3.731	3.730	3.729	3.727	3.726	3.724	3.723	3.722	3.720	3.719
2690	3.717	3.716	3.715	3.713	3.712	3.711	3.709	3.708	3.706	3.705
2700	3.704	3.702	3.701	3.700	3.698	3.697	3.695	3.694	3.693	3.691
2710	3.690	3.689	3.687	3.686	3.685	3.683	3.682	3.681	3.679	3.678
2720	3.676	3.675	3.674	3.672	3.671	3.670	3.668	3.667	3.666	3.664
2730	3.663	3.662	3.660	3.659	3.658	3.656	3.655	3.654	3.652	3.651
2740	3.650	3.648	3.647	3.646	3.644	3.643	3.642	3.640	3.639	3.638
2750	3.636	3.635	3.634	3.632	3.631	3.630	3.628	3.627	3.626	3.625
2760	3.623	3.622	3.621	3.619	3.618	3.617	3.615	3.614	3.613	3.611
2770	3.610	3.609	3.608	3.606	3.605	3.604	3.602	3.601	3.600	3.598
2780	3.597	3.596	3.595	3.593	3.592	3.591	3.589	3.588	3.587	3.586
2790	3.584	3.583	3.582	3.580	3.579	3.578	3.577	3.575	3.574	3.573
2800	3.571	3.570	3.569	3.568	3.566	3.565	3.564	3.563	3.561	3.560
2810	3.559	3.557	3.556	3.555	3.554	3.552	3.551	3.550	3.549	3.547
2820	3.546	3.545	3.544	3.542	3.541	3.540	3.539	3.537	3.536	3.535
2830	3.534	3.532	3.531	3.530	3.529	3.527	3.526	3.525	3.524	3.522
2840	3.521	3.520	3.519	3.517	3.516	3.515	3.514	3.512	3.511	3.510
2850	3.509	3.508	3.506	3.505	3.504	3.503	3.501	3.500	3.499	3.498
2860	3.497	3.495	3.494	3.493	3.492	3.490	3.489	3.488	3.487	3.486
2870	3.484	3.483	3.482	3.481	3.479	3.478	3.477	3.476	3.475	3.473
2880	3.472	3.471	3.470	3.469	3.467	3.466	3.465	3.464	3.463	3.461
2890	3.460	3.459	3.458	3.457	3.455	3.454	3.453	3.452	3.451	3.449
2900	3.448	3.447	3.446	3.445	3.444	3.442	3.441	3.440	3.439	3.438
2910	3.436	3.435	3.434	3.433	3.432	3.431	3.429	3.428	3.427	3.426
2920	3.425	3.423	3.422	3.421	3.420	3.419	3.418	3.416	3.415	3.414
2930	3.413	3.412	3.411	3.409	3.408	3.407	3.406	3.405	3.404	3.403
2940	3.401	3.400	3.399	3.398	3.397	3.396	3.394	3.393	3.392	3.391
2950	3.390	3.389	3.388	3.386	3.385	3.384	3.383	3.382	3.381	3.380

	0	1	2	3	4	5	6	7	8	9
$\tilde{\nu}$ (cm^{-1})					λ (μm)					
2960	3.378	3.377	3.376	3.375	3.374	3.373	3.372	3.370	3.369	3.368
2970	3.367	3.366	3.365	3.364	3.362	3.361	3.360	3.359	3.358	3.357
2980	3.356	3.355	3.353	3.352	3.351	3.350	3.349	3.348	3.347	3.346
2990	3.344	3.343	3.342	3.341	3.340	3.339	3.338	3.337	3.336	3.334
3000	3.333	3.332	3.331	3.330	3.329	3.328	3.327	3.326	3.324	3.323
3010	3.322	3.321	3.320	3.319	3.318	3.317	3.316	3.315	3.313	3.312
3020	3.311	3.310	3.309	3.308	3.307	3.306	3.305	3.304	3.303	3.301
3030	3.300	3.299	3.298	3.297	3.296	3.295	3.294	3.293	3.292	3.291
3040	3.289	3.288	3.287	3.286	3.285	3.284	3.283	3.282	3.281	3.280
3050	3.279	3.278	3.277	3.275	3.274	3.273	3.272	3.271	3.270	3.269
3060	3.268	3.267	3.266	3.265	3.264	3.263	3.262	3.261	3.259	3.258
3070	3.257	3.256	3.255	3.254	3.253	3.252	3.251	3.250	3.249	3.248
3080	3.247	3.246	3.245	3.244	3.243	3.241	3.240	3.239	3.238	3.237
3090	3.236	3.235	3.234	3.233	3.232	3.231	3.230	3.229	3.228	3.227
3100	3.226	3.225	3.224	3.223	3.222	3.221	3.220	3.219	3.218	3.216
3110	3.215	3.214	3.213	3.212	3.211	3.210	3.209	3.208	3.207	3.206
3120	3.205	3.204	3.203	3.202	3.201	3.200	3.199	3.198	3.197	3.196
3130	3 195	3.194	3.193	3.192	3.191	3.190	3.189	3.188	3.187	3.186
3140	3.185	3.184	3.183	3.182	3.181	3.180	3.179	3.178	3.177	3.176
3150	3.175	3.174	3.173	3.172	3.171	3.170	3.169	3.168	3.167	3.166
3160	3.165	3.164	3.163	3.162	3.161	3.160	3.159	3.158	3.157	3.156
3170	3.155	3.154	3.153	3.152	3.151	3.150	3.149	3.148	3.147	3.146
3180	3.145	3.144	3.143	3.142	3.141	3.140	3.139	3.138	3.137	3.136
3190	3.135	3.134	3.133	3.132	3.131	3.130	3.129	3.128	3.127	3.126
3200	3.125	3.124	3.123	3.122	3.121	3.120	3.119	3.118	3.117	3.116
3210	3.115	3.114	3.113	3.112	3.111	3.110	3.109	3.108	3.108	3.107
3220	3.106	3.105	3.104	3.103	3.102	3.101	3.100	3.099	3.098	3.097
3230	3.096	3.095	3.094	3.093	3.092	3.091	3.090	3.089	3.088	3.087
3240	3.086	3.085	3.085	3.084	3.083	3.082	3.081	3.080	3.079	3.078
3250	3.077	3.076	3.075	3.074	3.073	3.072	3.071	3.070	3.069	3.068
3260	3.067	3.067	3.066	3.065	3.064	3.063	3.062	3.061	3.060	3.059
3270	3.058	3.057	3.056	3.055	3.054	3.053	3.053	3.052	3.051	3.050
3280	3.049	3.048	3.047	3.046	3.045	3.044	3.043	3.042	3.041	3.040
3290	3.040	3.039	3.038	3.037	3.036	3.035	3.034	3.033	3.032	3.031
3300	3.030	3.029	3.028	3.028	3.027	3.026	3.025	3.024	3.023	3.022
3310	3.021	3.020	3.019	3.018	3.018	3.017	3.016	3.015	3.014	3.013
3320	3.012	3.011	3.010	3.009	3.008	3.008	3.007	3.006	3.005	3.004
3330	3.003	3.002	3.001	3.000	2.999	2.998	2.998	2.997	2.996	2.995
3340	2.994	2.993	2.992	2.991	2.990	2.990	2.989	2.988	2.987	2.986
3350	2.985	2.984	2.983	2.982	2.982	2.981	2.980	2.979	2.978	2.977
3360	2.976	2.975	2.974	2.974	2.973	2.972	2.971	2.970	2.969	2.968
3370	2.967	2.966	2.966	2.965	2.964	2.963	2.962	2.961	2.960	2.959
3380	2.959	2.958	2.957	2.956	2.955	2.954	2.953	2.952	2.952	2.951

	0	1	2	3	4	5	6	7	8	9
$\tilde{\nu}$ (cm^{-1})					λ (μm)					
3390	2.950	2.949	2.948	2.947	2.946	2.946	2.945	2.944	2.943	2.942
3400	2.941	2.940	2.939	2.939	2.938	2.937	2.936	2.935	2.934	2.933
3410	2.933	2.932	2.931	2.930	2.929	2.928	2.927	2.927	2.926	2.925
3420	2.924	2.923	2.922	2.921	2.921	2.920	2.919	2.918	2.917	2.916
3430	2.915	2.915	2.914	2.913	2.912	2.911	2.910	2.910	2.909	2.908
3440	2.907	2.906	2.905	2.904	2.904	2.903	2.902	2.901	2.900	2.899
3450	2.899	2.898	2.897	2.896	2.895	2.894	2.894	2.893	2.892	2.891
3460	2.890	2.889	2.889	2.888	2.887	2.886	2.885	2.884	2.884	2.883
3470	2.882	2.881	2.880	2.879	2.879	2.878	2.877	2.876	2.875	2.874
3480	2.874	2.873	2.872	2.871	2.870	2.869	2.869	2.868	2.867	2.866
3490	2.865	2.865	2.864	2.863	2.862	2.861	2.860	2.860	2.859	2.858
3500	2.857	2.856	2.856	2.855	2.854	2.853	2.852	2.851	2.851	2.850
3510	2.849	2.848	2.847	2.847	2.846	2.845	2.844	2.843	2.843	2.842
3520	2.841	2.840	2.839	2.838	2.838	2.837	2.836	2.835	2.834	2.834
3530	2.833	2.832	2.831	2.830	2.830	2.829	2.828	2.827	2.826	2.826
3540	2.825	2.824	2.823	2.822	2.822	2.821	2.820	2.819	2.818	2.818
3550	2.817	2.816	2.815	2.815	2.814	2.813	2.812	2.811	2.811	2.810
3560	2.809	2.808	2.807	2.807	2.806	2.805	2.804	2.803	2.803	2.802
3570	2.801	2.800	2.800	2.799	2.798	2.797	2.796	2.796	2.795	2.794
3580	2.793	2.793	2.792	2.791	2.790	2.789	2.789	2.788	2.787	2.786
3590	2.786	2.785	2.784	2.783	2.782	2.782	2.781	2.780	2.779	2.779
3600	2.778	2.777	2.776	2.775	2.775	2.774	2.773	2.772	2.772	2.771
3610	2.770	2.769	2.769	2.768	2.767	2.766	2.765	2.765	2.764	2.763
3620	2.762	2.762	2.761	2.760	2.759	2.759	2.758	2.757	2.756	2.756
3630	2.755	2.754	2.753	2.753	2.752	2.751	2.750	2.750	2.749	2.748
3640	2.747	2.746	2.746	2.745	2.744	2.743	2.743	2.742	2.741	2.740
3650	2.740	2.739	2.738	2.737	2.737	2.736	2.735	2.734	2.734	2.733
3660	2.732	2.731	2.731	2.730	2.729	2.729	2.728	2.727	2.726	2.726
3670	2.725	2.724	2.723	2.723	2.722	2.721	2.720	2.720	2.719	2.718
3680	2.717	2.717	2.716	2.715	2.714	2.714	2.713	2.712	2.711	2.711
3690	2.710	2.709	2.709	2.708	2.707	2.706	2.706	2.705	2.704	2.703
3700	2.703	2.702	2.701	2.701	2.700	2.699	2.698	2.698	2.697	2.696
3710	2.695	2.695	2.694	2.693	2.693	2.692	2.691	2.690	2.690	2.689
3720	2.688	2.687	2.687	2.686	2.685	2.685	2.684	2.683	2.682	2.682
3730	2.681	2.680	2.680	2.679	2.678	2.677	2.677	2.676	2.675	2.675
3740	2.674	2.673	2.672	2.672	2.671	2.670	2.670	2.669	2.668	2.667
3750	2.667	2.666	2.665	2.665	2.664	2.663	2.662	2.662	2.661	2.660
2760	2.660	2.659	2.658	2.657	2.657	2.656	2.555	2.655	2.654	2.653
3770	2.653	2.652	2.651	2.650	2.650	2.649	2.648	2.648	2.647	2.646
3780	2.646	2.645	2.644	2.643	2.643	2.642	2.641	2.641	2.640	2.639
3790	2.639	2.638	2.637	2.636	2.636	2.635	2.634	2.634	2.633	2.632
3800	2.632	2.631	2.630	2.630	2.629	2.628	2.627	2.627	2.626	2.625
3810	2.625	2.624	2.623	2.623	2.622	2.621	2.621	2.620	2.619	2.618

	0	1	2	3	4	5	6	7	8	9
$\tilde{\nu}$ (cm^{-1})					λ (μm)					
3820	2.618	2.617	2.616	2.616	2.615	2.614	2.614	2.613	2.612	2.612
3830	2.611	2.610	2.610	2.609	2.608	2.608	2.607	2.606	2.606	2.605
3840	2.604	2.603	2.603	2.602	2.601	2.601	2.600	2.599	2.599	2.598
3850	2.597	2.597	2.596	2.595	2.595	2.594	2.593	2.593	2.592	2.591
3860	2.591	2.590	2.589	2.589	2.588	2.587	2.587	2.586	2.585	2.585
3870	2.584	2.583	2.583	2.582	2.581	2.581	2.580	2.579	2.579	2.578
3880	2.577	2.577	2.576	2.575	2.575	2.574	2.573	2.573	2.572	2.571
3890	2.571	2.570	2.569	2.569	2.568	2.567	2.567	2.566	2.565	2.565
3900	2.564	2.563	2.563	2.562	2.561	2.561	2.560	2.560	2.559	2.558
3910	2.558	2.557	2.556	2.556	2.555	2.554	2.554	2.553	2.552	2.552
3920	2.551	2.550	2.550	2.549	2.548	2.548	2.547	2.546	2.546	2.545
3930	2.545	2.544	2.543	2.543	2.542	2.541	2.541	2.540	2.539	2.539
3940	2.538	2.537	2.537	2.536	2 535	2.535	2.534	2.534	2.533	2.532
3950	2.532	2.531	2.530	2.530	2.529	2.528	2.528	2.527	2.527	2.526
3960	2.525	2.525	2.524	2.523	2.523	2.522	2.521	2.521	2.520	2.520
3970	2.519	2.518	2.518	2.517	2.516	2.516	2.515	2.514	2.514	2.513
3980	2.513	2.512	2.511	2.511	2.510	2.509	2.509	2.508	2.508	2.507
3990	2.506	2.506	2.505	2.504	2.504	2.503	2.503	2.502	2.501	2.501

Recalculation of transmittance τ to absorbance D with correction for the background transmittance, τ_0

τ_0	100.0	99.5	99.0	98.5	98.0	97.5	97.0	96.5	96.0	95.5	95.0	τ_0
τ						10,000 . D						τ
100.0	0000											100.0
99.5	0022	0000										99.5
99.0	0044	0022	0000									99.0
98.5	0066	0044	0022	0000								98.5
98.0	0088	0066	0044	0022	0000							98.0
97.5	0110	0088	0066	0044	0022	0000						97.5
97.0	0132	0111	0089	0067	0045	0022	0000					97.0
96.5	0155	0133	0111	0089	0067	0045	0022	0000				96.5
96.0	0177	0156	0134	0112	0090	0067	0045	0023	0000			96.0
95.5	0200	0178	0156	0134	0112	0090	0068	0045	0023	0000		95.5
95.0	0223	0201	0179	0157	0135	0113	0090	0068	0045	0023	0000	95.0
94.5	0246	0224	0202	0180	0158	0136	0113	0091	0068	0046	0023	94.5
94.0	0269	0247	0225	0203	0181	0159	0136	0114	0091	0069	0046	94.0
93.5	0292	0270	0248	0226	0204	0182	0160	0137	0115	0092	0069	93.5
93.0	0315	0293	0272	0250	0227	0205	0183	0160	0138	0115	0092	93.0
92.5	0339	0317	0295	0273	0251	0229	0206	0184	0161	0139	0116	92.5
92.0	0362	0340	0318	0296	0274	0252	0230	0207	0185	0162	0139	92.0
91.5	0386	0364	0342	0320	0298	0276	0254	0231	0209	0186	0163	91.5
91.0	0410	0388	0366	0344	0322	0300	0277	0255	0232	0210	0187	91.0
90.5	0434	0412	0390	0368	0346	0324	0301	0279	0256	0234	0211	90.5
90.0	0458	0436	0414	0392	0370	0348	0325	0303	0280	0258	0235	90.0
89.5	0482	0460	0438	0416	0394	0372	0349	0327	0304	0282	0259	89.5
89.0	0506	0484	0462	0440	0418	0396	0374	0351	0329	0306	0283	89.0
88.5	0531	0509	0487	0465	0443	0421	0398	0376	0353	0331	0308	88.5
88.0	0555	0533	0512	0490	0467	0445	0423	0400	0378	0355	0332	88.0
87.5	0580	0558	0536	0514	0492	0470	0448	0425	0403	0380	0357	87.5
87.0	0605	0583	0561	0539	0517	0495	0473	0450	0428	0405	0382	87.0
86.5	0630	0608	0586	0564	0542	0520	0498	0475	0453	0430	0407	86.5
86.0	0655	0633	0611	0589	0567	0545	0523	0500	0478	0455	0432	86.0
85.5	0680	0659	0637	0615	0593	0570	0548	0526	0503	0480	0458	85.5
85.0	0706	0684	0662	0640	0618	0596	0574	0551	0529	0506	0483	85.0
84.5	0731	0710	0688	0666	0644	0621	0599	0577	0554	0531	0509	84.5
84.0	0757	0735	0714	0692	0669	0647	0625	0602	0580	0557	0534	84.0
83.5	0783	0761	0739	0717	0695	0673	0651	0628	0606	0583	0560	83.5
83.0	0809	0787	0766	0744	0721	0699	0677	0654	0632	0609	0586	83.0
82.5	0835	0814	0792	0770	0748	0726	0703	0681	0658	0635	0613	82.5
82.0	0862	0840	0818	0796	0774	0752	0730	0707	0685	0662	0639	82.0
81.5	0888	0867	0845	0823	0801	0778	0756	0734	0711	0688	0666	81.5
81.0	0915	0893	0871	0850	0827	0805	0783	0760	0738	0715	0692	81.0
80.5	0942	0920	0898	0876	0854	0832	0810	0787	0765	0742	0719	80.5

τ_0	100.0	99.5	99.0	98.5	98.0	97.5	97.0	96.5	96.0	95.5	95.0	τ_0
τ	10,000 . $\bar{D}$											τ
80.0	0969	0947	0925	0903	0881	0859	0837	0814	0792	0769	0746	80.0
79.5	0996	0975	0953	0931	0909	0886	0864	0842	0819	0796	0774	79.5
79.0	1024	1002	0980	0958	0936	0914	0891	0869	0846	0824	0801	79.0
78.5	1051	1030	1008	0986	0964	0941	0919	0897	0874	0851	0829	78.5
78.0	1079	1057	1035	1013	0991	0969	0947	0924	0902	0879	0856	78.0
77.5	1107	1085	1063	1041	1019	0997	0975	0952	0930	0907	0884	77.5
77.0	1135	1113	1091	1069	1047	1025	1003	0980	0958	0935	0912	77.0
76.5	1163	1142	1120	1098	1076	1053	1031	1009	0986	0963	0941	76.5
76.0	1192	1170	1148	1126	1104	1082	1060	1037	1015	0992	0969	76.0
75.5	1221	1199	1177	1155	1133	1111	1088	1066	1043	1021	0998	75.5
75.0	1249	1228	1206	1184	1162	1139	1117	1095	1072	1049	1027	75.0
74.5	1278	1257	1235	1213	1191	1168	1146	1124	1101	1078	1056	74.5
74.0	1308	1286	1264	1242	1220	1198	1175	1153	1130	1108	1085	74.0
73.5	1337	1315	1293	1271	1249	1227	1205	1182	1160	1137	1114	73.5
73.0	1367	1345	1323	1301	1279	1257	1234	1212	1189	1167	1144	73.0
72.5	1397	1375	1353	1331	1309	1287	1264	1242	1219	1197	1174	72.5
72.0	1427	1405	1383	1361	1339	1317	1294	1272	1249	1227	1204	72.0
71.5	1457	1435	1413	1391	1369	1347	1325	1302	1280	1257	1234	71.5
71.0	1487	1466	1444	1422	1400	1377	1355	1333	1310	1287	1265	71.0
70.5	1518	1496	1474	1452	1430	1408	1386	1363	1341	1318	1295	70.5
70.0	1549	1527	1505	1483	1461	1439	1417	1394	1372	1349	1326	70.0
69.5	1580	1558	1537	1515	1492	1470	1448	1425	1403	1380	1357	69.5
69.0	1612	1590	1568	1546	1524	1502	1479	1457	1434	1412	1389	69.0
68.5	1643	1621	1599	1577	1555	1533	1511	1488	1466	1443	1420	68.5
68.0	1675	1653	1631	1609	1587	1565	1543	1520	1498	1475	1452	68.0
67.5	1707	1685	1663	1641	1619	1597	1575	1552	1530	1507	1484	67.5
67.0	1739	1717	1696	1674	1652	1629	1607	1585	1562	1539	1516	67.0
66.5	1772	1750	1728	1706	1684	1662	1639	1617	1594	1572	1549	66.5
66.0	1805	1783	1761	1739	1717	1695	1672	1650	1627	1605	1582	66.0
65.5	1838	1816	1794	1772	1750	1728	1705	1683	1660	1638	1615	65.5
65.0	1871	1849	1827	1805	1783	1761	1739	1716	1694	1671	1648	65.0
64.5	1904	1883	1861	1839	1817	1794	1772	1750	1727	1704	1682	64.5
64.0	1938	1916	1895	1873	1850	1828	1806	1783	1761	1738	1715	64.0
63·5	1972	1950	1929	1907	1885	1862	1840	1818	1795	1772	1749	63.5
63.0	2007	1985	1963	1941	1919	1897	1874	1852	1829	1807	1784	63.0
62.5	2041	2019	1998	1976	1953	1931	1909	1886	1864	1841	1818	62.5
62.0	2076	2054	2032	2010	1988	1966	1944	1921	1899	1876	1853	62.0
61.5	2111	2089	2068	2046	2024	2001	1979	1957	1934	1911	1888	61.5
61.0	2147	2125	2103	2081	2059	2037	2014	1992	1969	1947	1924	61.0
60.5	2182	2161	2139	2117	2095	2072	2050	2028	2005	1982	1960	60.5
60.0	2218	2197	2175	2153	2131	2109	2086	2064	2041	2019	1996	60.0
59.5	2255	2233	2211	2189	2167	2145	2123	2100	2078	2055	2032	59.5
59.0	2291	2270	2248	2226	2204	2182	2159	2137	2114	2092	2069	59.0

τ_0	100.0	99.5	99.0	98.5	98.0	97.5	97.0	96.5	96.0	95.5	95.0	τ_0
τ	10,000 . D											τ
58.5	2328	2307	2285	2263	2241	2218	2196	2174	2151	2128	2106	58.5
58.0	2366	2344	2322	2300	2278	2256	2233	2211	2188	2166	2143	58.0
57.5	2403	2382	2360	2338	2316	2293	2271	2249	2226	2203	2181	57.5
57.0	2441	2419	2398	2376	2354	2331	2309	2287	2264	2241	2218	57.0
56.5	2480	2458	2436	2414	2392	2370	2347	2325	2302	2280	2257	56.5
56.0	2518	2496	2474	2452	2430	2408	2386	2363	2341	2318	2295	56.0
55.5	2557	2535	2513	2491	2469	2447	2425	2402	2380	2357	2334	55.5
55.0	2596	2575	2553	2531	2509	2486	2464	2442	2419	2396	2374	55.0
54.5	2636	2614	2592	2570	2548	2526	2504	2481	2459	2436	2413	54.5
54.0	2676	2654	2632	2610	2588	2566	2544	2521	2499	2476	2453	54.0
53.5	2716	2695	2673	2651	2629	2607	2584	2562	2539	2516	2494	53.5
53.0	2757	2735	2714	2692	2670	2647	2625	2603	2580	2557	2534	53.0
52.5	2798	2777	2755	2733	2711	2688	2666	2644	2621	2598	2576	52.5
52.0	2840	2818	2796	2774	2752	2730	2708	2685	2663	2640	2617	52.0
51.5	2882	2860	2838	2816	2794	2772	2750	2727	2705	2682	2659	51.5
51.0	2924	2903	2881	2859	2837	2814	2792	2770	2747	2724	2702	51.0
50.5	2967	2945	2923	2901	2879	2857	2835	2812	2790	2767	2744	50.5
50.0	3010	2989	2967	2945	2923	2900	2878	2856	2833	2810	2788	50.0
49.5	3054	3032	3010	2988	2966	2944	2922	2899	2877	2854	2831	49.5
49.0	3098	3076	3054	3032	3010	2988	2966	2943	2921	2898	2875	49.0
48.5	3143	3121	3099	3077	3055	3033	3010	2988	2965	2943	2920	48.5
48.0	3188	3166	3144	3122	3100	3078	3055	3033	3010	2988	2965	48.0
47.5	3233	3211	3189	3167	3145	3123	3101	3078	3056	3033	3010	47.5
47.0	3279	3257	3235	3213	3191	3169	3147	3124	3102	3079	3056	47.0
46.5	3325	3304	3282	3260	3238	3216	3193	3171	3148	3126	3103	46.5
46.0	3372	3351	3329	3307	3285	3262	3240	3218	3195	3172	3150	46.0
45.5	3420	3398	3376	3354	3332	3310	3288	3265	3243	3220	3197	45.5
45.0	3468	3446	3424	3402	3380	3358	3336	3313	3291	3268	3245	45.0
44.5	3516	3495	3473	3451	3429	3406	3384	3362	3339	3316	3294	44.5
44.0	3565	3544	3522	3500	3478	3456	3433	3411	3388	3366	3343	44.0
43.5	3615	3593	3571	3549	3527	3505	3483	3460	3438	3415	3392	43.5
43.0	3665	3644	3622	3600	3578	3555	3533	3511	3488	3465	3443	43.0
42.5	3716	3694	3672	3650	3628	3606	3584	3561	3539	3516	3493	42.5
42.0	3768	3746	3724	3702	3680	3658	3635	3613	3590	3568	3545	42.0
41.5	3820	3798	3776	3754	3732	3710	3687	3665	3642	3620	3597	41.5
41.0	3872	3850	3829	3807	3784	3762	3740	3717	3695	3672	3649	41.0
40.5	3925	3904	3882	3860	3838	3815	3793	3771	3748	3725	3703	40.5
40.0	3979	3958	3936	3914	3892	3869	3847	3825	3802	3779	3757	40.0
39.5	4034	4012	3990	3968	3946	3924	3902	3879	3857	3834	3811	39.5
39.0	4089	4068	4046	4024	4002	3979	3957	3935	3912	3889	3867	39.0
38.5	4145	4124	4102	4080	4058	4035	4013	3991	3968	3945	3923	38.5
38.0	4202	4180	4159	4137	4114	4092	4070	4047	4025	4002	3979	38.0
37.5	4260	4238	4216	4194	4172	4150	4127	4105	4082	4060	4037	37.5

τ_0	100.0	99.5	99.0	98.5	98.0	97.5	97.0	96.5	96.0	95.5	95.0	τ_0
τ						10,000 . D						τ
37.0	4318	4296	4274	4252	4230	4208	4186	4163	4141	4118	4095	37.0
36.5	4377	4355	4333	4311	4289	4267	4245	4222	4200	4177	4154	36.5
36.0	4437	4415	4393	4371	4349	4327	4305	4282	4260	4237	4214	36.0
35.5	4498	4476	4454	4432	4410	4388	4365	4343	4320	4298	4275	35.5
35.0	4559	4538	4516	4494	4472	4449	4427	4405	4382	4359	4337	35.0
34.5	4622	4600	4578	4556	4534	4512	4490	4467	4445	4422	4399	34.5
34.0	4685	4663	4642	4620	4597	4575	4553	4530	4508	4485	4462	34.0
33.5	4750	4728	4706	4684	4662	4640	4617	4595	4572	4550	4527	33.5
33.0	4815	4793	4771	4749	4727	4705	4683	4660	4638	4615	4592	33.0
32.5	4881	4859	4838	4816	4793	4771	4749	4726	4704	4681	4658	32.5
32.0	4948	4927	4905	4883	4861	4839	4816	4794	4771	4749	4726	32.0
31.5	5017	4995	4973	4951	4929	4907	4885	4862	4840	4817	4794	31.5
31.0	5086	5065	5043	5021	4999	4976	4954	4932	4909	4886	4864	31.0
30.5	5157	5135	5113	5091	5069	5047	5025	5002	4980	4957	4934	30.5
30.0	5229	5207	5185	5163	5141	5119	5097	5074	5051	5029	5006	30.0
29.5	5302	5280	5258	5236	5214	5192	5169	5147	5124	5102	5079	29.5
29.0	5376	5354	5332	5310	5288	5266	5244	5221	5199	5176	5153	29.0
28.5	5452	5430	5408	5386	5364	5342	5319	5297	5274	5252	5229	28.5
28.0	5528	5507	5485	5463	5441	5418	5396	5374	5351	5328	5306	28.0
27.5	5607	5585	5563	5541	5519	5497	5474	5452	5429	5407	5384	27.5
27.0	5686	5665	5643	5621	5599	5576	5554	5532	5509	5486	5464	27.0
26.5	5768	5746	5724	5702	5680	5658	5635	5613	5590	5568	5545	26.5
26.0	5850	5828	5807	5785	5763	5740	5718	5696	5673	5650	5628	26.0
25.5	5935	5913	5891	5869	5847	5825	5802	5780	5757	5735	5712	25.5
25.0	6021	5999	5977	5955	5933	5911	5888	5866	5843	5821	5798	25.0
24.5	6108	6087	6065	6043	6021	5998	5976	5954	5931	5908	5886	24.5
24.0	6198	6176	6154	6132	6110	6088	6066	6043	6021	5998	5975	24.0
23.5	6289	6268	6246	6224	6202	6179	6157	6135	6112	6089	6067	23.5
23.0	6383	6361	6339	6317	6295	6273	6250	6228	6205	6183	6160	23.0
22.5	6478	6456	6435	6413	6390	6368	6346	6323	6301	6278	6255	22.5
22.0	6576	6554	6532	6510	6488	6466	6443	6421	6398	6376	6353	22.0
21.5	6676	6654	6632	6610	6588	6566	6543	6521	6498	6476	6453	21.5
21.0	6778	6756	6734	6712	6690	6668	6646	6623	6601	6578	6555	21.0
20.5	6882	6861	6839	6817	6795	6773	6750	6728	6705	6682	6660	20.5
20.0	6990	6968	6946	6924	6902	6880	6857	6835	6812	6790	6767	20.0

τ_0	95.0	94.5	94.0	93.5	93.0	92.5	92.0	91.5	91.0	90.5	90.0	τ_0
τ						10,000 . D						τ
95.0	0000											95.0
94.5	0023	0000										94.5
94.0	0046	0023	0000									94.0
93.5	0069	0046	0023	0000								93.5
93.0	0092	0069	0046	0023	0000							93.0
92.5	0116	0093	0070	0047	0023	0000						92.5
92.0	0139	0116	0093	0070	0047	0024	0000					92.0
91.5	0163	0140	0117	0094	0071	0047	0024	0000				91.5
91.0	0187	0164	0141	0118	0094	0071	0047	0024	0000			91.0
90.5	0211	0188	0165	0142	0118	0095	0071	0048	0024	0000		90.5
90.0	0235	0212	0189	0166	0142	0119	0095	0072	0048	0024	0000	90.0
89.5	0259	0236	0213	0190	0167	0143	0120	0096	0072	0048	0024	89.5
89.0	0283	0260	0237	0214	0191	0168	0144	0120	0097	0073	0049	89.0
88.5	0308	0285	0262	0239	0215	0192	0168	0145	0121	0097	0073	88.5
88.0	0332	0309	0286	0263	0240	0217	0193	0169	0146	0122	0098	88.0
87.5	0357	0334	0311	0288	0265	0241	0218	0194	0170	0146	0122	87.5
87.0	0382	0359	0336	0313	0290	0266	0243	0219	0195	0171	0147	87.0
86.5	0407	0384	0361	0338	0315	0291	0268	0244	0220	0196	0172	86.5
86.0	0432	0409	0386	0363	0340	0316	0293	0269	0245	0221	0197	86.0
85.5	0458	0435	0412	0388	0365	0342	0318	0295	0271	0247	0223	85.5
85.0	0483	0460	0437	0414	0391	0367	0344	0320	0296	0272	0248	85.0
84.5	0509	0486	0463	0440	0416	0393	0369	0346	0322	0298	0274	84.5
84.0	0534	0512	0488	0465	0442	0419	0395	0371	0348	0324	0300	84.0
83.5	0560	0537	0514	0491	0468	0445	0421	0397	0374	0350	0326	83.5
83.0	0586	0564	0540	0517	0494	0471	0447	0423	0400	0376	0352	83.0
82.5	0613	0590	0567	0544	0520	0497	0473	0450	0426	0402	0378	82.5
82.0	0639	0616	0593	0570	0547	0523	0500	0476	0452	0428	0404	82.0
81.5	0666	0643	0620	0597	0573	0550	0526	0503	0479	0455	0431	81.5
81.0	0692	0669	0646	0623	0600	0577	0553	0529	0506	0482	0458	81.0
80.5	0719	0696	0673	0650	0627	0603	0580	0556	0532	0509	0484	80.5
80.0	0746	0723	0700	0677	0654	0631	0607	0583	0560	0536	0512	80.0
79.5	0774	0751	0728	0704	0681	0658	0634	0611	0587	0563	0539	79.5
79.0	0801	0778	0155	0732	0709	0685	0662	0638	0614	0590	0566	79.0
78.5	0829	0806	0783	0759	0736	0713	0689	0666	0642	0618	0594	78.5
78.0	0856	0833	0810	0787	0764	0740	0717	0693	0669	0646	0621	78.0
77.5	0884	0861	0838	0815	0792	0768	0745	0721	0697	0673	0649	77.5
77.0	0912	0889	0866	0843	0820	0797	0773	0749	0726	0702	0678	77.0
76.5	0941	0918	0895	0871	0848	0825	0801	0778	0754	0730	0706	76.5
76.0	0969	0946	0923	0900	0877	0853	0830	0806	0782	0758	0734	76.0
75.5	0998	0975	0952	0929	0905	0882	0858	0835	0811	0787	0763	75.5
75.0	1027	1004	0981	0958	0934	0911	0887	0864	0840	0816	0792	75.0
74.5	1056	1033	1010	0987	0963	0940	0916	0893	0869	0845	0821	74.5
74.0	1085	1062	1039	1016	1993	0969	0946	0922	0898	0874	0850	74.0

τ_0	95.0	94.5	94.0	93.5	93.0	92.5	92.0	91.5	91.0	90.5	90.0	τ_0
τ						10,000 . D						τ
73.5	1114	1091	1068	1045	1022	0999	0975	0951	0928	0904	0880	73.5
73.0	1144	1121	1098	1075	1052	1028	1005	0981	0957	0933	0909	73.0
72.5	1174	1151	1128	1105	1081	1058	1034	1011	0987	0963	0939	72.5
72.0	1204	1181	1158	1135	1112	1088	1065	1041	1017	0993	0969	72.0
71.5	1234	1211	1188	1165	1142	1118	1095	1071	1047	1023	0999	71.5
71.0	1265	1242	1219	1196	1172	1149	1125	1102	1078	1054	1030	71.0
70.5	1295	1272	1249	1226	1203	1180	1156	1132	1109	1085	1061	70.5
70.0	1326	1303	1280	1257	1234	1210	1187	1163	1139	1116	1091	70.0
69.5	1357	1334	1311	1288	1265	1242	1218	1194	1171	1147	1123	69.5
69.0	1389	1366	1343	1320	1296	1273	1249	1226	1202	1178	1154	69.0
68.5	1420	1397	1374	1351	1328	1305	1281	1257	1234	1210	1186	68.5
68.0	1452	1429	1406	1383	1360	1336	1313	1289	1265	1241	1217	68.0
67.5	1484	1461	1438	1415	1392	1368	1345	1321	1297	1273	1249	67.5
67.0	1516	1494	1471	1447	1424	1401	1377	1353	1330	1306	1282	67.0
66.5	1549	1526	1503	1480	1457	1433	1410	1386	1362	1338	1314	66.5
66.0	1582	1559	1536	1513	1489	1466	1442	1419	1395	1371	1347	66.0
65.5	1615	1592	1569	1546	1522	1499	1475	1452	1428	1404	1380	65.5
65.0	1648	1625	1602	1579	1556	1532	1509	1485	1461	1437	1413	65.0
64.5	1682	1659	1636	1613	1589	1566	1542	1519	1495	1471	1447	64.5
64.0	1715	1693	1669	1646	1623	1600	1576	1552	1529	1505	1481	64.0
63.5	1749	1727	1704	1680	1657	1634	1610	1586	1563	1539	1515	63.5
63.0	1784	1761	1738	1715	1691	1668	1644	1621	1597	1573	1549	63.0
62.5	1818	1796	1772	1749	1726	1703	1679	1655	1632	1608	1584	62.5
62.0	1853	1830	1807	1784	1761	1737	1714	1690	1666	1643	1619	62.0
61.5	1888	1866	1843	1819	1796	1773	1749	1725	1702	1678	1654	61.5
61.0	1924	1901	1878	1855	1832	1808	1785	1761	1737	1713	1689	61.0
60.5	1960	1937	1914	1891	1867	1844	1820	1797	1773	1749	1725	60.5
60.0	1996	1973	1950	1927	1903	1880	1856	1833	1809	1785	1761	60.0
59.5	2032	2009	1986	1963	1940	1916	1893	1869	1845	1821	1797	59.5
59.0	2069	2046	2023	2000	1976	1953	1929	1906	1882	1858	1834	59.0
58.5	2106	2083	2060	2037	2013	1990	1966	1943	1919	1895	1871	58.5
58.0	2143	2120	2097	2074	2051	2027	2004	1980	1956	1932	1908	58.0
57.5	2181	2158	2135	2111	2088	2065	2041	2018	1994	1970	1946	57.5
57.0	2218	2196	2173	2149	2126	2103	2079	2055	2032	2008	1984	57.0
56.5	2257	2234	2211	2188	2164	2141	2117	2094	2070	2046	2022	56.5
56.0	2295	2272	2249	2226	2203	2180	2156	2132	2109	2085	2061	56.0
55.5	2334	2311	2288	2265	2242	2218	2195	2171	2147	2124	2099	55.5
55.0	2374	2351	2328	2304	2281	2258	2234	2211	2187	2163	2139	55.0
54.5	2413	2390	2367	2344	2321	2297	2274	2250	2226	2203	2178	54.5
54.0	2453	2430	2407	2384	2361	2337	2314	2290	2266	2243	2218	54.0
53.5	2494	2471	2448	2425	2401	2378	2354	2331	2307	2283	2259	53.5
53.0	2534	2512	2489	2465	2442	2419	2395	2371	2348	2324	2300	53.0
52.5	2576	2553	2530	2507	2483	2460	2436	2413	2389	2365	2341	52.5

τ_0	95.0	94.5	94.0	93.5	93.0	92.5	92.0	91.5	91.0	90.5	90.0	τ_0
τ						10,000 . D						τ
52.0	2617	2594	2571	2548	2525	2501	2478	2454	2430	2406	2382	52.0
51.5	2659	2636	2613	2590	2567	2543	2520	2596	2472	2448	2424	51.5
51.0	2702	2679	2656	2632	2609	2586	2562	2539	2515	2491	2467	51.0
50.5	2744	2721	2698	2675	2652	2629	2605	2581	2557	2534	2510	50.5
50.0	2788	2765	2742	2718	2695	2672	2648	2625	2601	2577	2553	50.0
49.5	2831	2808	2785	2762	2739	2715	2692	2668	2644	2620	2596	49.5
49.0	2875	2852	2829	2806	2783	2759	2736	2712	2688	2665	2640	49.0
48.5	2920	2897	2874	2851	2827	2804	2780	2757	2733	2709	2685	48.5
48.0	2965	2942	2919	2896	2872	2849	2825	2802	2778	2754	2730	48.0
47.5	3010	2987	2964	2941	2918	2894	2871	2847	2823	2800	2775	47.5
47.0	3056	3033	3010	2987	2964	2940	2917	2893	2869	2846	2821	47.0
46.5	3103	3080	3057	3034	3010	2987	2963	2940	2916	2892	2868	46.5
46.0	3150	3127	2104	3081	3057	3034	3010	2987	2963	2939	2915	46.0
45.5	3197	3174	3151	3128	3105	3081	3058	3034	3010	2986	2962	45.5
45.0	3245	3222	3199	3176	3153	3129	3106	3082	3058	3034	3010	45.0
44.5	3294	3271	3248	3225	3201	3178	3154	3131	3107	3083	3059	44.5
44.0	3343	3320	3297	3274	3250	3227	3203	3180	3156	3132	3108	44.0
43.5	3392	3369	3346	3323	3300	3277	3253	3229	3206	3182	3158	43.5
43.0	3443	3420	3397	3373	3350	3327	3303	3280	3256	3232	3208	43.0
42.5	3493	3470	3447	3424	3401	3378	3354	3330	3307	3283	3259	42.5
42.0	3545	3522	3499	3476	3452	3429	3405	3382	3358	3334	3310	42.0
41.5	3597	3574	3551	3528	3504	3481	3457	3434	3410	3386	3362	41.5
41.0	3649	3626	3603	3580	3557	3534	3510	3486	3463	3439	3415	41.0
40.5	3703	3680	3657	3634	3610	3587	3563	3540	3516	3492	3468	40.5
40.0	3757	3734	3711	3688	3664	3641	3617	3594	3570	3546	3522	40.0
39.5	3811	3788	3765	3742	3719	3695	3672	3648	3624	3601	3576	39.5
39.0	3867	3844	3821	3797	3774	3751	3727	3704	3680	3656	3632	39.0
38.5	3923	3900	3877	3854	3830	3807	3783	3760	3736	3712	3688	38.5
38.0	3979	3956	3933	3910	3887	3864	3840	3816	3793	3769	3745	38.0
37.5	4037	4014	3991	3968	3945	3921	3898	3874	3850	3826	3802	37.5
37.0	4095	4072	4049	4026	4003	3979	3956	3932	3908	3884	3860	37.0
36.5	4154	4131	4108	4085	4062	4038	4015	3991	3967	3944	3919	36.5
36.0	4214	4191	4168	4145	4122	4098	4075	4051	4027	4003	3979	36.0
35.5	4275	4252	4229	4206	4183	4159	4136	4112	4088	4064	4040	35.5
35.0	4337	4314	4291	4267	4244	4221	4197	4174	4150	4126	4102	35.0
34.5	4399	4376	4353	4330	4307	4283	4260	4236	4212	4188	4164	34.5
34.0	4462	4440	4416	4393	4370	4347	4323	4299	4276	4252	4228	34.0
33.5	4527	4504	4481	4458	4434	4411	4387	4364	4340	4316	4292	33.5
33.0	4592	4569	4546	4523	4500	4476	4453	4429	4405	4381	4357	33.0
32.5	4658	4635	4612	4589	4566	4543	4519	4495	4472	4448	4424	32.5
32.0	4726	4703	4680	4657	4633	4610	4586	4563	4539	4515	4491	32.0
31.5	4794	4771	4748	4725	4702	4678	4655	4631	4607	4583	4559	31.5
31.0	4864	4841	4818	4794	4771	4748	4724	4701	4677	4653	4629	31.0

τ_0	95.0	94.5	94.0	93.5	93.0	92.5	92.0	91.5	91.0	90.5	90.0	τ_0
τ	10,000 . D											τ
30.5	4934	4911	4888	4865	4842	4818	4795	4771	4747	4723	4699	30.5
30.0	5006	4983	4960	4937	4914	4890	4867	4843	4819	4795	4771	30.0
29.5	5079	5056	5033	5010	4987	4963	4940	4916	4892	4868	4844	29.5
29.0	5153	5130	5107	5084	5061	5037	5014	4990	4966	4943	4918	29.0
28.5	5229	5206	5183	5160	5136	5113	5089	5066	5042	5018	4994	28.5
28.0	5306	5283	5260	5237	5213	5190	5166	5143	5119	5095	5071	28.0
27.5	5384	5361	5338	5315	5292	5268	5245	5221	5197	5173	5149	27.5
27.0	5464	5441	5418	5394	5371	5348	5324	5301	5277	5253	5229	27.0
26.5	5545	5522	5499	5476	5452	5429	5405	5382	5358	5334	5310	26.5
26.0	5628	5605	5582	5558	5535	5512	5488	5464	5441	5417	5393	26.0
25.5	5712	5689	5666	5643	5619	5596	5572	5549	5525	5501	5477	25.5
25.0	5798	5775	5152	5729	5705	5682	5658	5635	5611	5587	5563	25.0
24.5	5886	5863	5840	5816	5793	5770	5746	5723	5699	5675	5651	24.5
24.0	5975	5952	5929	5906	5883	5859	5836	5812	5788	5764	5740	24.0
23.5	6067	6044	6021	5997	5974	5951	5927	5904	5880	5856	5832	23.5
23.0	6160	6137	6114	6091	6068	6044	6021	5997	5973	5949	5925	23.0
22.5	6255	6232	6209	6186	6163	6140	6116	6092	6069	6045	6021	22.5
22.0	6353	6330	6307	6284	6261	6237	6214	6190	6166	6142	6118	22.0
21.5	6453	6430	6407	6384	6360	6337	6313	6290	6266	6242	6218	21.5
21.0	6555	6532	6509	6486	6463	6439	6416	6392	6368	6344	6320	21.0
20.5	6660	6637	6614	6591	6567	6544	6520	6497	6473	6449	6425	20.5
20.0	6767	6744	6721	6698	6675	6651	6628	6604	6580	6556	6532	20.0

τ_0	90.0	89.5	89.0	88.5	88.0	87.5	87.0	86.5	86.0	85.5	85.0	τ_0
τ						10,000 . D						τ
90.0	0000											90.0
89.5	0024	0000										89.5
89.0	0049	0024	0000									89.0
88.5	0073	0049	0024	0000								88.5
88.0	0098	0073	0049	0025	0000							88.0
87.5	0122	0098	0074	0049	0025	0000						87.5
87.0	0147	0123	0099	0074	0050	0025	0000					87.0
86.5	0172	0148	0124	0099	0075	0050	0025	0000				86.5
86.0	0197	0173	0149	0124	0100	0075	0050	0025	0000			86.0
85.5	0223	0199	0174	0150	0125	0100	0076	0050	0025	0000		85.5
85.0	0248	0224	0200	0175	0151	0126	0101	0076	0051	0025	0000	85.0
84.5	0274	0250	0225	0201	0176	0152	0127	0102	0076	0051	0026	84.5
84.0	0300	0275	0251	0227	0202	0177	0152	0127	0102	0077	0051	84.0
83.5	0326	0301	0277	0253	0228	0203	0178	0153	0128	0103	0077	83.5
83.0	0352	0327	0303	0279	0254	0229	0204	0179	0154	0129	0103	83.0
82.5	0378	0354	0329	0305	0280	0256	0231	0206	0180	0155	0130	82.5
82.0	0404	0380	0356	0331	0307	0282	0257	0232	0207	0182	0156	82.0
81.5	0431	0407	0382	0358	0333	0309	0284	0259	0233	0208	0183	81.5
81.0	0458	0433	0409	0385	0360	0335	0310	0285	0260	0235	0209	81.0
80.5	0484	0460	0436	0411	0387	0362	0337	0312	0287	0262	0236	80.5
80.0	0512	0487	0463	0439	0414	0389	0364	0339	0314	0289	0263	80.0
79.5	0539	0515	0490	0466	0441	0416	0392	0366	0341	0316	0291	79.5
79.0	0566	0542	0518	0493	0469	0444	0419	0394	0369	0343	0318	79.0
78.5	0594	0570	0545	0521	0496	0471	0446	0421	0396	0371	0345	78.5
78.0	0621	0597	0573	0548	0524	0499	0474	0449	0424	0399	0373	78.0
77.5	0649	0625	0601	0576	0552	0527	0502	0477	0452	0427	0401	77.5
77.0	0678	0653	0629	0605	0580	0555	0530	0505	0480	0455	0429	77.0
76.5	0706	0682	0657	0633	0608	0583	0559	0534	0508	0483	0458	76.5
76.0	0734	0710	0686	0661	0637	0612	0587	0562	0537	0512	0486	76.0
75.5	0763	0739	0714	0690	0665	0641	0616	0591	0566	0540	0515	75.5
75.0	0792	0768	0743	0719	0694	0669	0645	0620	0594	0569	0544	75.0
74.5	0821	0797	0772	0748	0723	0699	0674	0649	0623	0598	0573	74.5
74.0	0850	0826	0802	0777	0753	0728	0703	0678	0653	0627	0602	74.0
73.5	0880	0855	0831	0807	0782	0757	0732	0707	0682	0657	0631	73.5
73.0	0909	0885	0861	0836	0812	0787	0762	0737	0712	0686	0661	73.0
72.5	0939	0915	0891	0866	0841	0817	0792	0767	0742	0716	0691	72.5
72.0	0969	0945	0921	0896	0871	0847	0822	0797	0772	0746	0721	72.0
71.5	0999	0975	0951	0926	0902	0877	0852	0827	0802	0777	0751	71.5
71.0	1030	1006	0981	0957	0932	0907	0883	0858	0832	0807	0782	71.0
70.5	1061	1036	1012	0988	0963	0938	0913	0888	0863	0838	0812	70.5
70.0	1091	1067	1043	1018	0994	0969	0944	0919	0894	0869	0843	70.0
69.5	1123	1098	1074	1050	1025	1000	0975	0950	0925	0900	0874	69.5
69.0	1154	1130	1105	1081	1056	1032	1007	0982	0956	0931	0906	69.0

τ_0	90.0	89.5	89.0	88.5	88.0	87.5	87.0	86.5	86.0	85.5	85.0	τ_0
τ						10,000 . D						τ
68.5	1186	1161	1137	1113	1088	1063	1038	1013	0988	0963	0937	68.5
68.0	1217	1193	1169	1144	1120	1095	1070	1045	1020	0995	0969	68.0
67.5	1249	1225	1201	1176	1152	1127	1102	1077	1052	1027	1001	67.5
67.0	1282	1257	1233	1209	1184	1159	1134	1109	1084	1059	1033	67.0
66.5	1314	1290	1266	1241	1217	1192	1167	1142	1117	1091	1066	66.5
66.0	1347	1323	1298	1274	1249	1225	1200	1175	1150	1124	1099	66.0
65.5	1380	1356	1331	1307	1282	1258	1233	1208	1183	1157	1132	65.5
65.0	1413	1389	1365	1340	1316	1291	1266	1241	1216	1191	1165	65.0
64.5	1447	1423	1398	1374	1349	1324	1300	1275	1249	1224	1199	64.5
64.0	1481	1456	1432	1408	1383	1358	1333	1308	1283	1258	1232	64.0
63.5	1515	1490	1466	1442	1417	1392	1367	1342	1317	1292	1266	63.5
63.0	1549	1525	1500	1476	1451	1427	1402	1377	1352	1326	1301	63.0
62.5	1584	1559	1535	1511	1486	1461	1436	1411	1386	1361	1335	62.5
62.0	1619	1594	1570	1546	1521	1496	1471	1446	1421	1396	1370	62.0
61.5	1654	1629	1605	1581	1556	1531	1506	1481	1456	1431	1405	61.5
61.0	1689	1665	1541	1616	1592	1567	1542	1517	1492	1466	1441	61.0
60.5	1725	1701	1676	1652	1627	1603	1578	1553	1527	1502	1477	60.5
60.0	1761	1737	1712	1688	1663	1639	1614	1589	1563	1538	1513	60.0
59.5	1797	1773	1749	1724	1700	1675	1650	1625	1600	1574	1549	59.5
59.0	1834	1810	1785	1761	1736	1712	1687	1662	1636	1611	1586	59.0
58.5	1871	1847	1822	1798	1773	1749	1724	1699	1673	1648	1623	58.5
58.0	1908	1884	1860	1835	1811	1786	1761	1736	1711	1685	1660	58.0
57.5	1946	1922	1897	1873	1848	1823	1799	1773	1748	1723	1698	57.5
57.0	1984	1959	1935	1911	1886	1861	1836	1811	1786	1761	1735	57.0
56.5	2022	1998	1973	1949	1924	1900	1875	1850	1824	1799	1774	56.5
56.0	2061	2036	2012	1988	1963	1938	1913	1888	1863	1838	1812	56.0
55.5	2099	2075	2051	2027	2002	1977	1952	1927	1902	1877	1851	55.5
55.0	2139	2115	2090	2066	2041	2016	1992	1967	1941	1916	1891	55.0
54.5	2178	2154	2130	2105	2081	2056	2031	2006	1981	1956	1930	54.5
54.0	2218	2194	2170	2145	2121	2096	2071	2046	2021	1996	1970	54.0
53.5	2259	2235	2210	2186	2161	2137	2112	2087	2061	2036	2011	53.5
53.0	2300	2275	2251	2227	2202	2177	2152	2127	2102	2077	2051	53.0
52.5	2341	2317	2292	2268	2243	2218	2194	2169	2143	2118	2093	52.5
52.0	2382	2358	2334	2309	2285	2260	2235	2210	2185	2160	2134	52.0
51.5	2424	2400	2376	2351	2327	2302	2277	2252	2227	2202	2176	51.5
51.0	2467	2443	2418	2394	2369	2344	2319	2294	2269	2244	2218	51.0
50.5	2510	2485	2461	2437	2412	2387	2362	2337	2312	2287	2261	50.5
50.0	2553	2529	2504	2480	2455	2430	2405	2380	2355	2330	2304	50.0
49.5	2596	2572	2548	2523	2499	2474	2449	2424	2399	2374	2348	49.5
49.0	2640	2616	2592	2567	2543	2518	2493	2468	2443	2418	2392	49.0
48.5	2685	2661	2636	2612	2587	2563	2538	2513	2488	2462	2437	48.5
48.0	2730	2706	2681	2657	2632	2608	2583	2558	2533	2507	2482	48.0
47.5	2775	2651	2727	2702	2678	2653	2628	2603	2578	2553	2527	47.5

τ_0	90.0	89.5	89.0	88.5	88.0	87.5	87.0	86.5	86.0	85.5	85.0	τ_0
τ						10,000 . D						τ
47.0	2821	2797	2173	2748	2724	2699	2674	2649	2624	2599	2573	47.0
46.5	2868	2844	2819	2795	2770	2746	2721	2696	2670	2645	2620	46.5
46.0	2915	2891	2866	2842	2817	2792	2768	2743	2717	2692	2667	46.0
45.5	2962	2938	2914	2889	2865	2840	2815	2790	2765	2740	2714	45.5
45.0	3010	2986	2962	2937	2913	2888	2863	2838	2813	2788	2762	45.0
44.5	3059	3035	3010	2986	2961	2936	2912	2887	2861	2836	2811	44.5
44.0	3108	3084	3059	3035	3010	2986	2961	2936	2910	2885	2860	44.0
43.5	3158	3133	3109	3085	3060	3035	3010	2985	2960	2935	2909	43.5
43.0	3208	3184	3159	3135	3110	3085	3061	3035	3010	2985	2960	43.0
42.5	3259	3234	3210	3186	3161	3136	3111	3086	3061	3036	3010	42.5
42.0	3310	3286	3261	3237	3212	3188	3163	3138	3112	3087	3062	42.0
41.5	3362	3338	3313	3289	3264	3240	3215	3190	3165	3139	3114	41.5
41.0	3415	3390	3366	3342	3317	3292	3267	3242	3217	3192	3166	41.0
40.5	3468	3444	3419	3395	3370	3346	3321	3296	3270	3245	3220	40.5
40.0	3522	3498	3473	3449	3424	3399	3375	3350	3324	3299	3274	40.0
39.5	3576	3552	3528	3503	3479	3454	3429	3404	3379	3354	3328	39.5
39.0	3632	3608	3583	3559	3534	3509	3485	3460	3434	3409	3384	39.0
38.5	3688	3664	3639	3615	3590	3565	3541	3516	3490	3465	3440	38.5
38.0	3745	3720	3696	3672	3647	3622	3597	3572	3547	3522	3496	38.0
37.5	3802	3778	3754	3729	3705	3680	3655	3630	3605	3579	3554	37.5
37.0	3860	3836	3812	3787	3763	3738	3713	3688	3663	3638	3612	37.0
36.5	3919	3895	3871	3847	3822	3797	3772	3747	3722	3697	3671	36.5
36.0	3979	3955	3931	3906	3882	3857	3832	3807	3782	3757	3731	36.0
35.5	4040	4016	3992	3967	3943	3918	3893	3868	3843	3817	3792	35.5
35.0	4102	4078	4053	4029	4004	3979	3955	3929	3904	3879	3854	35.0
34.5	4164	4140	4116	4091	4067	4042	4017	3992	3967	3941	3916	34.5
34.0	4228	4203	4179	4155	4130	4105	4080	4055	4030	4005	3979	34.0
33.5	4292	4268	4243	4219	4194	4170	4145	4120	4095	4069	4044	33.5
33.0	4357	4333	4309	4284	4260	4235	4210	4185	4160	4135	4109	33.0
32.5	4424	4399	4375	4351	4326	4301	4276	4251	4226	4201	4175	32.5
32.0	4491	4467	4442	4418	4393	4369	4344	4319	4293	4268	4243	32.0
31.5	4559	4535	4511	4486	4462	4437	4412	4387	4362	4337	4311	31.5
31.0	4629	4605	4580	4556	4531	4506	4482	4457	4431	4406	4381	31.0
30.5	4699	4675	4651	4626	4602	4577	4552	4527	4502	4477	4451	30.5
30.0	4771	4747	4123	4698	4674	4649	4624	4599	4574	4548	4523	30.0
29.5	4844	4820	4796	4771	4747	4722	4697	4672	4647	4621	4596	29.5
29.0	4918	4894	4870	4845	4821	4796	4771	4746	4721	4696	4670	29.0
28.5	4994	4970	4945	4921	4896	4872	4847	4822	4797	4771	4746	28.5
28.0	5071	5047	5022	4998	4973	4948	4924	4899	4873	4848	4823	28.0
27.5	5149	5125	5101	5076	5051	5027	5002	4977	4952	4926	4901	27.5
27.0	5229	5205	5180	5256	5131	5106	5082	5057	5031	5006	4981	27.0
26.5	5310	5286	5261	5237	5212	5188	5163	5138	5113	5087	5062	26.5
26.0	5393	5368	5344	5320	5295	5270	5245	5220	5195	5170	5144	26.0

τ_0	90.0	89.5	89.0	88.5	88.0	87.5	87.0	86.5	86.0	85.5	85.0	τ_0
τ						10,000 . D						τ
25.5	5477	5453	5428	5404	5379	5355	5330	5305	5280	5254	5229	25.5
25.0	5563	5539	5514	5490	5465	5441	5416	5391	5366	5340	5315	25.0
24.5	5651	5627	5602	5578	5553	5528	5504	5478	5453	5428	5403	24.5
24.0	5740	5716	5692	5667	5643	5618	5593	5568	5543	5518	5492	24.0
23.5	5832	5808	5783	5759	5734	5709	5685	5659	5634	5609	5584	23.5
23.0	5925	5901	5877	5852	5828	5803	5778	5753	5728	5702	5677	23.0
22.5	6021	5996	5972	5948	5923	5898	5873	5848	5823	5798	5772	22.5
22.0	6118	6094	6070	6045	6021	5996	5971	5946	5921	5895	5870	22.0
21.5	6218	6194	6170	6145	6120	6096	6071	6046	6021	5995	5970	21.5
21.0	6320	6296	6272	6247	6223	6198	6173	6148	6123	6097	6072	21.0
20.5	6425	6401	6376	6352	6327	6303	6278	6253	6227	6202	6177	20.5
20.0	6532	6508	6484	6459	6435	6410	6385	6360	6335	6309	6284	20.0

τ_0	85.0	84.5	84.0	83.5	83.0	82.5	82.0	81.5	81.0	80.5	80.0	τ_0
τ						10,000 . D						τ
85.0	0000											85.0
84.5	0026	0000										84.5
84.0	0051	0026	0000									84.0
83.5	0077	0052	0026	0000								83.5
83.0	0103	0078	0052	0026	0000							83.0
82.5	0130	0104	0078	0052	0026	0000						82.5
82.0	0156	0130	0105	0079	0053	0026	0000					82.0
81.5	0183	0157	0131	0105	0079	0053	0027	0000				81.5
81.0	0209	0184	0158	0132	0106	0080	0053	0027	0000			81.0
80.5	0236	0211	0185	0159	0133	0107	0080	0054	0027	0000		80.5
80.0	0263	0238	0212	0186	0160	0134	0107	0081	0054	0027	0000	80.0
79.5	0291	0265	0239	0213	0187	0161	0134	0108	0081	0054	0027	79.5
79.0	0318	0292	0267	0241	0215	0188	0162	0135	0109	0082	0055	79.0
78.5	0345	0320	0294	0268	0242	0216	0189	0163	0136	0109	0082	78.5
78.0	0373	0348	0322	0296	0270	0244	0217	0191	0164	0137	0110	78.0
77.5	0401	0376	0350	0324	0298	0272	0245	0219	0192	0165	0138	77.5
77.0	0429	0404	0378	0352	0326	0300	0273	0247	0220	0193	0166	77.0
76.5	0458	0432	0406	0380	0354	0328	0302	0275	0248	0221	0194	76.5
76.0	0486	0460	0435	0409	0383	0356	0330	0303	0277	0250	0223	76.0
75.5	0515	0489	0463	0437	0411	0385	0359	0332	0305	0278	0251	75.5
75.0	0544	0518	0492	0466	0440	0414	0388	0361	0334	0307	0280	75.0
74.5	0573	0547	0521	0495	0469	0443	0417	0390	0363	0336	0309	74.5
74.0	0602	0576	0550	0525	0498	0472	0446	0419	0393	0366	0339	74.0
73.5	0631	0606	0580	0554	0528	0502	0475	0449	0422	0395	0368	73.5
73.0	0661	0635	0610	0584	0558	0531	0505	0478	0452	0425	0398	73.0
72.5	0691	0665	0639	0613	0587	0561	0535	0508	0481	0455	0428	72.5
72.0	0721	0695	0669	0644	0617	0591	0565	0538	0512	0485	0458	72.0
71.5	0751	0726	0700	0674	0648	0621	0595	0569	0542	0515	0488	71.5
71.0	0782	0756	0730	0704	0678	0652	0626	0599	0572	0545	0518	71.0
70.5	0812	0787	0761	0735	0709	0683	0656	0630	0603	0576	0549	70.5
70.0	0843	0818	0792	0766	0740	0714	0687	0661	0634	0607	0580	70.0
69.5	0874	0849	0823	0797	0771	0745	0718	0692	0665	0638	0611	69.5
69.0	0906	0880	0854	0828	0802	0776	0750	0723	0696	0669	0642	69.0
68.5	0937	0912	0886	0860	0834	0808	0781	0755	0728	0701	0674	68.5
68.0	0969	0943	0918	0892	0866	0839	0813	0786	0760	0733	0706	68.0
67.5	1001	0976	0950	0924	0898	0871	0845	0819	0792	0765	0738	67.5
67.0	1033	1008	0982	0956	0930	0904	0877	0851	0824	0797	0770	67.0
66.5	1066	1040	1015	0989	0963	0936	0910	0883	0857	0830	0803	66.5
66.0	1099	1073	1047	1021	0995	0969	0943	0916	0889	0863	0835	66.0
65.5	1132	1106	1080	1054	1028	1002	0976	0949	0922	0896	0868	65.5
65.0	1165	1139	1114	1088	1062	1035	1009	0982	0956	0929	0902	65.0
64.5	1199	1173	1147	1121	1095	1069	1043	1016	0989	0962	0935	64.5
64.0	1232	1207	1181	1155	1129	1103	1076	1050	1023	0996	0969	64.0

τ_0	85.0	84.5	84.0	83.5	83.0	82.5	82.0	81.5	81.0	80.5	80.0	τ_0
τ						10,000 . D						τ
63.5	1266	1241	1215	1189	1163	1137	1110	1084	1057	1030	1003	63.5
63.0	1301	1275	1249	1223	1197	1171	1145	1118	1091	1065	1037	63.0
62.5	1335	1310	1284	1258	1232	1206	1179	1153	1126	1099	1072	62.5
62.0	1370	1345	1319	1293	1267	1241	1214	1188	1161	1134	1107	62.0
61.5	1405	1380	1354	1328	1302	1276	1249	1223	1196	1169	1142	61.5
61.0	1441	1415	1389	1364	1337	1311	1285	1258	1232	1205	1178	61.0
60.5	1477	1451	1425	1399	1373	1347	1321	1294	1267	1240	1213	60.5
60.0	1513	1487	1461	1435	1409	1383	1357	1330	1303	1276	1249	60.0
59.5	1549	1523	1498	1472	1446	1419	1393	1366	1340	1313	1286	59.5
59.0	1586	1560	1534	1508	1482	1456	1430	1403	1376	1349	1322	59.0
58.5	1623	1597	1571	1545	1519	1493	1467	1440	1413	1386	1359	58.5
58.0	1660	1634	1609	1583	1556	1530	1504	1477	1451	1424	1397	58.0
57.5	1698	1672	1646	1620	1594	1568	1541	1515	1488	1461	1434	57.5
57.0	1735	1710	1684	1658	1632	1606	1579	1553	1526	1499	1472	57.0
56.5	1774	1748	1722	1696	1670	1644	1618	1591	1564	1537	1510	56.5
56.0	1812	1787	1761	1735	1709	1683	1656	1630	1603	1576	1549	56.0
55.5	1851	1826	1800	1774	1748	1722	1695	1669	1642	1615	1588	55.5
55.0	1891	1865	1839	1813	1787	1761	1735	1708	1681	1654	1627	55.0
54.5	1930	1905	1879	1853	1827	1801	1774	1748	1721	1694	1667	54.5
54.0	1970	1945	1919	1893	1867	1841	1814	1788	1761	1734	1707	54.0
53.5	2011	1985	1959	1933	1907	1881	1855	1828	1801	1774	1747	53.5
53.0	2051	2026	2000	1974	1948	1922	1895	1869	1842	1815	1788	53.0
52.5	2093	2067	2041	2015	1989	1963	1937	1910	1883	1856	1829	52.5
52.0	2134	2109	2083	2057	2031	2005	1978	1952	1925	1898	1871	52.0
51.5	2176	2150	2125	2099	2073	2046	2020	1994	1967	1940	1913	51.5
51.0	2218	2193	2167	2141	2115	2089	2062	2036	2009	1982	1955	51.0
50.5	2261	2236	2210	2184	2158	2132	2105	2079	2052	2025	1998	50.5
50.0	2304	2279	2253	2227	2201	2175	2148	2122	2095	2068	2041	50.0
49.5	2348	2323	2297	2271	2245	2218	2192	2166	2139	2112	2085	49.5
49.0	2392	2367	2341	2315	2289	2263	2236	2210	2183	2156	2129	49.0
48.5	2437	2411	2385	2359	2333	2307	2281	2254	2227	2201	2173	48.5
48.0	2482	2456	2430	2404	2378	2352	2326	2299	2272	2246	2218	48.0
47.5	2527	2502	2476	2450	2424	2398	2371	2345	2318	2291	2264	47.5
47.0	2573	2548	2522	2496	2470	2444	2417	2391	2364	2337	2310	47.0
46.5	2620	2594	2568	2542	2516	2490	2464	2437	2410	2383	2356	46.5
46.0	2667	2641	2615	2589	2563	2537	2511	2484	2457	2430	2403	46.0
45.5	2714	2688	2663	2637	2611	2584	2558	2531	2505	2478	2451	45.5
45.0	2762	2736	2711	2685	2659	2632	2606	2579	2553	2526	2499	45.0
44.5	2811	2785	2759	2733	2707	2681	2655	2628	2601	2574	2547	44.5
44.0	2860	2834	2808	2782	2756	2730	2704	2677	2650	2623	2596	44.0
43.5	2909	2884	2858	2832	2806	2780	2753	2727	2700	2673	2646	43.5
43.0	2960	2934	2908	2882	2856	2830	2803	2777	2750	2723	2696	43.0
42.5	3010	2985	2959	2933	2907	2881	2854	2828	2801	2774	2747	42.5

τ_0	85.0	84.5	84.0	83.5	83.0	82.5	82.0	81.5	81.0	80.5	80.0	τ_0
τ						10,000 . D						τ
42.0	3062	3036	3010	2984	2958	2932	2906	2879	2852	2825	2798	42.0
41.5	3114	3088	3062	3036	3010	2984	2958	2931	2904	2877	2850	41.5
41.0	3166	3141	3115	3089	3063	3037	3010	2984	2957	2930	2903	41.0
40.5	3220	3194	3168	3142	3116	3090	3064	3037	3010	2983	2956	40.5
40.0	3274	3248	3222	3196	3170	3144	3118	3091	3064	3037	3010	40.0
39.5	3328	3303	3277	3251	3225	3199	3172	3146	3119	3092	3065	39.5
39.0	3384	3358	3332	3306	3280	3254	3227	3201	3174	3147	3120	39.0
38.5	3440	3414	3388	3362	3336	3310	3284	3257	3230	3203	3176	38.5
38.0	3496	3471	3445	3419	3393	3367	3340	3314	3287	3260	3233	38.0
37.5	3554	3528	3502	3477	3450	3424	3398	3371	3345	3318	3291	37.5
37.0	3612	3587	3561	3535	3509	3483	3456	3430	3403	3376	3349	37.0
36.5	3671	3646	3620	3594	3568	3542	3515	3489	3462	3435	3408	36.5
36.0	3731	3706	3680	3654	3628	3602	3575	3549	3522	3495	3468	36.0
35.5	3792	3766	3141	3715	3688	3662	3636	3609	3583	3556	3529	35.5
35.0	3854	3828	3802	3776	3750	3724	3697	3671	3644	3617	3590	35.0
34.5	3916	3890	3865	3839	3813	3786	3760	3733	3707	3680	3653	34.5
34.0	3979	3954	3928	3902	3876	3850	3823	3797	3770	3743	3716	34.0
33.5	4044	4018	3992	3966	3940	3914	3888	3861	3834	3808	3780	33.5
33.0	4109	4083	4058	4032	4006	3979	3953	3926	3900	3873	3846	33.0
32.5	4175	4150	4124	4098	4072	4046	4019	3993	3966	3939	3912	32.5
32.0	4243	4217	4191	4165	4139	4113	4087	4060	4033	4006	3979	32.0
31.5	4311	4285	4260	4234	4208	4181	4155	4128	4102	4075	4048	31.5
31.0	4381	4355	4329	4303	4277	4251	4225	4198	4171	4144	4117	31.0
30.5	4451	4426	4400	4374	4348	4322	4295	4269	4242	4215	4188	30.5
30.0	4523	4497	4472	4446	4420	4393	4367	4340	4314	4287	4260	30.0
29.5	4596	4570	4545	4519	4493	4466	4440	4413	4387	4360	4333	29.5
29.0	4670	4645	4619	4593	4567	4541	4514	4488	4461	4434	4407	29.0
28.5	4746	4720	4694	4668	4642	4616	4590	4563	4536	4510	4482	28.5
28.0	4823	4797	4771	4745	4719	4693	4667	4640	4613	4586	4559	28.0
27.5	4901	4875	4849	4824	4797	4771	4745	4718	4692	4665	4638	27.5
27.0	4981	4955	4929	4903	4877	4851	4824	4798	4771	4744	4717	27.0
26.5	5062	5036	5010	4984	4958	4932	4906	4879	4852	4825	4798	26.5
26.0	5144	5119	5093	5067	5041	5015	4988	4962	4935	4908	4881	26.0
25.5	5229	5203	5177	5151	5125	5099	5073	5046	5019	4993	9465	25.5
25.0	5315	5289	5263	5237	5211	5185	5159	5132	5105	5079	5051	25.0
24.5	5403	5377	5351	5325	5299	5273	5246	5220	5193	5166	5139	24.5
24.0	5492	5466	5441	5415	5389	5362	5336	5309	5283	5256	5229	24.0
23.5	5584	5558	5532	5506	5480	5454	5427	5401	5374	5347	5320	23.5
23.0	5677	5651	5626	5600	5574	5547	5521	5494	5468	5441	5414	23.0
22.5	5772	5747	5721	5695	5669	5643	5616	5590	5563	5536	5509	22.5
22.0	5870	5844	5819	5793	5767	5740	5714	5687	5661	5634	5607	22.0
21.5	5970	5944	5918	5892	5866	5840	5814	5787	5760	5734	5707	21.5
21.0	6072	6046	6021	5995	5969	5942	5916	5889	5863	5836	5809	21.0

τ_0	85.0	84.5	84.0	83.5	83.0	82.5	82.0	81.5	81.0	80.5	80.0	τ_0
τ						10,000 . D						τ
20.5	6177	6151	6125	6099	6073	6047	6021	5994	5967	5940	5913	20.5
20.0	6284	6258	6232	6207	6180	6154	6128	6101	6075	6048	6021	20.0

Index

O

P

Q

R